Percid Fishes

Fish and Aquatic Resources Series

Series Editor: Professor Tony. J. Pitcher

Director, Fisheries Centre, University of British Columbia, Canada

The *Blackwell Science Fish and Aquatic Resources Series* is an initiative aimed at providing key books in this fast-moving field, published to a high international standard.

The Series includes books that review major themes and issues in the science of fishes and the interdisciplinary study of their exploitation in human fisheries. Volumes in the Series combine a broad geographical scope with in-depth focus on concepts, research frontiers and analytical frameworks. These books will be of interest to research workers in the biology, zoology, ichthyology, ecology, physiology of fish and the economics, anthropology, sociology and all aspects of fisheries. They will also appeal to non-specialists such as those with a commercial or industrial stake in fisheries.

It is the aim of the editorial team that books in the *Blackwell Science Fish and Aquatic Resources Series* should adhere to the highest academic standards through being fully peer reviewed and edited by specialists in the field. The Series books are produced by Blackwell Science in a prestigious and distinctive format. The Series Editor, Professor Tony J. Pitcher is an experienced international author, and founding editor of the leading journal in the field of fish and fisheries.

The Series Editor and Publisher at Blackwell Science, Nigel Balmforth, will be pleased to discuss suggestions, advise on scope, and provide evaluations of proposals for books intended for the Series. Please see contact details listed below.

Titles currently included in the Series

1. *Effects of Fishing on Marine Ecosystems and Communities* (S. Hall) 1999
2. *Salmonid Fishes* (Edited by Y. Altukhov *et al.*) 2000
3. *Percid Fishes* (J. Craig) 2000
4. *Fisheries Oceanography* (Edited by P. Harrison & T. Parsons) 2000
5. *Sustainable Fishery Systems* (A. Charles) 2000
6. *Krill* (Edited by I. Everson) 2000

For further information concerning books in the series, please contact:

Nigel Balmforth, Professional Division, Blackwell Science, Osney Mead, Oxford OX2 0EL, UK
Tel: +44 (0) 1865 206206; Fax +44 (0) 1865 721205
e-mail: nigel.balmforth@blacksci.co.uk

Percid Fishes
Systematics, Ecology and Exploitation

John F. Craig

Dunscore, Scotland

Blackwell
Science

Editorial Offices:
Osney Mead, Oxford OX2 0EL
25 John Street, London WC1N 2BL
23 Ainslie Place, Edinburgh EH3 6AJ
350 Main Street, Malden
MA 02148 5018, USA
54 University Street, Carlton
Victoria 3053, Australia
10, rue Casimir Delavigne
75006 Paris, France

Other Editorial Offices:

Blackwell Wissenschafts-Verlag GmbH
Kurfürstendamm 57
10707 Berlin, Germany

Blackwell Science KK
MG Kodenmacho Building
7–10 Kodenmacho Nihombashi
Chuo-ku, Tokyo 104, Japan

First published 2000

Set in 10/13pt Times
by DP Photosetting, Aylesbury, Bucks

DISTRIBUTORS

Marston Book Services Ltd
PO Box 269
Abingdon
Oxon OX14 4YN
(*Orders:* Tel: 01235 465500
Fax: 01235 465555)

USA
Blackwell Science, Inc.
Commerce Place
350 Main Street
Malden, MA 02148 5018
(*Orders:* Tel: 800 759 6102
781 388 8250
Fax: 781 388 8255)

Canada
Login Brothers Book Company
324 Saulteaux Crescent
Winnipeg, Manitoba R3J 3T2
(*Orders:* Tel: 204 837-2987
Fax: 204 837-3116)

Australia
Blackwell Science Pty Ltd
54 University Street
Carlton, Victoria 3053
(*Orders:* Tel: 03 9347 0300
Fax: 03 9347 5001)

A catalogue record for this title is available from the British Library

ISBN 0-632-05616-9

Library of Congress
Cataloging-in-Publication Data
Craig, John F.
Percid fishes: systematics, ecology, and exploitation/John F. Craig.
p. cm.
Includes bibliographical references (p.).
ISBN 0-632-05616-9 (pbk.)
1. Percidae. 2. Fish-culture. I. Title.
QL638.P4 C73 2000
597′.75—dc21 99-059648

For Nicholas, Stephen, Alastair and Rachel

Contents

Series Foreword

Fish researchers (a.k.a. fish freaks) like to explain, to the bemused bystander, how fish have evolved an astonishing array of adaptations, so much so that it can be difficult for them to comprehend why anyone would study anything else. Yet, at the same time, fish are among the last wild creatures on our planet that are hunted by humans for food. As a consequence, few today would fail to recognise that the reconciliation of exploitation with the conservation of biodiversity provides a major challenge to our current knowledge and expertise. Even evaluating the trade-offs that are needed is a difficult task. Moreover, solving this pivotal issue calls for a multidisciplinary conflation of fish physiology, biology and ecology with social sciences such as economics and anthropology in order to probe new frontiers of applied science. The *Blackwell Science Series* on *Fish and Aquatic Resources* is an initiative aimed at providing key, peer-reviewed texts in this fast-moving field.

While bony fish stem from a great radiation that followed the invention of the swimbladder in the Cretaceous period 100 million years ago, some fish groups, such as the sharks, lungfish and sturgeons are more ancient beasts. Survivors from earlier eras may be more widespread than we think: the deep sea coelacanths, formerly known only from the Indian Ocean, have recently turned up in Indonesia. Also, these fishes may be more effectively adapted to specialised niches than their ancient body plans would suggest. For example, rays and angel sharks have perfected the art of the ambush predator, while most cartilaginous fishes can detect electric discharges in the nerves of their prey.

Bony fish themselves have evolved into an amazing array of habitats and niches. As well as the open sea, there are fish in lakes, ponds, rivers and rock pools; in deserts, forests, mountains, the great deeps of the sea, and the extreme cold of the Antarctic; in warm waters of high alkalinity or of low oxygen; and in habitats like estuaries or mudflats, where their physiology is challenged by continuous change. Air-breathing climbing perch (regularly found up trees), walking catfish and mangrove mudskippers are currently repeating the land invasion of their Carboniferous ancestors. We can marvel at high-speed swimming adaptations in the fins, tails, gills and muscles of marlins, sailfish and warm-blooded tunas; gliding and flapping flight in several groups of fish; swinging, protusible jaws providing suction-assisted feeding that have evolved in parallel in groupers, carps and cods; parental care in mouth-brooding cichlids; the birth of live young in many sharks, tooth carps, rockfish and blennies; immense migrations in salmon, shads and tunas; and even the so-called four-eyed fish, with eyes divided into upper air and lower water-adapted sections.

In addition to food, recreation (and inspiration for us fish freaks), it has moreover recently been realised that fish are essential components of aquatic ecosystems that provide vital services to human communities. But, sadly, virtually all sectors of the stunning biodiversity of fishes are at risk from human activities. In fresh water for example, the largest mass extinction event since the end of the dinosaurs has occurred as the introduced Nile perch in Lake Victoria eliminated over 100 species of endemic haplochromine fish. But, at the same time, precious food and income from the Nile perch fishery was created in a miserably poor region. In the oceans, we have barely begun to understand the profound changes that have accompanied a vast expansion of human fishing over the past 100 years.

John Craig's *Percid Fishes*, the third volume in the Blackwell Science Series on *Fish and Aquatic Resources* is a comprehensive monograph summarising the biology, ecology, exploitation and economic importance of all groups of this remarkable assemblage.

From over 9300 member species of the Perciformes, a principally marine order of advanced bony fish, the Percidae, comprising the perch, ruffe, percarina, zander, apron and darter groups, are an ancient Northern hemisphere family with about 189 freshwater species, dating back to a bass-like ancestor 38 million years ago near the cataclysmic start of the Oligocene period. Many freshwater habitats are almost evanescent in geological time and exhibit spatial connections that may form and reform, so the percid group's subsequent evolution has been driven by the complex series of Northern hemisphere glaciations, so much so that some consider the group polyphyletic. An early radiation in European freshwater appears to have been suppressed when cyprinids evolved, but in North America the darters have carried a high diversity through to the present day. In general, percids live in acid, oxygen-rich freshwaters, but one species, the Caspian pike-perch, lives in salt water, and several species in the Baltic and the Danube delta regularly invade brackish waters.

Most of the book is concerned with perch and zander. A chapter by Lawrence Page reviews the North American darters (Etheostomatinae), and ruffe, percarina, asprete and zingel are covered in a self-contained chapter at the end of the book. John Craig has written chapters that thoroughly survey percid morphology, anatomy, growth, mortality, reproduction, development, feeding, ionic balance, swimming, migrations, parasites, diseases, population dynamics, community structure, biomanipulation and fisheries. A chapter on aquaculture by Patrick Kestemont and Charles Mélard provides a bonus.

For me, highlights of the book are descriptions of the sturdy and flexible perch skeleton, the protrusible jaw suction feeding mechanism and the advanced closed swimbladder. The optic area of the perch brain (the mid-brain tectum) has 'on' and 'off' cell responses that parallel the similar but more famous neurones in the fore-brain of mammals; one set is tuned to respond maximally to the three vertical dark stripes of the perch flanks. Perch don't swim fast, but are extremely manoeuvrable.

After stalking and chasing, perch seize their prey sideways and turn them before swallowing – a system that reduces the effect of anti-predator spines. Another mechanism behind perch success is that they appear to have a flexible life history response to crowded conditions induced by productive environments; they reduce growth rate and mature at earlier ages and smaller size, resulting in populations that have almost constant biomass. In addition, percids exhibit an opportunistic diet; although larger percids are characteristically piscivorous, both perch and zander will eat invertebrates and plankton if they have to and chitinase is produced by the pancreas. Perch shoal to catch prey, and although cooperative hunting has not been unequivocally demonstrated, my hunch is that a reward awaits a clever graduate student who can devise the right controls and experiments.

Archaeological work shows that perch and zander fished along the Baltic shores in 6000 BP were much bigger than today's fish (no surprise there!). Today, Finland catches the most percids, followed by Kazakhstan, Russia and Turkey. Walleye and yellow perch are also caught and marketed in Canada and the USA. Such fisheries provide a great product: I have greatly enjoyed a German perch (barsch) dish with salt bacon and celery washed down with one of those appley dry Rieslings that the Germans never seem to export. Craig discusses the issues that face these freshwater fisheries in their endeavours to be sustainable.

Towards the end of the book, the reader will find four interesting case histories of percid ecology to compare: Oneida lake, New York; Western Lake Erie, Canada; Lake Ijssel, Netherlands; and Windermere, United Kingdom, where the author spent the early part of his career. Indeed, there is much fascinating detail in this book, and John Craig has assembled over 1250 references to the primary literature with which to support his lucid prose. I am confident that this important book will provide a valuable source of ideas and reference for percid fishes for many years to come.

Professor Tony J. Pitcher
Editor, Blackwell Science Fish and Aquatic Resources Series
Director, Fisheries Centre, University of British Columbia, Vancouver, Canada

Preface

This monograph provides a comprehensive coverage of percid biology. This very important family of fishes inhabits the temperate Northern hemisphere and provides a renewable resource for commercial and recreational fisheries and suitable candidates for aquaculture.

It is over fifteen years ago that I started to assemble information which formed the basis of my book *The biology of perch and related fish*, published in 1987. The number of publications on percids since that time has been considerable, although it is interesting to note that gains in knowledge have not been in proportion to the prolific output. Indeed, the foundations in all aspects of percid taxonomy, morphology, physiology, ecology and fisheries management were well established by the mid 1980s.

The most important and novel research in the last ten years includes molecular genetics and the use of DNA in stock identification. The techniques have been used to trace the ancestry and evolution of the family. There have been rapid developments in aquaculture and rearing of several members of this group. Knowledge has been gained in reproductive physiology and constraints in culture have been identified and overcome. Fishes are not domesticated like farm animals so their culture requires special attention. Patrick Kestemont and Charles Mélard have written a detailed account of this technology. This area of development is very pertinent as percid capture fisheries continue to decline.

There is now greater appreciation of the role of percids in the context of communities. A greater understanding of the effects of predation and both inter- and intra-competition has enabled managers to make improved predictions about their stocks and how they should be cared for. The application of bioenergetic and ecosystem models, which are becoming more sophisticated and more realistic, has aided the formulation of fishery policy and conservation. A quantitative approach is necessary to predict and manage the effects of varying harvest rates, climate change, pollutants and stocking with both indigenous and exotic fishes. Models rely on good, accurate data usually collected over an extended period. A number of percid populations have been monitored on a long-term basis which has provided very valuable and necessary information. However, the direct causative factors controlling population dynamics, in particular recruitment, are still poorly understood. In addition, the role played by percids in the ecosystem and how they can be used in biomanipulation needs futher investigation. They are unpredictable in how they respond to natural and human induced perturbations.

There is a need to conserve our present day fish fauna which have an important function in many food webs. The darters, which are confined to North America but exhibit extensive diversity, are a case in point. Larry Page has provided a comprehensive chapter on this colourful group. There is also a need to understand other members of the family such as zingels and the asprete which are rare and threatened by extinction. The diversity of fishes continues to be threatened not only by adverse abiotic factors affecting habitat but also by elements such as introductions of exotics. The recent introduction of the ruffe into the North American Great Lakes and into water bodies outside its normal range in Europe has provoked interest and attention to a fish that was in the past considered to be somewhat innocuous.

I am privileged to have been able to work with these attractive fish for over thirty years and on two continents. As a biologist I continue to find them fascinating and worthy of further study.

John F. Craig

Acknowledgements

I am very grateful to the following for all their valuable help and willingness to provide information and assistance: John Babaluk, Nigel Balmforth, Kieth Brickley, Jean Collins, Stephen Craig, Sue Grant, Charmaine Johnson, Amira Mohamed, Maurizio Perotti, Ian Pettman and Linda Temperosa.

My wife, Hilary, provided the necessary driving force. She also undertook the very arduous task of compiling the bibliography. I cannot thank her enough.

Chapter 1
Taxonomy and Distribution

1.1 Introduction

The perch family, the Percidae, belongs to the order Perciformes and suborder Percoidei. The order Perciformes, the largest group of vertebrates, encompasses 18 suborders of fish, 148 families about 1496 genera and about 9293 species (Nelson 1994). The members of this huge group of spiny-rayed fish have a world-wide distribution and are especially common in tropical and subtropical seas and are less abundant in temperate waters. They normally inhabit marine inshore areas although 1922 species (mainly cichlids) occur only in freshwater. Individuals of about 2185 species enter freshwater for at least part of their life cycle (Nelson 1994). The members of the Perciformes are very varied in size and shape but in general have the following common characteristics. They usually have laterally compressed bodies and most have ctenoid scales (tooth-like processes are present on their free edge) although cycloid scales are found in some species: e.g. the scales of *Epinephelus* Bloch (Serranidae) above the lateral line are nearly all ctenoid, and those below it are cycloid. The Perciformes are physoclists (the duct connecting the swim bladder to the gut is closed). There are usually two dorsal fins but never an adipose fin. The first dorsal fin is supported by spines and the second by soft rays. The anal fin has two or three spines and a series of soft rays. The caudal fin has 17 or fewer principal rays, 15 are branched. Pelvic fins, when present, are jugular or thoracic in position, each with one spine and five or fewer soft rays. The pelvic girdle is often attached to the cleithra. The pectoral fins are elevated on the sides of the fish with the base of the fins more or less vertically oriented. The skull is elongate and well ossified. The upper jaw is bordered by the premaxillae and both jaws contain well-developed teeth. The post-temporal is usually forked and articulates with the skull; the orbitosphenoid and the hypocoracoid are absent. There are four to seven branchiostegal rays and gill rakers are toothed. Perciformes have 24 or more vertebrae and inter-muscular bones are absent.

The order contains a widely developed assemblage of fishes and may be paraphyletic, consisting of distinct lines which evolved in parallel to a perciform grade of organisation (Nelson 1994). A thorough and detailed examination of the classification and distribution of the Perciformes is provided by Johnson (1993) and Johnson & Patterson (1993).

1.2 The Percoidei

The largest suborder of the Perciformes, the Percoidei, contains 71 families (Table 1.1), about 528 genera and 2860 species. Nelson (1994) considers it to be the basic group of the Perciformes from which others evolved. Approximately 338 (12%) species live in freshwater. There is no general agreement on how the percoid families might be interrelated. Some series of families appear to be more similar to one another than to most other percoid families (e.g. Carangidae, Nematistiidae and Coryphaenidae, or Lutjanidae, Haemulidae and Sparidae) but there are no sharp morphological discontinuities demarcating such groups, nor is it clear whether the similarities are the result of families being related or from evolutionary convergence.

Table 1.1 Families of the suborder Percoidi (After Nelson 1994.)

Family Name		Common Name
(1)	Pomatomidae	Bluefishes
(2)	Rachycentridae	Cobia
(3)	Coryphaenidae	Dolphins
(4)	Lobotidae	Tripletails
(5)	Echeneididae	Remoras
(6)	Caritsiidae	Manefishes
(7)	Bramidae	Pomfrets
(8)	Nematistiidae	Roosterfish
(9)	Centrarchidae	Sunfishes
(10)	Percidae	Perches
(11)	Percichthyidae	Temperate basses
(12)	Centropomidae	Snooks, glassfishes and perches
(13)	Serranidae	Sea basses
(14)	Grammidae	Basslets
(15)	Pempheridae	Sweepers
(16)	Apogonidae	Cardinalfishes
(17)	Lutjanidae	Snappers
(18)	Malacanthidae	Tilefishes
(19)	Haemulidae	Grunts
(20)	Gerreidae	Mojarras
(21)	Inermiidae	Bonnetmouths
(22)	Cirrhitidae	Hawkfishes
(23)	Chaetodontidae	Butterflyfishes
(24)	Mullidae	Goatfishes
(25)	Sciaenidae	Drums or croakers
(26)	Carangidae	Jacks and pompanos
(27)	Sparidae	Porgies
(28)	Centracanthidae	*Centracanthus*, *Spicara*
(29)	Coracinidae	Galjoen fishes
(30)	Lactariidae	False trevallies
(31)	Priacanthidae	Bigeyes
(32)	Pseudochromidae	Dottybacks

Cont.

Table 1.1 Cont.

Family Name	Common Name
(33) Pomacanthidae	Angelfishes
(34) Monodactylidae	Moonfishes or fingerfishes
(35) Leiognathidae	Ponyfishes, slimys or slipmouths
(36) Teraponidae	Grunters or tigerperches
(37) Plesiopidae	Roundheads
(38) Menidae	Moonfish
(39) Bathyclupeidae	*Bathyclupea*
(40) Oplegnathidae	Knifejaws
(41) Pentacerotidae	Armourheads
(42) Toxotidae	Archerfishes
(43) Nandidae	Leaffishes
(44) Nemipteridae	Threadfin breams
(45) Banjosidae	*Banjos*
(46) Cepolidae	Bandfishes
(47) Kuhliidae	Aholeholes
(48) Glaucosomatidae	Pearl perch, *Glaucosoma*
(49) Sillaginidae	Smelt-whitings
(50) Cheilodactylidae	Morwongs
(51) Latrididae	Trumpeters
(52) Kyphosidae	Sea chubbs
(53) Arripidae	Australian salmon
(54) Leptobramidae	Beachsalmon
(55) Chironemidae	Kelpfishes
(56) Aplodactylidae	Sea carp; also called marble fishes
(57) Enoplasidae	Oldwife
(58) Dinolestidae	*Dinolestes*
(59) Emmelichthyidae	Rovers
(60) Lethrinidae	Scavengers or emperors
(61) Chandidae	Asiatic glassfish
(62) Moronidae	Temperate basses
(63) Acropomatidae	Temperate ocean-basses
(64) Ostracoberycidae	*Ostracoberyx*
(65) Callanthiidae	*Callanthias*
(66) Notograptidae	*Notograpus*
(67) Opistognathidae	Jawfishes
(68) Dinopercidae	*Centrarchops* and *Dinoperca*
(69) Epigonidae	Deepwater cardinalfishes
(70) Polynemidae	Threadfins
(71) Drepanidae	*Drepene*

1.3 Percidae taxonomy and distribution

The family Percidae, unlike most of the 71 percoid families, is confined to freshwater, in the temperate and subarctic regions of North America and Eurasia (Nelson 1994). Some introductions of species of the family have been made into the Southern hemisphere. Percid fossils occur as far back as the Oligocene in both northern landmasses (Romer 1966). The existence in eastern North America, of the subfamily Etheostomatinae Agassiz (four genera, *c.* 173 species) (Chapter 12) is further evidence of a long, independent history of the family in North America. Similarly, the existence of the endemic European genera *Zingel* (three species) and *Romanichthys* (one species) of the subfamily Luciopercinae Jordan and Evermann and *Gymnocephalus* (four species) and *Percarina* (one species) of the subfamily Percinae Bonaparte (Song *et al.* 1998), is evidence of a long, independent period of evolution there.

Song *et al.* (1998) classified the Percidae based on mitochondrial cytochrome b DNA sequence data. Their classification (Fig. 1.1) is similar to that of Collette and Bănărescu (1977) although the latter grouped Percini (*Perca*, *Gymnocephalus* and *Percarina*) and Etheostomatini (*Percina*, *Crystallaria*, *Ammocrypta* and *Etheostoma*) into the subfamily Percinae. The Percinae have the anterior-most interhaemal bone greatly enlarged (Fig. 1.2). Also, the lateral line usually does not extend onto the caudal fin and the anal spines are large and well developed. In the Luciopercinae, the anterior-most interhaemal bone is not larger than the other interhaemal bones, the lateral line does extend onto the caudal fin and the anal spines are not well developed (Collette & Bănărescu 1977). Song *et al.* (1998) believed that similarities between Etheostomatinae and some members of Luciopercinae (*Zingel* and *Romanichthys*) have resulted from convergent evolution as these fish have independently invaded benthic rheophilic habitats. Wiley (1992) had previously claimed that the Etheostomatinae and the small luciopercines were monophyletic.

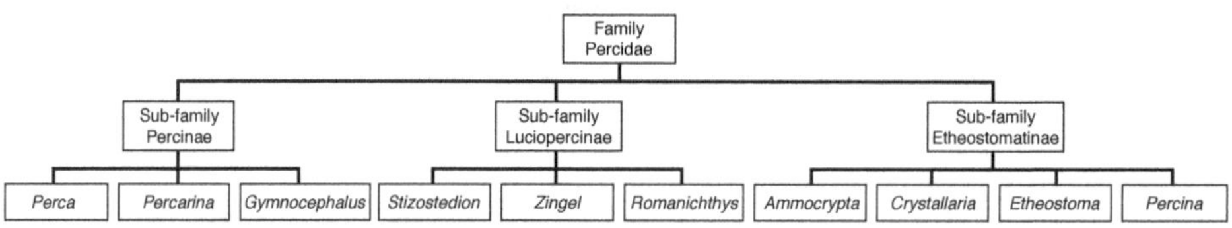

Fig. 1.1 Taxonomic relationships of the Percidae (based on Song *et al.* 1998).

There are three species of *Perca* Linnaeus. *P. fluviatilis* Linnaeus (Eurasian perch, referred to hereafter as perch; Russian = okun) is widely distributed throughout Eurasia but also has been introduced into South Africa, Australia and New Zealand. *Perca flavescens* (Mitchill) (yellow perch) is a North American perch and *P. schrenki* Kessler (Balkhush perch; Russian = Balkhushkii okun) is found only in eastern Kazakh (Russia) (Fig. 1.3).

There are four species of *Gymnocephalus* Bloch: *G. cernuus* (Linnaeus) (ruffe; Russian = ërsh), *G. schraetser* (Linnaeus) (striped ruffe; Russian = polosatyi ërsh), *G. acerinus* (Güdenstadt) (Don ruffe; Russian = Donskoiërsh) and *G. baloni* Holčik & Hensel (Balon's ruffe; Slovak – hrebenačka Balonava) (see Holčik & Hensel 1974). The distribution of these four species of ruffe is shown in Figure 1.4.

The one species of *Percarina* Nordmann, *P. demidoffi* Nordmann (percarina;

(a) (b)

Fig. 1.2 Radiographs of (a) four species of Percinae from top to bottom: *Percarina demidoffi*, *Perca schrenki*, *Gymnocephalus cernuus* and *G. schraetser*, showing the enlarged anterior interhaemal bones, and (b) four species of Luciopercinae, from top to bottom: *Romanichthys vascanicola*, *Zingel zingel*, *Stizostedion canadense* and *S. volgensis*, with no enlargement of the interhaemal bones. (After Collette 1963.)

Russian = perkarina) is distributed in the Sea of Azov and parts of the Black Sea (Fig. 1.5).

There are four genera of darters which are distributed solely in North America (Fig. 1.6). These are *Percina* Haldeman (40 described species), *Crystallaria* Jordan and Gilbert (one species), *Ammocrypta* Jordon (six species) and *Etheostoma* Rafinesque (126 described species).

Stizostedion Rafinesque has five species: three in Europe, *S. lucioperca* (Linnaeus) (zander; Russian = sudak), *S. marina* (Cuvier) (sea pikeperch; Russian = morskoi sudak) and *S. volgensis* (Gmelin) (Volga pikeperch; Russian = bersh) and two in North America, *S. canadense* (Smith) (sauger) and *S. vitreum* (Mitchill) (walleye) (Fig. 1.7).

There are three *Zingel* Cloquet species: *Z. zingel* (Linnaeus) (zingel; Russian = tschop), *Z. streber* Siebold (streber; Russian = malyi tschop) and *Z. asper* (Linnaeus) (French = apron). Zingel and streber are found in the Danube and Vardar Rivers and apron is distributed in the Rhône River (Fig. 1.8). Zingel's close relative *Romanichthys* Dumitrescu, Bănărescu and Stoica, is monotypic and represented by *R. valsanicola* Dumitrescu, Bănărescu and Stoica (asprete) which was found in the Arges and Riul-Doamnei Rivers but is extinct there (Stanescu 1971) and is now only found in the Vilsan River (Fig. 1.8).

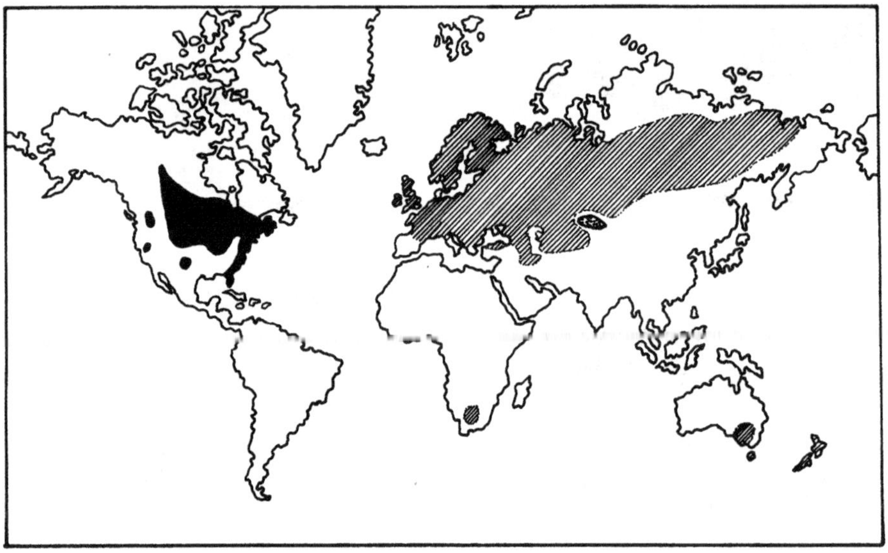

Fig. 1.3 The distribution of *Perca* species. ■ = *Perca flavescens*, ▨ = *P. fluviatilis* and ⬚ = *P. schrenki.*

1.4 Some general characteristics of the percids

Percids are hardy species tolerating a wide range of environmental conditions. Water temperature limits their distribution through its effects on metabolic processes. The upper limit for *Perca* species is a summer air isotherm of about 31°C which is reached on the edge of their range in Italy, Greece, the USA and Australia. The lower limits of the temperature range may influence reproductive success rather than the survival of the adults. The northern distribution limit of walleye to about the summer air isotherm of 13°C probably is more a reflection of temperatures required for maturation and spawning success on a seasonal cyclic basis than limitations to metabolism. Perch and yellow perch eggs and young stages can tolerate low pH and have intermediate requirements in terms of temperature and oxygen concentrations when compared to other temperate freshwater fishes. High production, growth and survival of percids require appropriate food such as zooplankton, littoral invertebrates and prey fish, at the correct stage in the life cycle. Perch species are very successful animals within the limits of tolerable environmental conditions. They have relatively high fecundity, fairly unspecific spawning requirements and can expand rapidly in numbers when they are introduced to new bodies of water such as reservoirs and canals. The same criteria apply to *Stizostedion* species although zander appears to tolerate higher summer maxima than walleye; it may also be better adapted to spawning than walleye. Nest building and guarding leads to better fertilisation and hatching rate and to lower mortality of eggs and

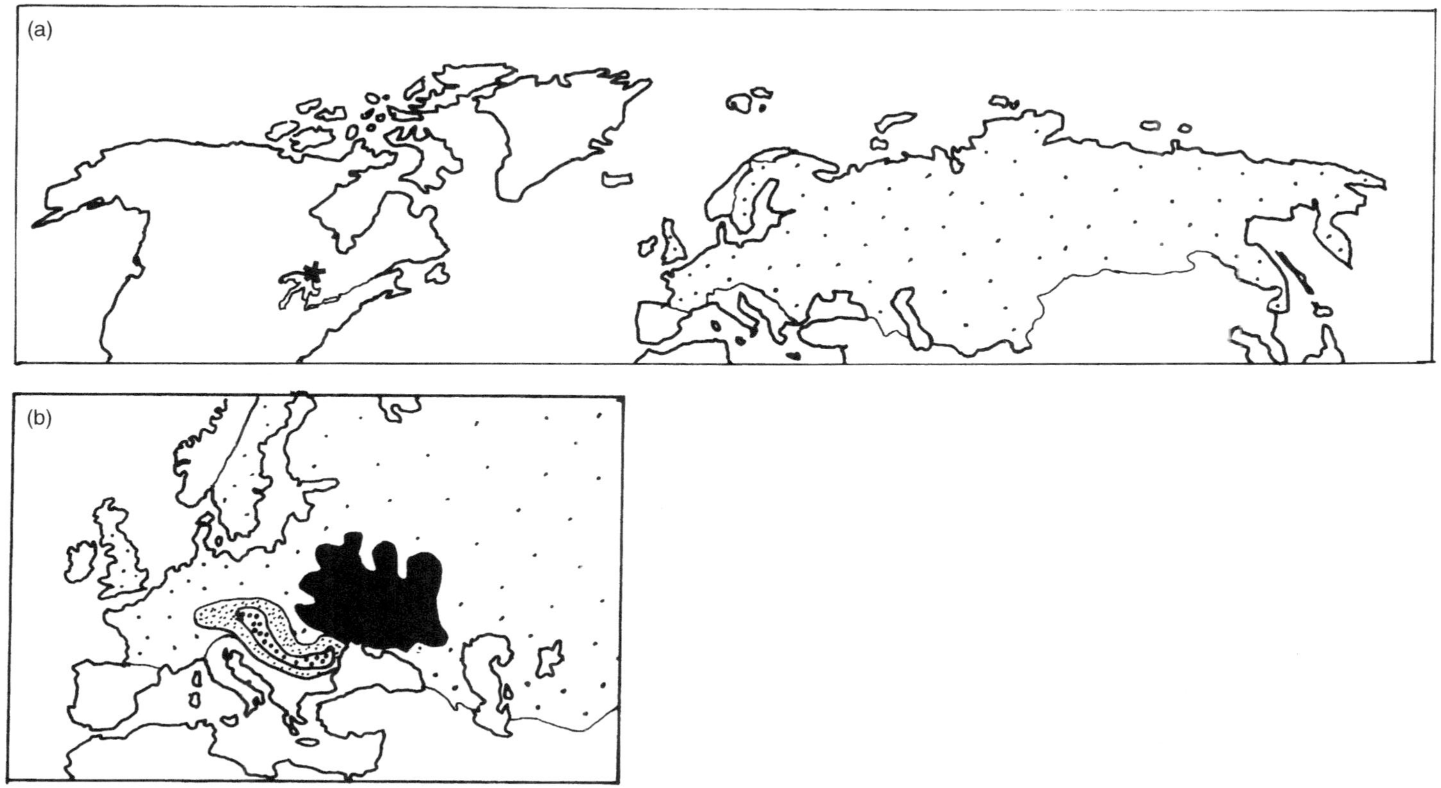

Fig. 1.4 The distribution of *Gymnocephalus* species. [dotted box] = *Gymnocephalus cernuus*. It has been introduced into Lakes Superior and Huron (★). [fine-stippled box] = *Gymnocephalus schraetser*, [coarse-stippled box] = *G. baloni* and *G. schraetser* combined and [black box] = *G. acerinus*.

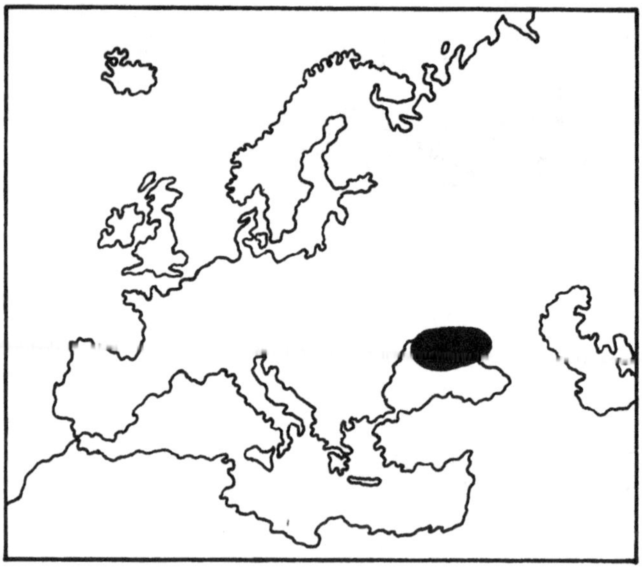

Fig. 1.5 The distribution of *Percarina demidoffi* ■.

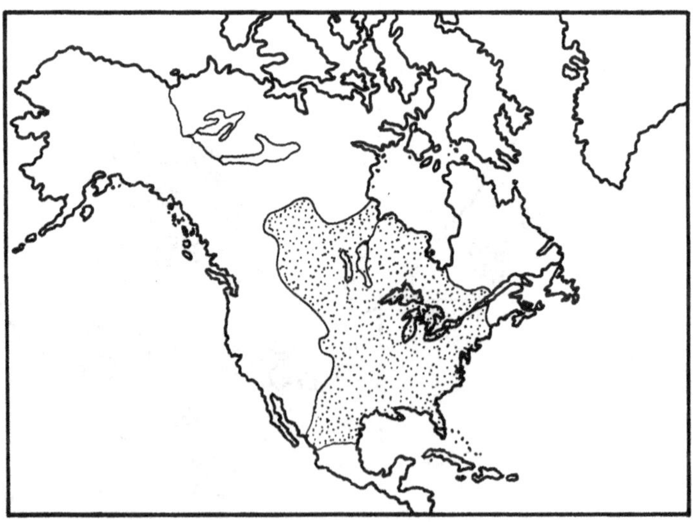

Fig. 1.6 The distribution of the *Etheostomatinae* ⁚⁚.

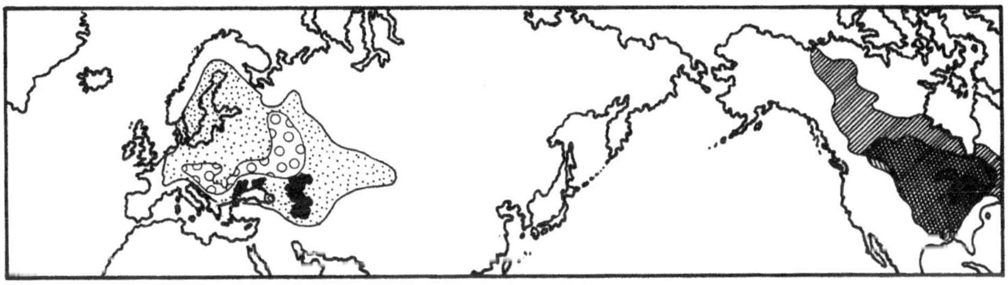

Fig. 1.7 The distribution of *Stizostedion species.* [stippled] = *Stizostedion lucioperca,* [solid black] = *S. marina,* [circles] = *S. volgensis,* [hatched] = *S. canadense* and [hatched] = *S. vitreum.*

larvae. In addition the zander is more fecund than the walleye. These factors probably enable zander to spawn successfully under stressed conditions and to flourish in areas where it has been introduced. Both zander and walleye are particularly successful in lakes with low mean depth, where the water is mixed and there is no thermal stratification. *Stizostedion* species also may compete successfully with other predators because their activities are usually confined to twilight and night.

The plastic nature of percid adaptation to the environment undoubtedly has enabled them to expand within their niche both naturally and as a result of human influence. Specialisation often leads to the end of an evolutionary line. The amount of DNA in teleost haploid cells ranges from 0.4 to 4.4×10^{-12} g. It tends to decline the more specialised the fish becomes (Hinegardner 1968; Hinegardner & Rosen

Fig. 1.8 The distribution of *Zingel* and *Romanichthys* species. [hatched] = *Zingel zingel,* [hatched] *Z. streber,* [solid black] = *Z. asper* and ★ = *R. valsanicola.*

1972). The Percoidei are a fairly specialised group of fish but within the group, members of the Percidae family are of a generalised form. Thus *P. flavescens* has a DNA content of 1.2×10^{-12} g compared to a more specialised species such as *Chaetodon ocellatus* Bloch (Chaetodontidae, butterfly fishes) which has a DNA content of 0.9×10^{-12} g. Specialisation usually means the loss of structures and possibly an evolutionary 'dead-end'. The generalised form of the percids makes them radially adaptive. However, they have certainly not been as successful as Cyprinidae. Up to the Miocene period the Percidae probably played a major role in the fish fauna in Europe. But the appearance of cyprinids at this time may have prevented diversification of the percids. This was not the case in North America where only one fairly primitive cyprinid appeared during the Miocene. Thus, some groups of percids were able to diversify leading to the present day extensive Etheostomatinae fauna (see Chapter 12).

A Perch ancestor probably evolved from an anadromous offshoot of the Serranidae during the Cenozoic. The basic form of the family Percidae is found in Europe. Fossils of *P. fluviatilis* have been discovered in Miocene deposits in the former USSR (Lebedev 1952; Yakovlev 1960), Pliocene deposits in the former USSR, Austria and Belgium (Newton 1908; Weinfurter 1950; Sychevskaya & Devyatkin 1960) and Pleistocene deposits in the former USSR and Germany (Weiler 1933; Gripp & Beyle 1937; Svetovidov & Dorofeeva 1963). Fossils of *P. flavescens* have been found from Pleistocene deposits in Oklahoma, USA. These deposits contain fossils of animals which inhabited the Great Plains at that time (Smith 1954).

It is generally accepted that *Perca* and *Stizostedion* species invaded North America by a north Pacific route across the Bering land bridge (Yakovlev 1961; Collette & Bănărescu 1977; Billington *et al.* 1990, 1991; Song 1995). A number of scientists believed that the route was from Europe across the Atlantic. Cĭhăr (1975) suggested that perch moved from Europe to North America during the late Pleistocene glacial Würm period, 13–15,000 BP, through brackish water along the foot of the receding ice sheet. The Rocky Mountains in the west and the Mackenzie Mountains in the northwest stopped their advance in North America. The spread eastward in Asia was halted by the Kolyma and Anadyr Mountains. Cĭhăr (1975) drew these conclusions from the fact that perch are not found in north Siberian rivers (the furthest east is the Kolyma River, Russia) nor the yellow perch in North American rivers running into the Bering Sea and Pacific Ocean. Recently (nineteenth century), yellow perch have been introduced by humans into water basins flowing into the Pacific Ocean.

Perca fluviatilis and *P. flavescens* are very alike in morphology, physiology and behaviour and they are limited in their distribution by similar factors such as salinity, current speed and oxygen concentration. There has been much controversy over their taxonomy. Mitchill (1818), Gunther (1859), Steindachner (1878), Day (1880; 1886), Sterba (1962), Scott and Crossman (1973) and Collette and Bănărescu (1977) have classified *P. fluviatilis* and *P. flavescens* as separate species. Jordon and Gilbert (1877), Smith (1892), Berg (1905; 1949), Pokrovskii (1951) and Cĭhăr (1975) have made *P. flavescens* a subspecies of *P. fluviatilis*, *P. fluviatilis flavescens*. Thorpe (1977) compared many aspects of the morphology and biology of the two animals and found them very similar. Collette and Bănărescu (1977) have pointed out that there is a difference in the position of the predorsal bone. In *P. fluviatilis* the bone is anterior to the first neural spine but in *P. flavescens* the bone is inserted between the first and second neural spines. MacCrimmon *et al.* (1989) found a specimen of yellow perch that was intermediate between these two types. The fish had a predorsal bone that was anterior to the first neural spine (as in *P. fluviatilis*) but the pterygiophore supporting the first dorsal spine extended between the second and third neural spines (as is typical in *P. flavescens*). However, Song (1995) compared the entire nucleotide sequences of cytochrome b gene and found a large genetic divergence between *P. fluviatilis* and *P. flavescens*. These two species probably became separated a long time ago, about eight million years BP. This would support the theory that *Perca* invaded North America from Siberia. They may have done this at a similar time to *Stizostedion*. The divergence times between the two North American species, *S. canadense* and *S. vitreum*, between two European species, *S. lucioperca* and *S. volgensis* and between the North American and two European species were calculated by Billington *et al.* from allozyme and mitochondrial DNA data (1990, 1992) (Table 1.2). The evidence supports the theory that *Stizostedion* ancestors invaded North America during the Pliocene via the Bering land bridge *c.* ten million years BP. *Stizostedion* was probably well established in North America by the mid Pliocene.

Table 1.2 Approximate divergence times (million years BP ± S.E.) for North American and European *Stizostedion* species based on allozyme and mtDNA data. Studies are from: (i) Billington *et al.* (1991) and (ii) Billington *et al.* (1990). (After Billington *et al.* 1991, with kind permission of Academic Press.)

	Allozyme	mtDNA
North America (i)	1.58 ± 1.14 (26 loci)	4.07 ± 0.73
North America (ii)	3.12 ± 1.33 (36 loci)	4.06 ± 0.73
	2.85 ± 1.22 (39 loci)	
European (i)	6.02 ± 2.32 (26 loci)	1.98 ± 0.47
North American & European (i)	13.28 ± 3.78 (26 loci)	7.42 ± 1.01
North American & European (ii)	10.59 ± 2.74 (36 loci)	7.86 ± 1.18

The Balkhash perch, *P. schrenki*, has developed in isolation since the Tertiary period, unaffected by glacial activity. However, when *P. schrenki* has been introduced into waters containing *P. fluviatilis*, hybridisation has occurred and the resulting populations exhibit convergence of morphological characteristics from the two species (Dukravets & Biryukov 1976; Dukravets & Mamilov 1993).

Stizostedion vitreum has been represented by two subspecies, *S. vitreum vitreum* Mitchill and the blue pike, *S. vitreum glaucum* (Hubbs 1926). The blue pike inhabited Lake Erie and Lake Ontario but is now extinct (McAllister 1970). As well as competition from stocked exotic fish and selective fishing, the gene pool of the blue pike may have become mixed with that of *S. vitreum vitreum* by hybridisation and the subspecies lost (Regier *et al.* 1969). The environmental requirements of walleye are similar to those of yellow perch although the walleye extends further north than the yellow perch. Their niches are closely connected and probably these species have evolved together. Thus yellow perch is the chief food of walleye in many water-bodies and predation may play a part in controlling recruitment both for the predator and prey.

Stizostedion lucioperca originated in the Elbe River and the drainage basins of the Black, Baltic, Caspian and Aral Seas and in the Maritsa River which flows into the Aegean Sea (Sakovich 1986). It has spread throughout Europe as a result of human introductions. It also now occurs in western Turkey (Aksiray 1961) and Morocco (Brunet 1957). Many of the introductions have been recent and well monitored. The zander did not appear in French waters until the 1900s. In 1910 it inhabited the Rhine River and the canal connecting the Rhine and Rhône Rivers (Gagne 1977). By 1915 it was found in the Doubs River, by 1920 in the Sâone River and by 1932 in the Rhône basin (Vivier 1951). In 1948 it inhabited the brackish waters of Etang du Vaccares, France, where it has now become very abundant. The zander spread through numerous introductions into canals, rivers and ponds. Today it is found in all river basins and a large number of lakes in France.

The zander was first introduced into England in 1878 (Sachs 1878). Several introductions were made and by 1910 a breeding population was established (Fitter 1959). The zander became established in the Great Ouse Relief Channel, England, in 1963 from the original, introduced, Woburn stocks (Wheeler & Maitland 1973; Linfield & Rickards 1979). The zander has spread uncontrollably from this open water system to many of the canals and rivers of East Anglia.

Stizostedion volgensis is an eastern European species while *S. marina*, a partially marine species, is found in the Black and Caspian Seas and the lower reaches of the River Bug and River Dnieper. Both species have probably evolved in partial isolation.

Within a percid species there are a number of races and stocks but the extensive variation in morphological characteristics within even a population makes it often difficult to differentiate between these stocks. Some attempts have been made in this direction, for example in comparing scale shape and brain mass to body mass ratio between populations. The latter was shown to vary between three populations of perch in the former USSR (Yakovleva *et al.* 1976). For example, two stocks of perch were identified in one lake by separation of their blood proteins (Kirsipuu 1967). This would be consistent with the theory that two forms of perch live in the same body of water, a predatory deep-water form and a littoral form living off invertebrates and small fish. Care must be taken to ensure that blood protein comparisons have a genetic basis (Kirsipuu 1971). Examination of 19 enzymatic loci using starch gel electrophoresis failed to show any differences in samples of yellow perch from Green Bay, Lake Michigan or nearby Keyes Lake, USA (Leary &

Booke 1982). All of the loci were effectively monomorphic. Significant differences were found at four loci between the pooled Lake Michigan yellow perch and a sample from Lake Champlain, Vermont. However, the reasons for the differences are confusing and do not give any indication of the evolution of these stocks. In a study of yellow perch from 13 localities chosen from throughout the species distribution in North America, Todd and Hatcher (1993) measured the variation in enzymes in muscle and liver using starch-gel electrophoresis. Western populations were less variable than eastern ones and the distribution of alleles indicated that the two groups were derived from Mississippi and Atlantic glacial refugia. Postglacial dispersal between the two populations appeared to be limited.

Starch gel electrophoresis of muscle myogens demonstrated a polymorphism of three patterns in walleye, A, B and AB from lakes in central Canada (Uthe *et al.* 1966; Uthe & Ryder 1970). The authors related the distribution of these patterns within populations to the glacial isolation proposed for walleye, one Atlantic and one Mississippi with a later mixing as might have occurred in the Great Lakes. All three forms of polymorphism occur in the Great Lakes. However, the extent of walleye sampling was not extensive enough to derive definite conclusions about continental-scale zoogeographic events, although it is supported by work on other animals.

The most extensive work on stock identification and attempts to resolve post-Pleistocene re-colonisation of large percids in North America, based on variation in mitochondrial DNA (mtDNA), has been carried out by Billington (1996). Forty two walleye mtDNA haplotypes were identified which could be divided into five groups showing distinct geographical distribution. In contrast, sauger showed very little variation with only four recognised haplotypes. Thirteen yellow perch mtDNA haplotypes were identified with one haplotype predominating in all populations examined. From this work, Billington (1996) concluded that mtDNA was useful for stock identification in walleye but had limitations in application for sauger and yellow perch. In a previous study Clayton *et al.* (1974) had found differences between walleye populations in two headwater lakes in Saskatchewan compared to other western Canadian populations based on the frequency of the MdhB-3 allele. The authors considered that there were insufficient data to draw any definite conclusions from these population differences. However, the presence of dissimilar genetic groups has been used in determining the success of stocking walleye into waters which contain indigenous populations (Schweigert *et al.* 1977; Murphy *et al.* 1983; Mathias *et al.* 1992). The differences between the frequencies of MDH alleles can be used as 'genetic tags' in the absence of other suitable marks or morphological discriminators. Todd and Haas (1993) were able to differentiate walleye from Lake St Clair and the western basin of Lake Erie using a combination of marking with tags and allozyme markers.

Hybridisation between walleye and sauger has been recorded (Billington *et al.* 1988) (see Chapter 11) and they probably diverged in recent time. Lynch *et al.* (1982) have bred 'saugeye' from female walleye crossed with male sauger. These saugeye grew and developed satisfactorily and by the end of the third year they had reached an average length of 350 to 400 mm. Other attempts at hybridisation between different species have met with less success. In cases where the embryo developed, such as a cross between *P. fluviatilis* and *S. lucioperca*, the larvae died after hatching (Kammerer 1907; Balon 1956). In a cross between *P. fluviatilis* and *Gymnocephalus schraetser* embryos developed but did not hatch (Kammerer 1907). Hubbs (1971) tried to cross *P. flavescens* with a number of percids including *S. vitreum*, *Percina sciera* (Swain) and *Etheostoma spectabile* (Agassiz) but the progeny failed to develop or died after hatching. Closely related species can breed (*P. fluviatilis* × *P. flavescens*, see Chapter 11) and produce hybrids but this is not possible between more widely separated species. Success with crosses is mainly achieved between species of the same genera.

Chapter 2
Morphology and Anatomy

2.1 Introduction

Fundamental to an understanding of the biology of fishes is an appreciation of their anatomy and morphology. This appreciation is essential in the study of taxonomy, systematics and form and function. Anatomically and morphologically the percids are advanced perciform teleosts, well adapted, through their response to external stimuli and in maintaining internal homeostasis, to life in freshwater.

2.2 External anatomy

The great variability in colour of *Perca* and *Stizostedion* species depends on the habitat in which they live. *Perca* species are among the most attractively coloured fish in temperate fresh waters. In shallow areas where light penetration is good *Perca* species tend to be darkly coloured whereas in poorly lit areas without vegetation they are lightly coloured. Sometimes carotinoids, derived from crustaceans in the diet, make the perch deeply reddish-yellow. Usually the dorsal surface is bright green to olive and this colour is extended down the sides in seven tapering bars (Plates 1 & 2; Fig. 2.1). The sides of the body are yellow to yellow green and the ventral surface grey to white. The eyes are green to yellow in colour, as are the dorsal and caudal fins. The first spine of the dorsal fin is often black and there is black colouration on the membrane between spines one and two and on the membranes between the last four or five spines. The pectoral fins are amber and transparent and the pelvic and anal fins are silver white to yellow and opaque. The body of the Balkhush perch is usually light coloured and the dark transverse bands are absent (Berg 1965). Some perch have been found with entirely black bodies.

All species of *Stizostedion* are highly variable in colour. Their eyes are always silvery due to the reflective nature of the tapetum lucidum. The zander is coloured brownish to brownish-black on the dorsal surface and this extends onto the flanks in eight to twelve characteristic bands which are most apparent in juveniles. The sides of the fish are greenish to blue-grey and the ventral surface is a silvery to dirty white. During spawning this surface becomes marble blue in the males and pure white in the females (Holčik & Mihalik 1970). Both dorsal fins have irregular rows of black

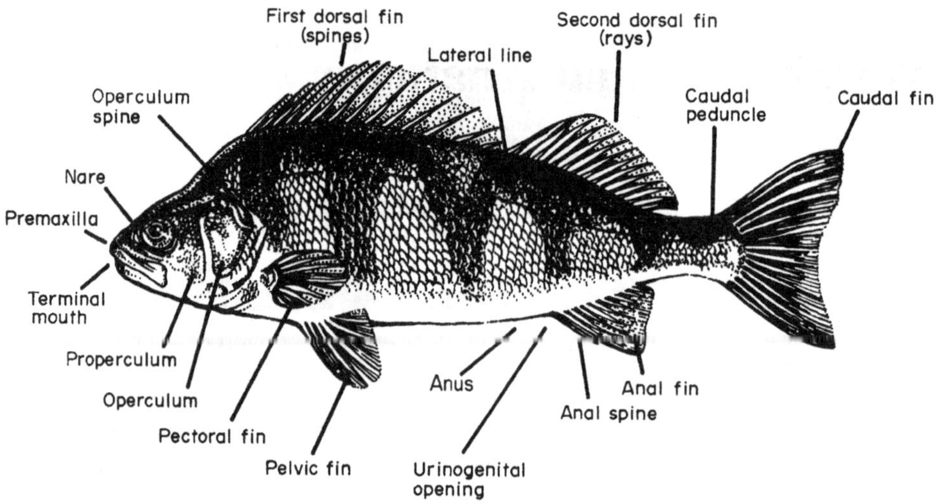

Fig. 2.1 The external features of perch.

spots on the membranes between spines (Fig. 2.2a) and the caudal fin is spotted in transverse rows. Other fins are yellowish grey, the pectoral and pelvic fins being almost transparent. The head has numerous shining blue to brass coloured zones on a marbled brown to grey-green background. Colouration of the Volga pikeperch is similar to the zander but the sea pikeperch is lighter than the other two species (Berg 1965) (Fig. 2.2b). The back of the sea pikeperch is light grey with 12 to 13 transverse dark bars on the body. Quite often these bars are indistinct. The first dorsal fin is usually without spots, dark grey or black in colour. The other fins are grey.

The walleye is paler in turbid waters but in clear waters it is more vividly marked (Scott & Crossman 1973). The body is olive-brown to golden-brown to yellow. The dorsal surface is darker and the flanks paler with golden flecks on the scales. The ventral surface is yellow-white to milk-white. The first dorsal fin is dusky, clear or vaguely speckled and has a characteristic black blotch at the base of the last few membranes (Fig. 2.2c). The second dorsal and caudal fins have small dots of pigment in regular rows. The lower lobe of the caudal fin and the tip of the anal fin are milk-white in colour. The pectoral fins are olive with black dots at the base and the pelvic fins are yellow or orange-yellow.

The sauger is sandy to dull brown. The dorsal surface is usually brown with three or four brown stripes expanding onto the flanks. There are also several large round spots on the sides (Fig. 2.2d). The ventral surface is milk-white. The first dorsal fin has dark pigment on the edge of the membranes and rows of distinct spots below this. There is no black blotch at the base of the posterior membranes as in the walleye. The second dorsal fin of the sauger has two bands of dark spots on the membranes and the caudal fin also has rows of dark brown bars. The lower lobe of

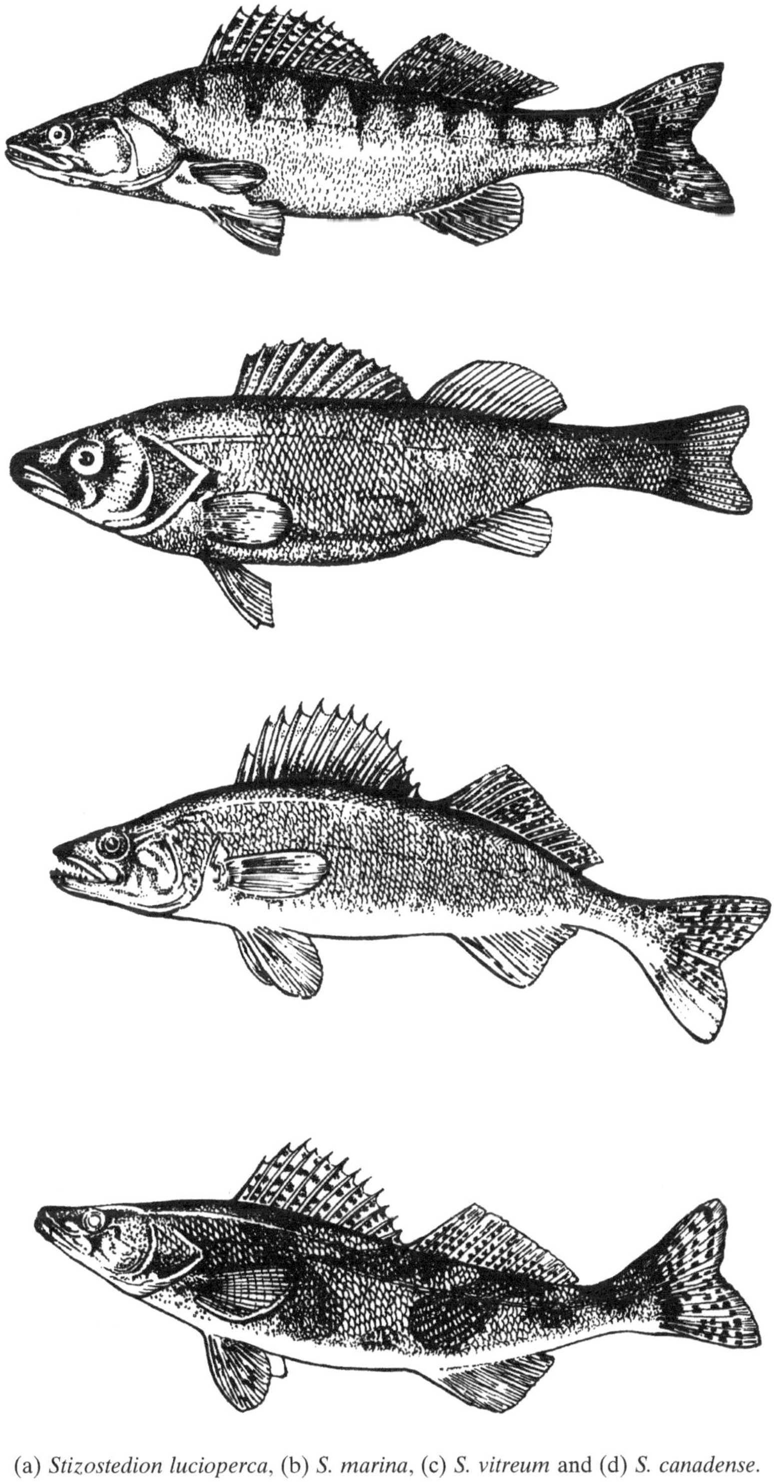

Fig. 2.2 (a) *Stizostedion lucioperca*, (b) *S. marina*, (c) *S. vitreum* and (d) *S. canadense*.

the caudal fin sometimes has a white tip, as does the anal fin, which is white with small spots of pigment. The pelvic fins are also white and speckled with pigment. The pectoral fins are almost transparent and amber with a dark spot at the base of each fin.

The skin of percids is similar to other teleosts consisting of a thin epidermis and a thicker dermis (Greenwood 1975; Whitear 1986). The epidermis, derived from the ectoderm, is composed entirely of living cells. Male perch have thicker skin than females for most of the year (Lindesjöö 1994). Regardless of sex, the epidermis is thickest in the prespawning period while the dermis is thickest after spawning. Ctenoid scales are distinguished on the basis of their surface sculpture but their origin and development are similar to cycloid scales. Ctenoid scales are so called because of the numerous teeth or cteni at the posterior end. These cteni may number up to 70 in zander (Linfield & Rickards 1979). Percidae do not produce a lot of mucous so the body often feels rough to touch due to the presence of these cteni on the scales. The number of scales along the lateral line varies between genera and species (Coburn & Gaglione 1992) (Table 2.1).

Table 2.1 Numbers of scales along the lateral line of species of *Perca* and *Stizostedion*.

Perca	Scales	*Stizostedion*	Scales
P. schrenki	41–54	*S. volgensis*	70–83
P. flavescens	51–61	*S. marina*	75–88
P. fluviatilis	56–77	*S. lucioperca*	80–97
		S. canadense	82–100
		S. vitreum	83–104

2.3 The skeleton

The skeleton of *Perca* species is a compact, sturdy structure (Fig. 2.3a). The outer layer of the skull is made up of dermal bone. The inner skull and other skeletal structures are subdermal in origin and their formation is usually preceded by the development of cartilage (endochondral bones). Osse (1969) has reviewed the functional morphology of the head of perch. The following brief description is based on this work and that of Chiasson (1966) for yellow perch. A summary of the bones of the skull (neurocranium) is given in Table 2.2. The elongated nasal bones are located anteriorly in the olfactory region (Fig. 2.3b). The paired frontal bones form most of the anterior roof of the skull and also the dorsal border of the orbits. The lateral ethmoids are attached to the sides of the frontals and form the anterior edge of the orbits. The lacrimal bones also border the orbit anteriorly and the suborbital bones (four on each side of the head) extend from them around the orbit to the frontal and dermosphenotic bones. In the dorsal median, the large supraoccipital

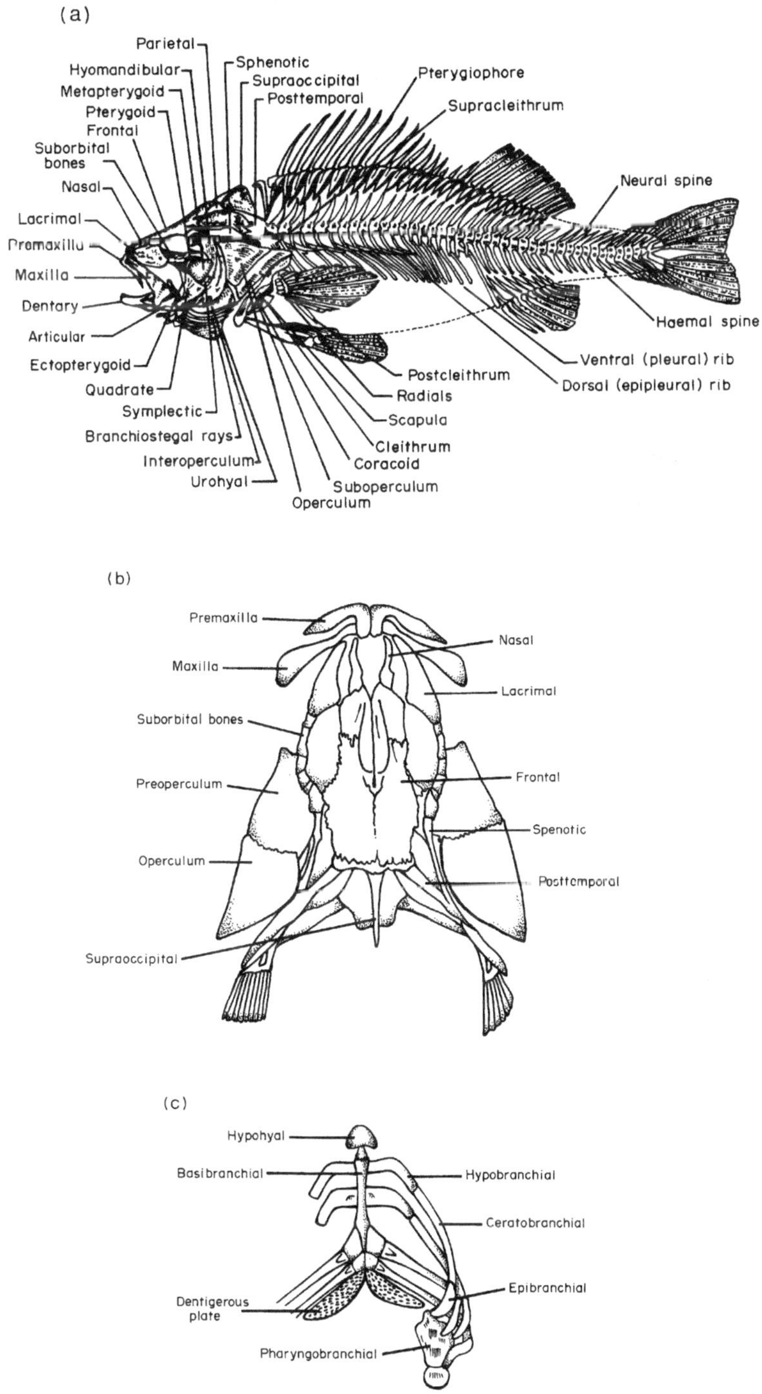

Fig. 2.3 The skeleton of perch: (a) lateral view, (b) dorsal view of the skull and (c) branchial system.

Table 2.2 Bones of the skull or cranium of perch. Dermal bones are in italics and endochondral bones are in normal print.

Neurocranium				Branchiocranium		
Olfactory region	Orbital region	Otic region	Basicranial region	Oromandibular region	Hyoid region	Branchial region
Nasal	*Frontal*	*Parietal*	Parasphenoid	Quadrate	Hyomandibular	Epibranchial
Lateral ethmoid	*Lacrimal*	*Epiotic*	Basioccipital	Articular	Symplectic	Hypobranchial*
Mesethmoid	*Suborbital*	*Scale bone*		*Premaxilla**	*Preopercular*	Ceratobranchial
*Vomer**	*Dermosphenotic*	*Post temporal*		*Maxillar*	*Opercular*	Pharyngobranchial complex*
		Pterotic		*Angular*	*Interopercular*	
		Exocipital		*Dentary**	*Subopercular*	
		Supraoccipital		Articular	Ceratohyal	
		Epiotic		*Palatine**	Epihyal	
		Opisthotic		*Ectopterygoid**	Interhyal	
		Sphenotic		*Metapterygoid*	*Branchiostegal ray* (6)	
		Prootic		*Pterygoid*	*Urohyal*	

* Bones bearing teeth.

crest lies posteriorly and the mesethmoid lies anteriorly. The parietal bones are small and lie posteriorly to the frontals. They are separated in the midline by the dorsal spine of the supraoccipital. In the palate of the mouth, the narrow median parasphenoid extends from the vomer to the basioccipital (see Fig. 2.4d). The narrow palatines also articulate with the vomer posteriorly and laterally and extend back to meet the ectopterygoids. The pterygoid bones form most of the roof of the mouth. The premaxillae and maxillae are both paired bones of the upper jaw. The premaxillae are curved bones which are connected posteriorly and ventrally to the maxillae and the mesethmoid. The maxillae also articulate with the vomer and lacrimal bones. Each side of the lower jaw consists of three bones, the dentary, the articular and the angular which are connected together by cartilage and connective tissue. The articular forms the caudal portion of the lower jaw and at its thickened posterior end it articulates with the quadrate. The articular bone is developed from part of Meckel's cartilage (the primary lower jaw). The angular bone is situated in the posterior-ventral angle of the lower jaw and forms the retro-articular process. The jaw is suspended from the neurocranium by means of the suspensorium. The suspensorium consists of a number of bones forming a V-shaped complex. The bones comprise the palatine, ectopterygoid, metapterygoid, symplectic, pre-opercular, hyomandibular, quadrate and pterygoid. The hyomandibular is located between the quadrate and cranium. It is connected to the cranium by means of the anterior-dorsal ball and socket joint (formed with the splenotic and prootic) and a dorsal ridge-like articulating surface formed partly with the pterotic. The epiotic, opisthotic, prootic and sphenotic contain semicircular canals (see Section 2.8 and Fig. 2.9). The hyomandibular is also jointed with the opercular posteriorly and the interhyal bone at its most ventro-posterior tip. Teeth, villiform and in bands, are located on the premaxilla, palatine, ectopterygoid, dentary and vomer bones (Table 2.2). *Perca* species have no canines. In *Stizostedion* teeth are located on the same bones and some canines are present (except in adult Volga pikeperch).

The opercular series consists of the preopercular, the opercular, the subopercular and the interopercular. The vertical limb of the L-shaped preopercular is firmly connected to the hyomandibular along most of its length and forms part of the suspensorium. The horizontal limb of the bone bears seven spines and the posterior vertical part is serrated. The opercular is a large flat triangular bone. Anteriorly it is connected by skin to the preopercular and in the ventral posterior region the opercular is joined to the subopercular by connective tissue. Dorsally it lies close to the pectoral girdle. The subopercular is a flat elongated bone which forms the border of the gill flap. It is connected to the interopercular by connective tissue.

The hyoid arch is located behind the lower jaw in the floor of the buccal cavity. It bears the six branchiostegal rays. The ceratohyals form the main part of the apparatus and are separated in the anterior midline by the hypohyal. The interhyal suspends the apparatus from the hyomandibular and symplectic.

The five branchial arches form the gill apparatus which is dorsally attached to the ventral side of the skull and ventrally to the copulae. Except for the fifth, each branchial arch bears two gills. The bones are covered on the medial sides with two rows of bony plates each plate bearing numerous recurved teeth. The second and third branchial arches are complete (hypobranchial, ceratobranchial, epibranchial and pharyngobranchial) but the first lacks a pharyngobranchial, the fourth a hypobranchial and the fifth consists only of a ceratobranchial which forms the dentigerous plate (Fig. 2.3c). The pharyngobranchials lie close together in the posterior region of the buccal cavity forming a complex. They are also covered with recurved teeth.

The pectoral girdle forms the posterior section of the opercular chamber. It consists of five pairs of bones: the post temporal, the supracleithrum, the cleithrum, the scapula and the coracoid which join in the ventral midline. The cleithrum is the largest bone of the girdle and of the head region. Anteriorly it is attached to the supracleithrum which in turn articulates with the skull at the post temporal. The pelvic girdle is simple in structure; it is made up of paired pelvic plates or basipterygia which are fused in the midline. The bones are connected to the cleithra by ligaments. The median fins are dermal in origin and are made up of fin rays each of which is supported by a pterygiophore (except the caudal fin) (Fig. 2.3a). There are some variations in the number of spines and rays found in the fins of different species of *Perca* and *Stizostedion* (Table 2.3). The anterior pterygiophores are fused together in the anal fin to form a base for the first two fin rays. The internal spine of this fused structure is connected to the ribs of the vertebral column and the first true haemal arch is just behind this spine.

The vertebral column of *Perca* species is made up of 36–40 vertebrae. The notochord passes through the notochordal canal in the middle of the centrum and the spinal cord passes through the neural arch. The ribs are of two types, ventral (pleural) and dorsal (epipleural) (Fig. 2.3a). The dorsal ribs are attached to the posterior surfaces of the ventral ribs by ligaments but the last seven ventral ribs do not have dorsal ribs attached to them. In the tail region the vertebrae contain haemal arches through which blood vessels pass. The base of the caudal fin consists of compressed neural spines called epurals. The last vertebrae form a plate called the urostyle below which lies a modified haemal arch called the hypural.

2.4 Musculature

The head and body muscles of perch contain four main fibre types: two red and two white (Akster & Osse 1978). These muscle types can be distinguished by ATPase activity, pH stability of ATPase and reactions with specific antisera. White and red muscles are associated with anaerobic and aerobic metabolism respectively. White muscles in the head are only active during rigorous and rapid movements such as eating and coughing. Each type of red muscle can be divided further into two subtypes. In the head, red muscles with a comparatively high ATPase (energy requiring enzymes) activity are functional during rapid respiratory movements. Red

Table 2.3 Number of spines and rays found in the dorsal and anal fins of *Perca* and *Stizostedion* species. (After Scott & Crossman 1973 and Berg 1965.)

	P. schrenki	*P. fluviatilis*	*P. flavescens*	*S. marina*	*S. volgensis*	*S. lucioperca*	*S. vitreum*	*S. canadense*
1st Dorsal spines	12–13	12–13	13–15	13–14	12–16	13–17	12–16	13–15
2nd Dorsal spines	2–3	1–3	1–2	2–4	1–2	1–3	1–2	1–2
2nd Dorsal rays	12–13	13–15	12–15	15–18	20–22	19–27	17–22	16–21
Anal spines	2	2	2	2–3	2	2–3	2	2
Anal rays	7–9	7–10	6–8	10–22	9–10	11–13	11–14	11–14

muscles with low ATPase are associated with little movement and quiet respiration. The myotomes consist mainly of white muscle (an adaptation to sudden movement such as darting at prey) and there is no extensive development of red muscle. Red muscle in the body of teleosts is usually associated with sustained swimmers and in percids may function in the less strenuous aspects of normal activity such as maintaining position or searching behaviour. Akster (1981) has determined that there is a decrease in contraction velocity of muscle fibres from white through to deep red which is in agreement with the activities oulined. Gill *et al.* (1989) described three distinct (red, pink and white) fibre types in the myotomal muscles of yellow perch. White fibres predominated in the myotomal musculature. The proportion of red fibres increased from the anterior to the caudal regions of the body. The caudal region may experience greater flexure during swimming.

Cranial muscles can be divided into jaw muscles, suspensorium muscles, opercular muscles, ventral head muscles and muscles of the branchial arches (Fig. 2.4a-d). The muscles of the branchial arches are part of the pharyngeal jaw apparatus (Lauder 1983). The adductor mandibulae is the most complicated muscle in *Perca* species. It is a large triangular muscle covering most of the suspensorium and is divided into four parts (Fig. 2.4a). The head muscles are important in changing the shape of the head when the fish is capturing prey (see Chapter 5). Muscles of the paired fins comprise both adductors and abductors. The pectoral fin muscles originate on the scapula and coracoid bones and insert on the fin radials, the adductors on the anterior faces of the bones and the abductors medially. The pelvic fin muscles originate on the pelvic plates and insert on the pelvic fin radials, the adductors on the medial region of the bones and the abductors on the outer surfaces. The median fin movements are also controlled by two sets of muscles. The first set is superficial, is found in the skin and originates in the connective tissue and insert on the base of the fin spines. The second group of muscles runs between the pterygiophores. The most anterior of these muscles is attached to the posterior edge of the supraoccipital bones. The caudal fin is controlled by the trunk myomeres. The body musculature is made up of W-shaped myotomes with the two bottom points of the W facing to the posterior (Chiasson 1966).

2.5 The gills

The pharynx wall is pierced by five pairs of gill slits (see Fig. 2.7a), the first between the last branchiostegal rays and the first branchial arches, the last between the last branchial arches and the dentigerous plate and the rest between the branchial arches. The gills are attached to the four branchial arches and are covered laterally by the opercular shield. Each gill is composed of two hemibranchs, one anterior and one posterior making up a holobranch (Hughes 1984). Each hemibranch is made up of gill filaments (supported by gill rays) which in turn support lamellae which increase the surface area of the gill. Niimi and Morgan (1980) found that the first two gill arches in walleye contained both the highest number of filaments and lamellae and the largest gill and lamellae surface areas. The gill surface area increased significantly with body mass (Fig. 2.5). The epithelium of the filaments contain mucous, chloride and rodlet cells (Leino 1982; Matei 1984). The distribution and quantity of these cells may vary depending on the environment in which the fish lives (Matei 1984). The branchial arches bear gill rakers on their inner anterior surfaces. The numbers of these gill rakers varies according to species and there is considerable variation in the number within species. These comb-like rakers prevent material passing from the mouth to the delicate lamellae of the filaments and thereby aid in feeding on small particles.

2.6 The gut

The cavities of percids include the peritoneal cavity containing the gut, glands, kidneys, gonads and swim bladder, and the pericardial cavity containing the heart. The gut of a percid (Fig. 2.6) is typical of a carnivorous fish. The oesophagus is short and connects the pharynx with the stomach. The stomach ends posteriorly as a blind sac and the intestine joins the pyloric region of the stomach in a dorsal position. The first section of the intestine is S-shaped, the bottom of the S running to the straight intestine which ends at the anus. At the anterior end of the intestine are the intestinal (pyloric) caecae. There are three of these caecae in *Perca* species, three in walleye and Volga pikeperch, four to nine in zander and five to seven in sea pikeperch. The anterior dorsal region of the intestine receives four to six bile ducts from the gall bladder which in turn is fed by a system of tubules from the liver. The spleen is positioned posteriorly on the dorsal surface of the stomach and the pancreas is located on the ventral side of the anterior part of the intestine. The swim bladder is found dorsally in the peritoneal cavity below the kidneys. It is connected to the oesophagus by the pneumatic duct which is closed in adult percids. The swim bladder acts as a hydrostatic organ.

The gut wall is made up of an inner mucosa layer and an outer submucosa layer. The mucosa layer contains various secretory and absorptive cells and the submucosa is made up of blood vessels, nerves, muscle fibres and connective tissue (Kapoor *et al.* 1975). The mucosa of the oesophagus contains basal epithelial, surface epithelial and mucous cells. The muscle of the submucosa is composed of an inner layer of longitudinal striated muscle fibres surrounded by an outer layer of circular muscle containing both striated and smooth fibres (Hirji 1983). Towards the stomach the longitudinal muscles cease and the muscle layer is represented solely by circular muscle. The stomach is lined by circular muscle with an outer narrow layer of longitudinal muscle. An inner circular and an outer longitudinal muscle layer surround the intestine. Noaillac-Depeyre and Gas (1978) have described the cytology of the gastric mucosa, occurring in the stomach. Cell types include mucous, mucous neck, oxytic and endocrine. The endocrine cells found between the mucoid

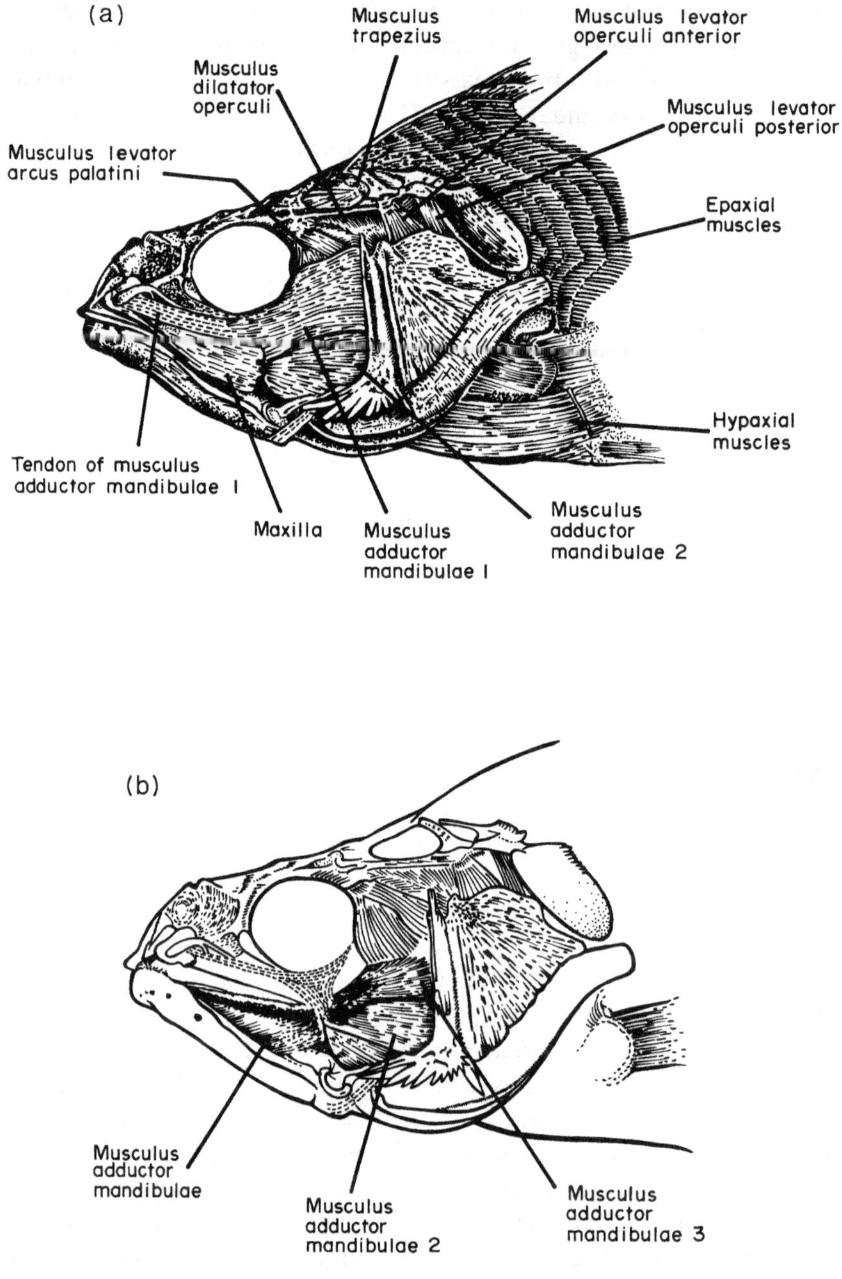

Fig. 2.4 The head musculature of perch: (a) lateral view of the left superficial head muscles, (b) lateral view of the head muscles with the musculus adductor mandibulae removed, (c) lateral view of the aponeurotic system of the left musculus adductor mandibulae and (d) ventral view of the skull showing the adductor muscles (lower jaw, hyoid and branchial arches have been removed). (After Osse 1969.)

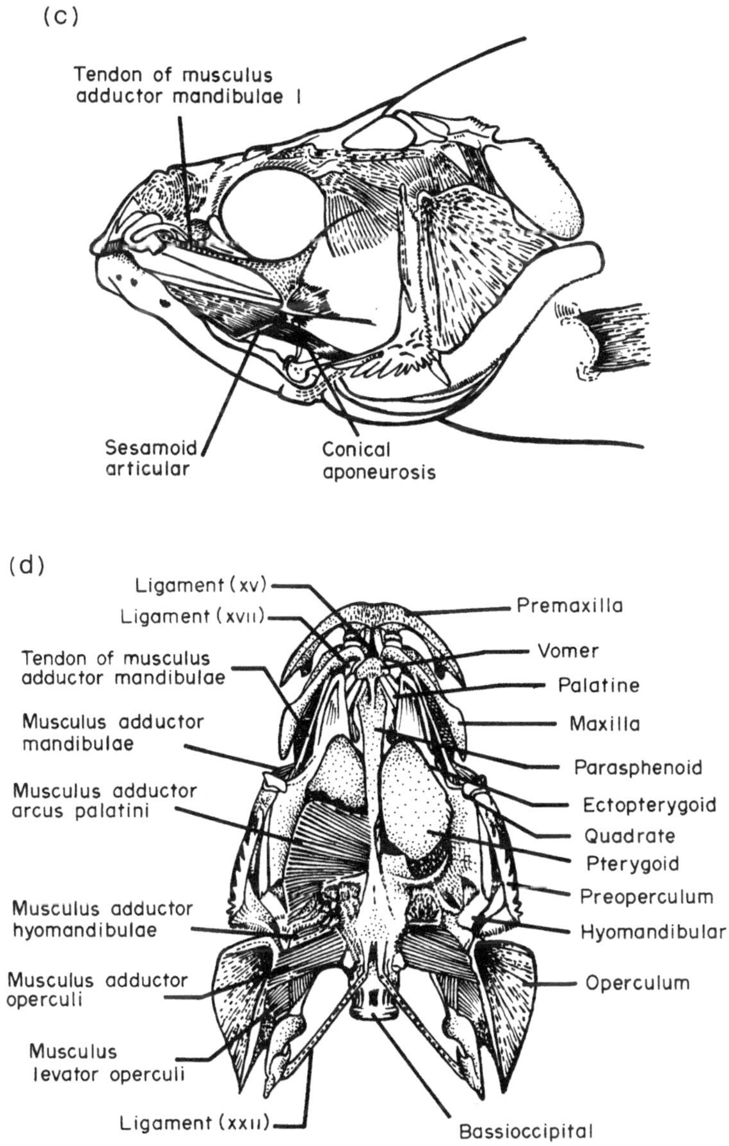

cells are of three types distinguished by the type of secretory granules they produce (Noaillac-Depeyre & Gas 1982). The intestine can be divided up into three segments based on the nature of the absorptive epithelium (Noaillac-Depeyre & Gas 1979). Enterocytes concerned with lipid absorption occur in the proximal section while those in the middle section undertake pinocytosis of exogenous proteins. The distal segment contains absorptive cells with short microvilli.

Rodlet cells are pear-shaped secretory cells found in the epithelium of various organs in teleosts. In the gut of perch rodlet cells are found in the mucosa of the

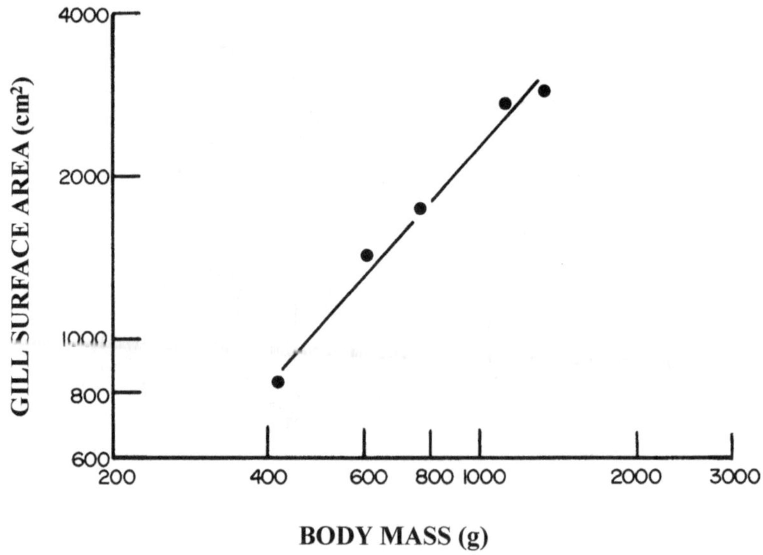

Fig. 2.5 The relationship between the gill surface area and body mass of walleye. (After Niimi & Morgan 1980.)

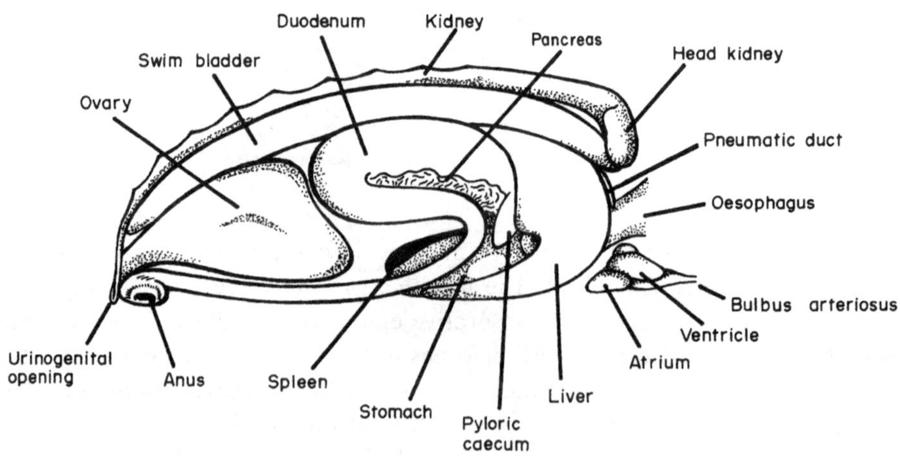

Fig. 2.6 Lateral view of the body cavity of yellow perch.

stomach, intestinal caecae and anterior region of the intestine (Hirji & Courtney 1979).

2.7 Circulatory system

The heart is located in the pericardial cavity and consists of four chambers: the sinus venosus, atrium, ventricle and bulbus arteriosus (Fig. 2.6 and see also Fig. 2.7c). Blood passes from the bulbus arteriosus to the ventral aorta and is distributed to the gills via the afferent branchial arteries (Fig. 2.7a). These arteries divide up to form capillaries within the gill filaments. Oxygenated blood is then passed to the dorsal aorta from the efferent branchial arteries. The internal carotid arteries, serving the brain, extend from the first efferent branchial arches. The dorsal aorta passes along the trunk of the fish and through the haemal arches into the tail where it becomes the caudal artery. The dorsal aorta and caudal artery serve the viscera (Fig. 2.7b) and body muscles with oxygenated blood. The anterior, cranial cardinal veins drain blood from the head and the two posterior, caudal cardinal veins from the muscle, backbone and kidneys. These veins empty into the two common cardinal veins (duct of Cuvier) which, with the hepatic veins (two pairs), drain into the sinus venosus (Fig. 2.7c). The liver receives blood from the gut via the hepatic portal vein.

In a study of the haematology of perch, zander and ruffe from the River Danube (Table 2.4) Ivanc *et al.* (1996) found significant differences between species which they related to their distribution and ecology especially the ability of perch and ruffe to live in areas with low oxygen levels.

2.8 Excretory and reproductive systems

The kidneys of percids are paired and elongated and lie pressed against the dorsal body wall in the peritoneal cavity (Fig. 2.6). The kidneys are separated anteriorly by the dorsal aorta but posteriorly they are fused in the midline. Each kidney is in two parts, the head kidney at the anterior end and the trunk kidney posterior to it. The head kidney is not directly concerned with ion or water regulation. It contains lymphoid, hematopoietic, interrenal and suprarenal tissue. The trunk region of the kidney is directly involved with excretion and osmotic regulation (see Chapter 6). Distributed throughout the trunk region are clusters of nephrons each cluster fed by an artery and drained by a vein. The nephron consists of (a) a renal corpuscle containing a well-vascularised glomerulus, (b) a ciliated neck region, (c) an initial proximal segment with prominent brush border and numerous prominent lysosomes, (d) a second proximal segment with numerous mitochondria but less well developed brush border, (e) a narrow ciliated intermediate segment, (f) a distal segment with elongate mitochondria and (g) a collecting duct (for detail, see Fig. 6.3) (Hickman & Trump 1969). The collecting ducts (nephronic ducts) empty into

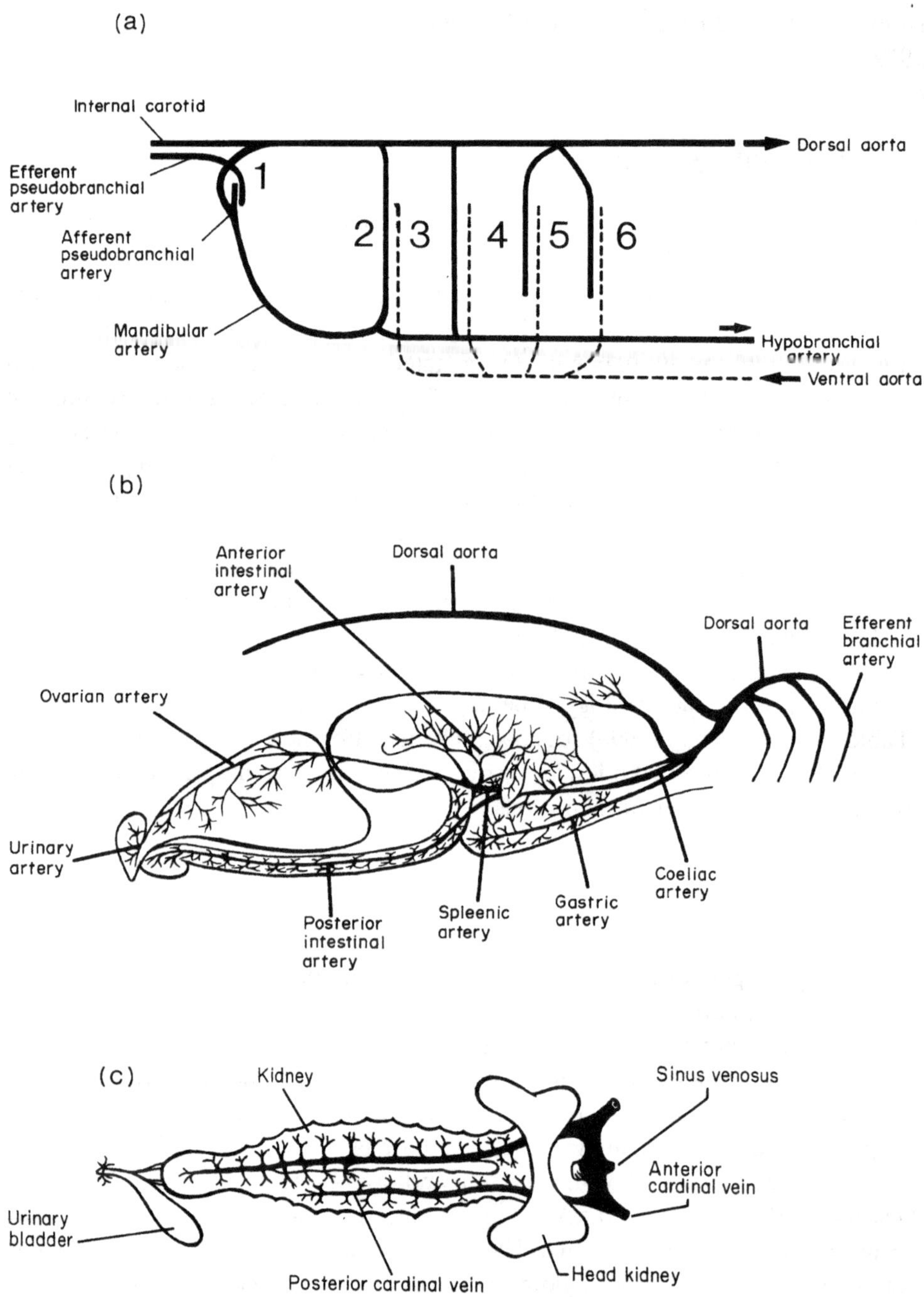

Fig. 2.7 The circulatory system of perch: (a) the gills (the branchial slits are 1, spiracle, 2, hyoidean and 3–6 aortic), (b) the viscera and (c) kidneys of yellow perch.

Table 2.4 Mean value, standard deviation and range of body mass, total length, red blood cell count (RBC), haemoglobin concentration, haematocrit (Hct), mean corpuscular volume (MCV), mean corpuscular haemoglobin (MCH) and mean corpuscular haemoglobin concentration (MCHC) for *Stizostedion lucioperca*, *Perca fluviatilis* and *Gymnocephalus cernuus*. The season caught and the number of fish examined are also given. Letters a, b and c denote species in which the mean of the parameter differs significantly ($p < 0.05$). (After Ivanc *et al.* 1996, with kind permission from Schweizerbart'sche Verlasbuchhandlung.)

Species and season	No.		Body mass (g)	Total body length (mm)	RBC (count l^{-1})	Haemoglobin concentration ($g\,l^{-1}$)	Hct ($l\,l^{-1}$)	MCV (fl)	MCH ($pg\,l^{-1}$)	MCHHC ($g\,Hb\,l^{-1}$)
S. lucioperca	12	Mean	462	355	1.821E + 12	56.99 b	0.430 c	240.72 c	31.66 b,c	134.74 b,c
		SD	377	90	2.587E + 11	5.04	0.050	42.13	3.41	23.78
Autumn		Range	128–1258	265–520	1.513E + 12–2.200E + 12	49.60–66.19	0.321–0.492	145.91–307.50	25.89–37.92	113.36–206.20
P. fluviatilis	10	Mean	109	186	1.630E + 12 c	79.69 a,c	0.431 c	265.19 c	49.19 a,c	185.72 a,c
Autumn		SD	92	44	1.179E + 11	7.42	0.041	29.73	6.21	14.32
		Range	26–325	130–265	1.467E + 12–1.835E + 12	68.35–94.38	0.389–0.487	228.83–317.91	39.54–59.92	155.71–203.44
G. cernuus	7	Mean	29	125	2.002E + 12 b	55.97 b	0.376 a,b	171.47 a,b	27.00 a,b	158.09 a,b
		SD	12	18	3.055E + 11	7.65	0.024	17.61	2.38	12.44
Spring		Range	15–48	108–165	1.400E + 12–2.30 E + 12	42.30–63.45	0.354–0.425	155.18–208.13	23.85–31.07	142.63–174.63

branches of the ureter. Each ureter (Wolffian duct) runs ventrally in the trunk kidney and opens into the urinary bladder (Fig. 2.7c). In the female perch the urinary bladder is fused with the oviduct to form a urogenital sinus. There is no cloaca in adult perch. Externally there is a urogenital papilla which in the male is an obvious extension of the urinary bladder. Excretions are passed through a urinary pore in the papilla.

The testes of the male are paired and are located in the midline of the peritoneal cavity between the gut and swim bladder. They are united by a septum for about one third of their length at their posterior end. The testes have a number of longitudinal folds and the vasa deferentia run from these folds and eventually join at the posterior midline. The single duct from the two testes empties at the base of the urinary bladder. Spermatozoa are produced in a zone of the testes composed of numerous tubules. The tubules consist of a layer of germinal epithelial cells and an outer layer of connective tissue. The spermatic artery lies in a dorsal groove of the testes. The ovaries of *Perca* and *Stizostedion* species are also paired but during development in the former these ovaries fuse together to form a single ovary (Fig. 2.6). This ovary is located in a similar position to the testes. At the anterior end the ovary is blunt and rounded but posteriorly it is tapered. In *Stizostedion* species the paired ovaries unite posteriorly. Often the right ovary is larger than the left. The ovary and oviduct are enveloped in a fold from the peritoneum. The ovary wall consists of three parts: an outer epithelial layer (peritoneum), a layer of connective tissue, and an inner germinal epithelium which gives rise to the ova. There are transverse blood vessels throughout the length of the ovaries of *Perca* and *Stizostedion* species which becomes more conspicuous with maturity. Hermaphrodites have been reported for both *Perca* (Chevey 1922; Turner 1927; Brunelli & Rizzo 1928; Jellyman 1976) and *Stizostedion* (Dence 1938) species.

The diploid chromosome number $2n = 48$ has been determined for many of the Eurasian and North American percids (Danzmann 1979; Mayr *et al.* 1987; Ráb *et al.* 1987).

2.9 The nervous system

The nervous system of teleosts consists of the central (brain and spinal cord), the peripheral (cranial and spinal nerves) and the autonomic (parasympathetic and sympathetic) nervous systems and sensory organs (nose, ear, eye and lateral line). The brain is divided up into the forebrain (prosencephalon), the midbrain (mesencephalon) and the hindbrain (rhombencephalon). The forebrain consists of the telencephalon and diencephalon. The paired olfactory bulbs are distinct posteriorly from the telencephalon but anteriorly they blend with the telencephalon to form the olfactory nerves (Fig. 2.8a). The telencephalon contains the two most anterior ventricles. The diencephalon consists of the epithalmus, thalmus (divided into dorsal thalmus and ventral thalmus) and the hypothalmus. The floor of the

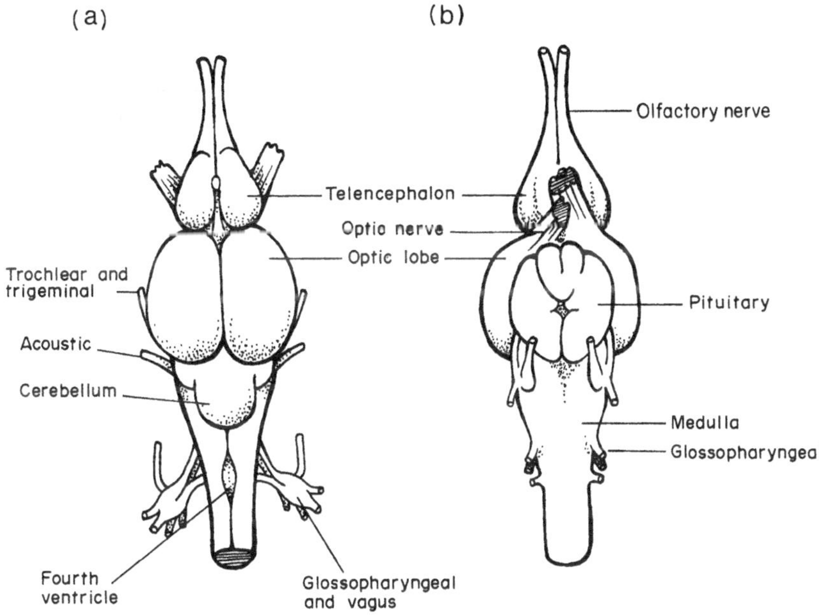

Fig. 2.8 The brain of yellow perch: (a) dorsal and (b) ventral views.

diencephalon is formed by the hypothalmus which is connected to the well developed pituitary (Fig. 2.8b). The thalmus of the diencephalon is joined to the telencephalon, hypothalmus and mid- and hindbrains (Dodd & Kerr 1963). The epithalmus is made up of the pineal organ and the habenular nuclei. The midbrain is composed of the optic lobes and is divided into the optic tectum and tegmentum. The optic tectum forms the roof of the third ventricle and the tegmentum the base of the ventricle. A channel runs forward in the midline to connect with the telencephalon ventricles and posteriorly through the aqueduct of Sylvias to the fourth ventricle. The hindbrain consists of the cerebellum and medullar oblongata. The latter merges into the spinal cord. The fourth ventricle is located dorsally in the rhombencephalon. The cranial nerves and their functions are listed in Table 2.5; a detailed description of perch gill innervation is given by Dunel-Erb *et al.* 1993. The spinal cord runs the length of the body of the fish through the neural arches of the vertebrae and ends in the urophysis, a neuroendocrine organ which has an osmoregulatory function. The spinal nerves are paired and pass from the spinal cord through foraminae in the vertebrae. The first three spinal nerves run to the muscles of the pectoral and pelvic girdles. The first spinal nerve also has a dorsal branch, which serves the back of the cranium, and a ventro-anterior branch which innervates the hyoid muscles (Chiasson 1966).

Percids like other teleosts have a ganglionated sympathetic nerve chain. This is found medial to the ribs within the body cavity extending caudally to the tail. The ganglia, two per spinal segment, are connected with the spinal nerves by both grey

Table 2.5 Cranial nerves of *Perca flavescens*. (After Chiasson 1966.)

Nerve	Name	Origin	Function	Serving
O	Nervus terminalis.	Ventral border of olfactory bulb.	Sensory	Sense endings of snout and olfactory epithelium.
I	Olfactory.	Anterior border of olfactory bulb.	Sensory.	Nasal epithelium.
II	Optic.	Optic lobes, ventral to telencephalon.	Sensory.	Retina of eye.
III	Oculomotor.	Posterior ventral end of medulla, between optic lobes and cerebellum.	Motor.	Superior and inferior rectus and inferior oblique muscles.
IV	Trochlear.	Dorso-lateral anterior surface of medulla.	Motor.	Superior oblique muscles.
V	Trigeminal.	Anterior lateral border of medulla oblongata.	Motor and sensory.	Jaw muscles, tactile endings in skin of the head.
VI	Abducens.	Anterior ventral end of medulla oblongata.	Motor.	Lateral rectus muscle and rectractor bulbi muscles in part.
VII	Facial portion of Acoustico-facialis.	Lateral border of medulla oblongata just posterior to the Trigeminal and with Acoustic.	Motor and sensory.	Skin of head, taste endings and lateral line organs.
VIII	Acoustic portion of Acoustico-facialis.	With Facial from lateral border of medulla oblongata.	Sensory.	Inner ear and lateral line organ.
IX	Glossopharyngeal.	Posterior lateral border of medulla with Vagus.	Motor and sensory.	Gill muscles and lateral line organ.
X	Vagus.	Posterior lateral border of medulla with Glossopharyngeal.	Motor and sensory.	Gills, heart, lateral line organ and anterior alimentary tract.

and white rami communicantes after the dorsal and ventral roots of the nerve have joined. Both afferent and efferent sympathetic fibres pass through the ventral root. The major part of the parasympathetic system is represented by the vagus nerve and to a lesser extent by the oculomotor.

Chemoreception is achieved by taste, olfaction and an overall 'body sense' (Hara 1971). Taste is carried out through receptors in the mouth, pharynx, gill cavity and gill arches. The olfactory organs of yellow perch, innervated by the first cranial nerve, are oval and rosette shaped and nearly fill two cartilaginous chambers just under the rostrodorsal surface of the head (Lyons 1983). The rosette consists of a number of neuroepithelial lamellae radiating from a midline raphe. There are 14 to 16 of these primary lamellae. The olfactory epithelium is developed best in the proximal portions of the lamellae and consists of basal, sensory and ciliated sustentacular cells. The distal section of the lamellae are made up of mucous cells and undifferentiated epithelium.

The ear of percids, as in all fishes, consists of the labyrinth or inner ear only (Fig. 2.9) (Lowenstein 1971). The anterior vertical semicircular canal, utriculus and sacculus are located inside the cranial cavity and the horizontal and posterior vertical semicircular canals are embedded in the skeleton (prootic, opisthotic and epiotic). Branches of nerve VIII (Table 2.5) serve the laybrinth. The semicircular canals are filled with labyrinthine fluid. At one end the canals enlarge to form the spherical ampullae which contain a sensory apparatus. This apparatus consists of the crista ampullaris which is covered by an epithelium containing sensory cells. Otoliths are found in the utriculus, sacculus and lagena. The otoliths, composed of calcium salts, are heavier than the surrounding endolymph and lie above a sensory epithelium. When the epithelium deviates from the horizontal plane the otolith slides along it and activates the sensory cells. The otolith from the sacculus is used to determine the age of many species of fishes but has not been used extensively for percids. The labyrinth functions as a sound receptor (in conjunction with the swim bladder), as a gravity receptor, as a receptor for angular accelerations, and in the maintenance and regulation of muscle tone. The hearing thresholds of perch are approximately in the range of 1–1000 Hz (Enger *et al.* 1993).

The eyes of species of *Perca* and *Stizostedion* are located within the orbits and moved by the inferior, superior, lateral and medial recto muscles and the superior

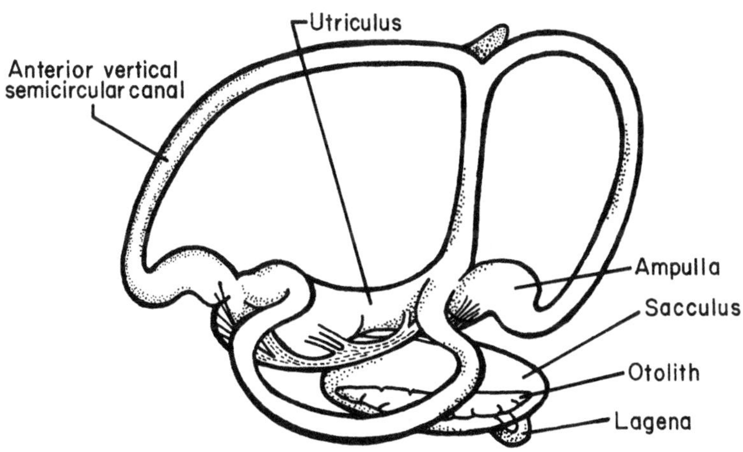

Fig. 2.9 The ear of perch.

and inferior oblique muscles. The eyeball is covered by the sclera, which is thin and transparent in the anterior region to form the cornea. The choroid layer, which forms the heavily pigmented iris proximal to the cornea, is inside the sclera coat. The lens is spherical and suspended from the anterior edge of the retina by a suspensory ligament. The visual acuity of perch increases exponentially with age, mainly as a result of lens development and increase in focal length (Guma'a 1982; Miller *et al.* 1993). The pupil of the perch is egg-shaped (the point directed forward). The eyeball is filled with vitreous humour. The retina lines the inside of the eyeball in the posterior region. The organisational structure of the retina varies between *Perca* and *Stizostedion* species; the former is active during the day while the latter is mainly nocturnal. Before metamorphosis the retina of perch contains only cones; rods appear at or after this stage (Guma'a 1982). The adult fish has a retina of the duplex type made up of eight layers of cells and membranes (Blaxter 1968). The distribution of cones in the retina indicates the importance of acute vision in perch and yellow perch. The cones are arranged in a regular square mosaic consisting of identical twin cones surrounding central single cones at the corner positions. The absorbance maximum for the identical twin cones is at 620 nm while the absorbance of the pigment at the central single cone is 535 nm (Loew & Wahl 1991) The retinal cone photoreceptors of larval and juvenile yellow perch contain a visual pigment with a peak absorbance near 400 nm (near-ultra violet) which disappears in adults (Loew *et al.* 1993). The disappearance coincides with a change in behaviour from feeding on zooplankton near the surface to a dermersal omnivorous habit. The highest density of cones is in a temporal region of the retina. The cones of walleye and zander are arranged for increased sensitivity rather than visual acuity and are organised in a predominantly row-like pattern with single, double and triple cone types. Atypical retinomotor movements were observed in a localised area of the walleye retina (Wahl 1994). The twisting resulted in a local, cyclic change of the cone mosaic from the predominately row type to the square type. Only single cones are present in zander <8 mm (Ahlbert 1969). There are two forms of cone in the retina of adult walleye, midwave and longwave with maximum sensitivity at 533 and 605 nm respectively (Burkhardt & Hassin 1983). The rods of *Perca* species are relatively large and evenly spaced while those of *Stizostedion* species are relatively small and grouped (Ali *et al.* 1977). The tapetum lucidum is a distinctive feature of the eye of *Stizostedion* species. Moore (1944) first described it in detail for North American species and Ali and Anctil (1968; 1977) and Zyznar and Ali (1975) have carried out further investigations. The tapetum lucidum is a reflective layer found in the pigmented epithelium layer of the retina. The reflecting material in the cells is a reduced pteridine, 7, 8-dihydroxanthopterin. In dark-adapted eyes the rods are in groups optically isolated from each other by the tapetal processes. Light can be reflected back and forth between the tapetal processes with additional absorption by the rods after each reflection (Zyznar & Ali 1975). European species of *Stizostedion* have a relatively smaller effective area for reflection by the tapetum lucidum than North American species. Aggregation of photoreceptor cells in the retina of juvenile walleye into groups of 20–30, macroreceptors, occurs at *c.* 60 mm TL. These macroreceptors are thought to increase acuity at low light levels and provide walleye with the ability to feed in turbid water (Vandenbyllaardt *et al.* 1991).

The lateral line system is of considerable importance as a sensory organ to teleosts in general (Disler 1950; Flock 1971; Disler & Smirnov 1977). Its main function is that of a tactile receptor for the detection of water displacements. It acts as a near field receptor operating at low frequencies and can be considered as an acoustic sense organ. Sensory hairs from a receptor cell are attached to a cupular that rests upon a sensory epithelium. The sensory cells or neuromasts may either be at the bottom of the canal sunk below the skin (these are particularly concentrated in the head region) or as epidermal organs situated at the skin surface with the cupula extending into the water. The yellow perch has a lateral line composed of neuromasts enclosed in narrow channels freely open to the water, typical of active diurnal, shallow water fishes. The ruffe has a lateral line similar to deep-sea fishes, with very wide canals, large neuromasts and membranes covering the canal openings (Janssen 1997; see Chapter 13). The lateral line runs from behind the head to the base of the caudal fin and extends over the scale portion of the fin in *Stizostedion* species. In members of both *Perca* and *Stizostedion* it is curved upwards following the line of the back. The acoustic nerve VIII acts as the peripheral sensory supply to the lateral line (Pearson 1936). It is also served by nerves VII, IX and X (Table 2.5).

2.10 The endocrine system

The endocrine system of teleosts is discussed in Hoar and Randall (1969) and Baker and Wigham (1979) give an account of endocrine control of metabolic processes. Endocrine glands of teleosts include the pituitary, pancreas, thyroid, adrenal, ultimobranchial, pineal body, urophysis, ovary and testes. Through a feedback system the large pituitary gland, found attached to the hypothalmus below the brain, controls other endocrine glands and with its close contact with the nervous system is probably the most important gland. The pituitary consists of two parts: the neurohypothesis, formed from a down growth of the floor of the diencephalon, and the adenohypophysis, originating embryonically from epidermal tissue in the buccal cavity. The pancreas is concentrated into islets bearing three types of endocrine cells. Type A produce glucagon, type B insulin and type D, somatostatin (Epple 1969; Matty 1985; Evans 1993). Other hormones are probably also produced by the pancreas.

The thyroid gland reacts to levels of pituitary thyrotrophic hormone (TSH). It accumulates iodine and produces the tyrosine-derived hormone, thyroxin. Thyroxin is manufactured by secretory follicles, which are scattered in connective tissue throughout the pharyngeal region. The adrenal gland is closely associated with the kidney and is made up of interrenal, suprarenal (chromaffin tissue) and stannius corpuscle components. The interrenal organ produces steroids. The interrenal and suprarenal glands are found in the head kidney and develop from coelomic epithelium. The interrenal tissue of zander is located as two clusters around each posterior cardinal vein. Suprarenal tissue is also found near the posterior cardinal vein and its branches, usually where there is no interrenal tissue. With a few minor exceptions, the same system is found in perch and other percids (Mezhnin 1972). Stannius corpuscles are developed from outgrowths of the walls of Wolffian ducts. Three to five corpuscles lie in the zander trunk kidney close to its connection with the head kidney. In the Volga pikeperch, three corpuscles lie in the caudal region of the trunk kidney and in perch two corpuscles lie in a similar region (Mezhnin 1979). The adrenal tissue undergoes changes with the reproductive cycle in perch (Mezhnin 1977) although its main function is water and ion balance. The ultimobranchial gland found in the last gill arch is responsible for calcium regulation. Hormones involved in the reproductive cycle have been described by Dabrowski *et al.* (1996) (see Chapters 4 and 11).

Chapter 3
Age, Growth and Mortality

3.1 Introduction

Percids are temperate mesotherms living in a wide range of environmental conditions but in particular shallow and moderately productive freshwaters. They are able to withstand low temperatures including those near freezing. They may adapt to these conditions by increasing heart rate (Sephton & Driedzic 1991; Driedzic *et al.* 1996). They are also able to tolerate high temperatures. Growth studies of fish, in the form of measuring increases in length or mass, are common in the literature but the underlying processes of growth are complex. Growth starts from division of the fertilised oocyte and continues through the absorption of the yolk sac, exogenous feeding and in many fishes throughout life. Several stages occur between feeding and eventual tissue deposition and these stages are under physiological control which in turn are influenced by internal and external factors (Bakanov *et al.* 1987; see Chapter 5). An understanding of percid growth is essential in these economically important fish both in the management of natural populations and in aquaculture production.

3.2 Internal and external factors

Rising water temperatures and increasing day lengths may stimulate increased activity of the pituitary gland which produces both thyrotropic and gonadotropic hormones. Pituitary growth, anabolic steroid and thyroid hormones, and possibly insulin, stimulate growth (Donaldson *et al.* 1979). Swift and Pickford (1965) studied the seasonal effects on growth of the hormones secreted by the perch pituitary gland. An assay technique was used in which male *Fundulus heteroclitus* (Linnaeus) were injected with a hormone preparation. In the winter months when there was little growth, the level of the hormone was low but with the onset of growth in the spring the amount of hormone increased ten-fold to a peak in June. It then declined to a minimum in August, which was well below the winter minimum value. Growth responses to the growth hormone produced by α cells in the pars distalis of the pituitary gland are dependent on temperature which is the most important external controlling factor. Not only does temperature affect hormone production but also it affects all body processes by its direct influence on enzymes. Essentially, it controls growth, feeding, reproduction, activity, survival and distribution. Hokanson (1977a) has given an account of the temperature requirements of percids. Pike, *Esox lucius* Linnaeus, having a physiological optimum temperature of 26°C close to that of percids, also belong to this group of temperate mesotherms. Other temperate fish are either stenotherms, for example the rainbow trout, *Oncorhynchus mykiss* Richardson, with a physiological optimum temperature of 17°C, or eurytherms, for example the carp, *Cyprinus carpio* Linnaeus, with a physiological optimum temperature of 32°C. The ability of the major *Perca* and *Stizostedion* species to tolerate high temperatures is reduced in the order zander, perch, yellow perch, walleye and sauger (Table 3.1). The upper lethal temperatures derived for percids show some variation in the literature due to the methods and the size of fish used in their derivation. The critical thermal maximum is obtained by gradually increasing the temperature until death and is usually several degrees above the incipient lethal temperature which is determined from sudden exposure of the fish to increased temperature (Fry 1947). Thermal tolerance is minimal at embryogenesis, maximal during fry development and then gradually declines in adults (Lapkin *et al.* 1981).

Table 3.1 Physiological optima, incipient lethal and critical thermal maxima (°C) for species of *Stizostedion* and *Perca* (*not determined). (After Hokanson 1977a.)

Species	Physiological optimum (°C)	Incipient lethal (°C)	Critical thermal maximum (°C)
S. canadense	22.0	30.4	*
S. vitreum	22.6	31.6	*
P. flavescens	24.7	29.2–34.0	33.4
P. fluviatilis	25.4	31.4–33.5	35.5–35.9
S. lucioperca	27.0	34.3–35.0	37.0

Willemsen (1977a) found that perch and zander congregated near the cooling water discharge of the Flevo power plant on Lake Ijssel in the Netherlands. Experiments showed that the fish were able to feed at high temperatures there although in the case of perch they were near to the upper lethal limits. Willemsen recorded only a mortality of five per cent at 29°C and eight per cent at 30.5°C for perch and no mortalities for zander below 32°C. Although these fish were able to feed and survive at such temperatures it is probable that they were in a state of stress and more vulnerable to disease. Weatherley (1963) found that interrenal tissue in perch kidney underwent atrophy when the fish were exposed to temperatures reaching 30–31°C. The interrenal tissue functions in a similar way to the adrenocortical tissue of mammals in opposing stress. Studies using telemetry have shown that percids, as a rule, do not make use of these artificial warmed areas in their environment (Ross & Siniff 1980; 1982; Maclean *et al.* 1982). In Lake Monona,

USA, maximum temperatures in the outflow from the power plant reached 35°C in the summer and 14°C in the winter. The unheated parts of the lake rarely exceeded 29°C in the summer (Neill & Magnuson 1974). Yellow perch avoided the heated effluent and those caught had a body temperature <29°C. There is evidence that at extremes of temperature the avoidance of the temperature overrides other factors such as food availability or hunger of the fish (Neill & Magnuson 1974). Percids thus show the ability to select out preferred temperatures for growth and reproduction. This ability to thermoregulate has been demonstrated experimentally by Reynolds and Casterline (1979). In their experiment yellow perch exhibited diel patterns of behaviour selecting a maximum of 16.7°C at predawn and 23.8°C at dusk. The overall mean for 24 h was 20.2°C. Although these experiments were held in April under natural photoperiod and the yellow perch would be unlikely to encounter such temperatures at this time, the selected temperature values are close to the physiological optimum for yellow perch (Table 3.1).

Light has an important influence on the body mechanisms and organisation in nearly all organisms. It is probably, apart from temperature, ration size and size of fish, the most important factor governing metabolic activity, feeding, growth and reproduction (Ryder 1977; Karås 1990; Wieser & Medgyesy 1991). The effect of light on hormones in teleosts has not been well studied and there is little information for percids. Seasonal differences in the mobilisation or deposition of food stores depend on the nutritional status of the fish and the action of individual hormones such as growth hormone, thyroxin and insulin. It is also important to have synergistic co-operation between hormones which is, theoretically, brought about as a result of environmental influences (Baker & Wigham 1979). Hormones such as prolactin, cortisol and thyroxin fluctuate diurnally. Additionally, peaks of activity for these hormones are brought in and out of phase at different times of the year. Hormone levels may increase in the blood plasma in response to photoperiod as well as to temperature.

Certain cells of the body may only become sensitive to external stimuli during specific phases in a cycle. Light cannot have a direct effect on hormone production, as does temperature but must act through a transducing process. In the hypothalmus–pituitary complex gonadotrophic hormones are secreted in the mesoadenohypophysis and prolactin and adrenocorticotropin in the proadenohypophysis. Thyrotropin and somatotropin secreting cells occur in both. Pituitary activity is controlled by neurosecretory cells in the hypothalmus with major concentrations in the nucleus lateralis tuberis and the nucleus preopticus (Peter & Nagahama 1974). These cells are neurally controlled by appropriate receptors for environmental stimuli (Scott 1979). Part of this complex is the pineal body or epiphysis which contains photoreceptive cells. Thus the pineal body may be affected by light conditions and act as a neuroendocrine transducer for photoperiod information.

3.3 Age

Perca and *Stizostedion* species have been aged by a variety of calcified body structures. The use of opercular bones for ageing perch was well established by Le Cren (1947). A broad opaque zone on the bone represents summer growth (Fig. 3.1). As growth diminishes with the onset of autumn and then winter, this zone gradually becomes transparent. The narrow transparent zone abruptly changes to the opaque zone in late spring. If growth is interrupted in the summer, a false check may be laid down but this usually starts and ends sharply. Before an ageing method can be adopted it is essential that it is validated (Bagenal & Tesch 1978). Craig (1974b) examined monthly samples of perch opercular bones throughout the year, from Slapton Ley, England, and counted the number of bones with checks on the outer edge (Fig. 3.2). The checks on the bone were found to be annual and the majority were laid down in May when perch growth in the lake resumed. Ageing of walleye by use of the opercular bone has been verified by applying tetracycline antibiotic (Babaluk & Campbell 1987; Babaluk *et al.* 1993) (Fig. 3.3).

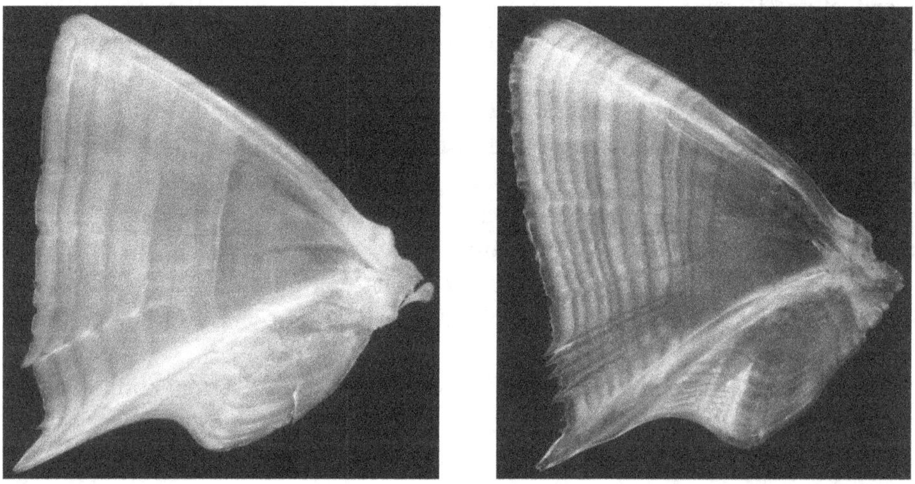

Fig. 3.1 Opercula of perch from Windermere: (a) seven years old (1968 year-class) and (b) 16 years old (1959 year-class).

Although opercular bones have been considered by many biologists to be satisfactory ageing structures for *Perca* and *Stizostedion* species other methods have been used including the examination of cleithra, otoliths (Fig. 3.4) and vertebrae (Campbell & Babaluk 1979; Belanger & Hogler 1982; Schmitt & Hubert 1982; Erickson 1983; Schlueter 1989; Brooks *et al.* 1994).

The use of structures such as scales and fins (Fig. 3.5) do not require sacrificing the fish (Carlander 1961; Campbell & Babaluk 1979; Sterns 1981; Belanger & Hogler 1982; Abdu-Nabi 1983). Al-Absy and Carlander (1988) recommended the

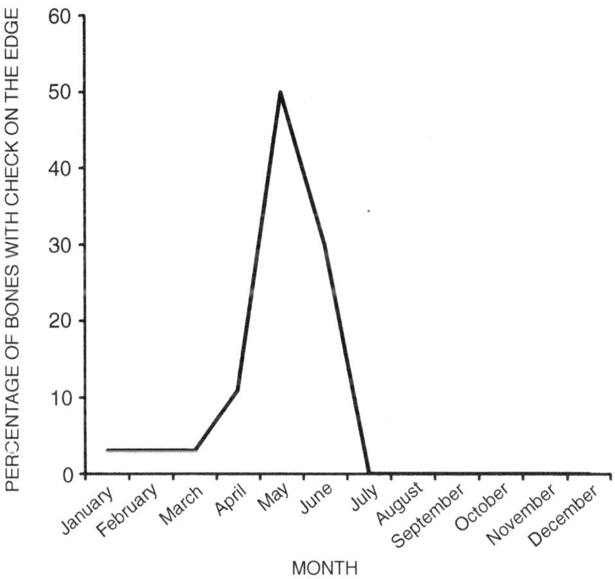

Fig. 3.2 The monthly frequency (%) of Slapton Ley perch with a growth check on the edge of the operculum. (After Craig 1974b.)

standardisation of the body site used for the removal of scales for ageing of yellow perch. They suggested a site below the lateral line at the tip of the pectoral fin when the fin was pressed to the body. The scales here were considered to have the least variability in scale size and lowest frequency in regeneration.

Walleye were found by Glenn and Mathias (1985) to first develop scales in the caudal peduncle region of the body at *c.* 30 mm TL. The authors found that scale growth and circulus formation was directly related to body growth, with maximum deposition occurring in late July and early August. However, as the fish get older, scales become more difficult and less precise for readers to assign an age. Slow growth or reabsorption of previously deposited scale tissue frequently leads to an underestimation of age based on scales. Heidinger and Clodfelter (1987) obtained 100% agreement between walleye aged from the otolith and their known age (0–4 yr) but only 75% when scale ageing was compared to the known ages. Robillard & Marsden (1996) also obtained greater precision ageing walleye with otoliths rather than scales, especially for walleye > 150 mm.

3.4 Growth

Percid fish show extensive plasticity in their growth, the variations caused by differences in abiotic and biotic factors. These variations can be seen in comparisons between mean growth of different populations, between mean growth of year

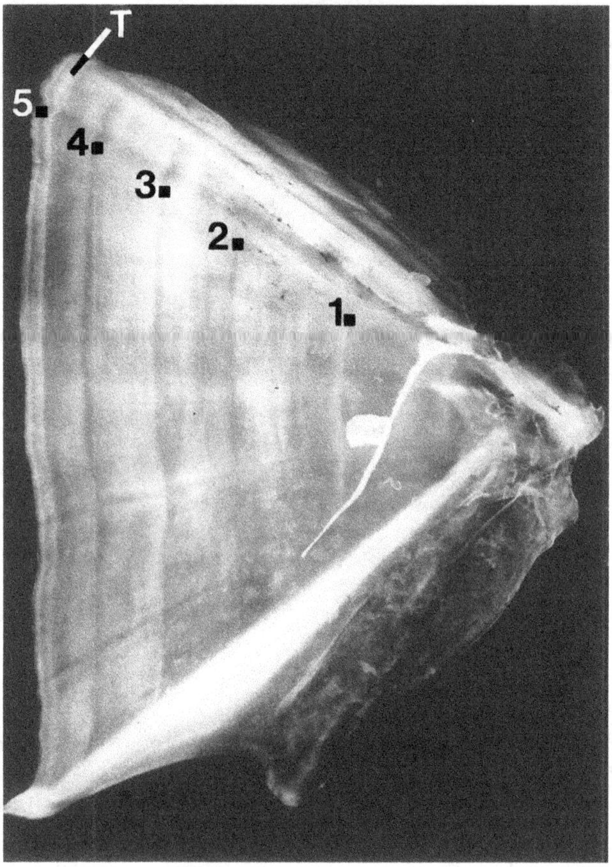

Fig. 3.3 The operculum of a walleye five plus years old, injected with tetracycline in 1978 and recaptured in 1979. Annual checks and the tetracycline mark, T, are indicated. The bone is photographed under reflected, ultraviolet light. (Photograph: J. Babaluk.)

classes within the same population and between individual growth from the same year class. Growth dynamics of larval and juvenile percids are more sensitive to external factors such as temperature and prey availability than adults (Post 1990).

Perch and yellow perch hatch at a length of about 5 mm (range 4–7 mm) (Konstantinov 1957; Hokanson & Kleiner 1974), zander at about 4 mm (Woynarovich 1960), sauger at 4–6 mm and walleye slightly larger at 7 mm (range 6-9 mm) (Scott & Crossman 1973). The growth in length of both genera is usually sigmoidal in the first year of life (Fig. 3.6) (Kubečka 1994) although between year variations can be considerable.

Species of *Perca* and *Stizostedion* populations in the Northern hemisphere grow more slowly the further north they live. This is a reflection of reduction in mean annual temperatures with latitude (Fig. 3.7) (Colby & Nepszy 1981). Population viability relies on the ability of fish to complete a minimum amount of growth during

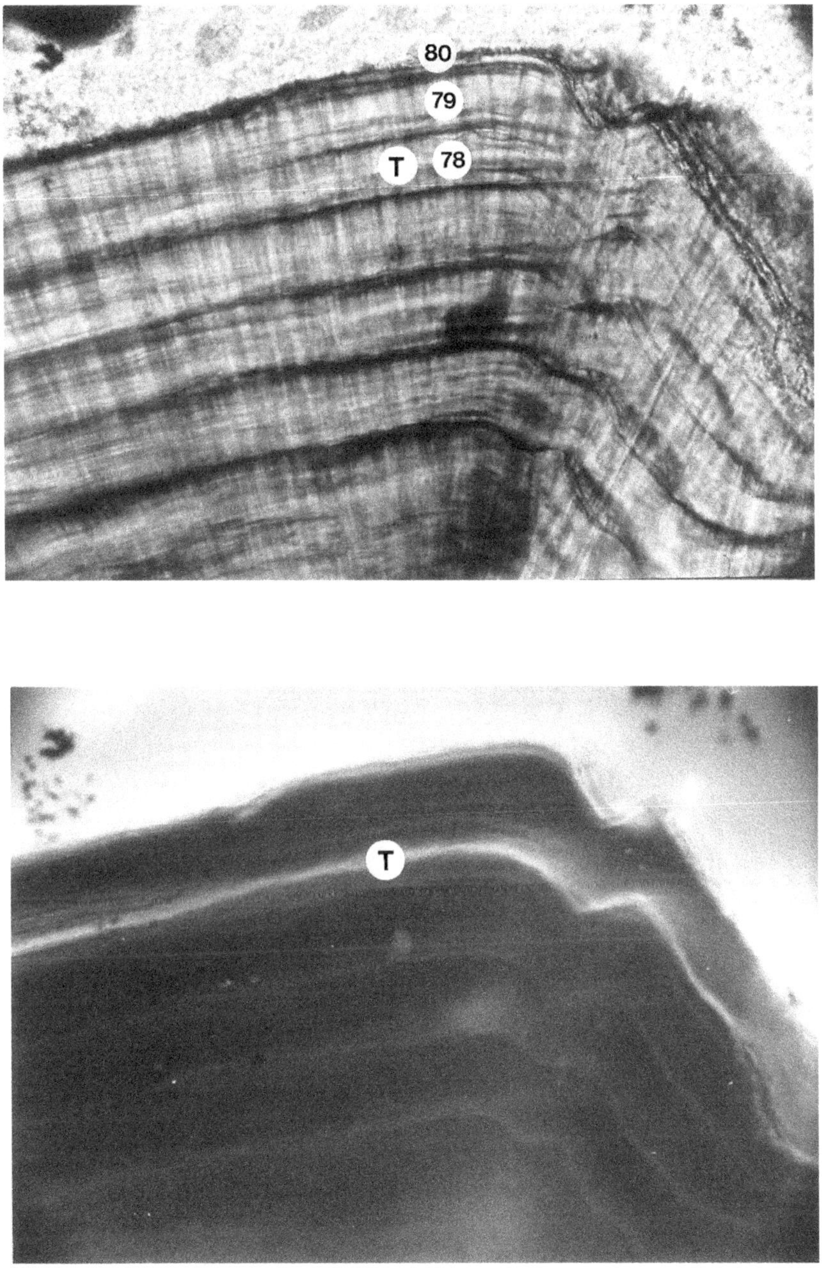

Fig. 3.4 A transverse, dorso-ventral cross-section through the otolith of a walleye seven plus years old, injected with tetracycline in 1978 and recaptured in 1980. The 1978–1980 checks and the tetracyline mark, T, are indicated. The upper photograph is with transmitted, white light and the bottom with reflected, ultraviolet light. (Photograph: J. Babaluk.)

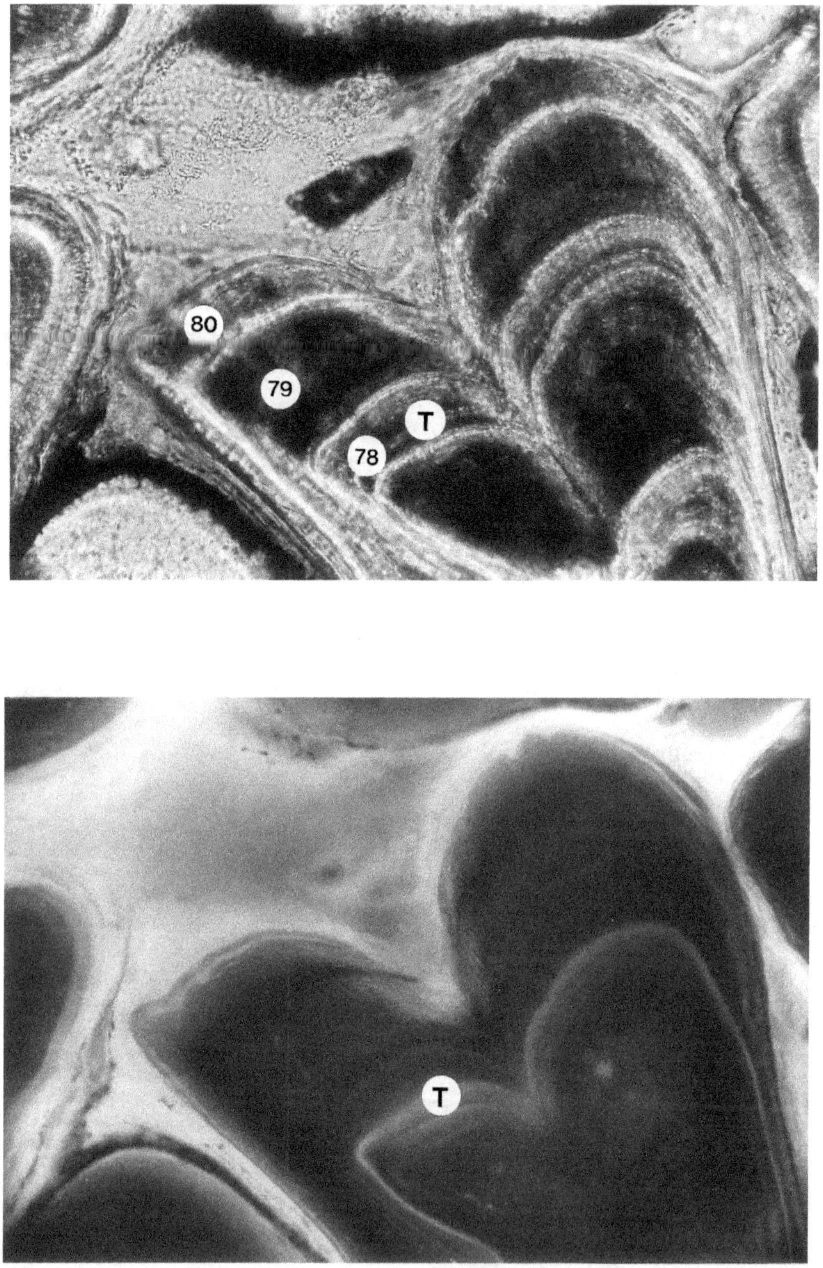

Fig. 3.5 A cross-section through the pelvic fin rays of a walleye five plus years old, injected with tetracycline in 1978 and recaptured in 1980. The 1978–1980 checks and the tetracycline mark, T, are indicated. The upper photograph is with transmitted, white light and the bottom with reflected, ultraviolet light. (Photograph: J. Babaluk.)

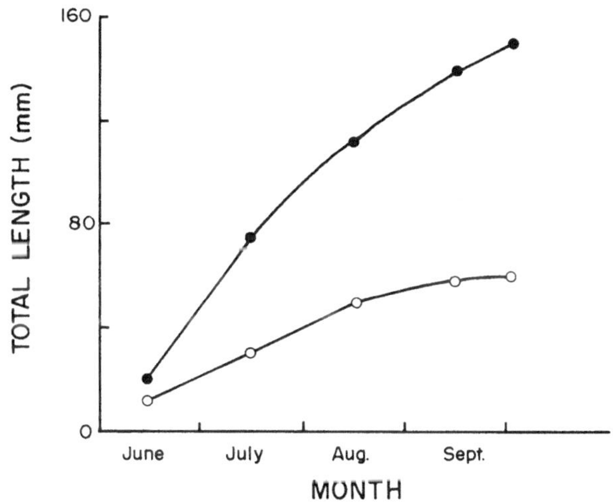

Fig. 3.6 Monthly lengths of young-of-the-year Windermere perch (○) and Oneida Lake walleye (●). (After Guma'a 1978a and Colby *et al.* 1979.)

their first year. From south to north across the distribution of percids, this ability is increasingly restricted as the growing season shortens and the starvation period lengthens. Shuter and Post (1990) show this constraint is sufficient to explain the present northern limits of perch and yellow perch. Percids such as yellow perch are very sensitive to starvation during the winter months (Sullivan 1986) although Radke & Eckmann (1999) found that overwinter mortality caused by starvation was rare in perch from central European lakes. Sensitivity to extremes of temperature and its effect on growth of a year class is probably most acute during the first year (Pereira *et al.* 1995)

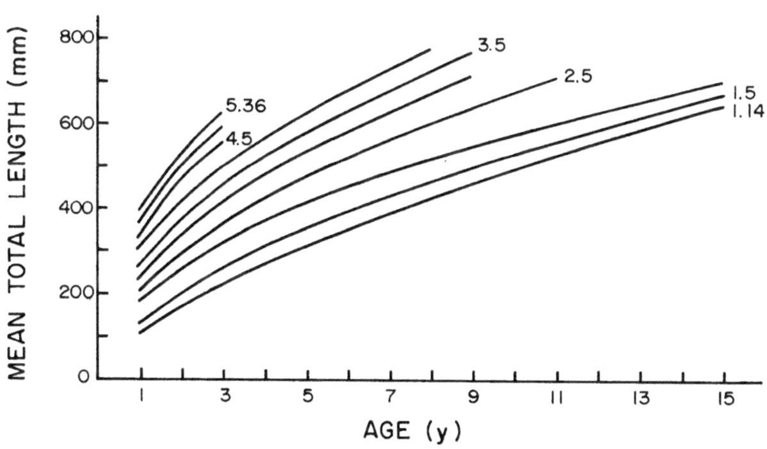

Fig. 3.7 Mean total length at age of female walleye over their range represented by growth degree days, >5°C × 10^{-3}, given adjacent to the growth curves. (After Colby & Nepszy 1981.)

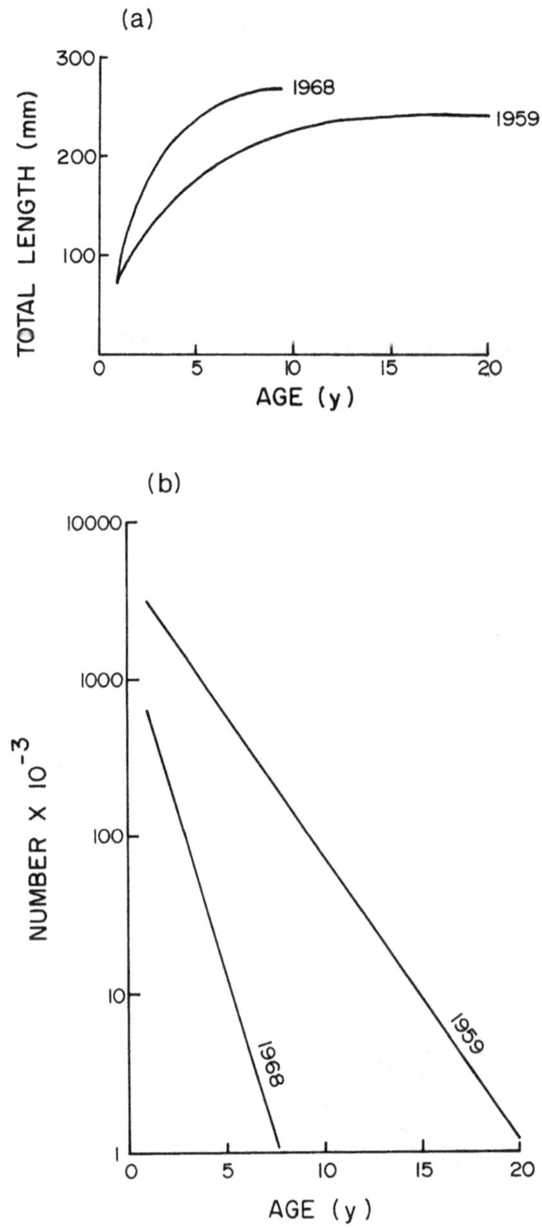

Fig. 3.8 Windermere perch (a) mean lengths and (b) catch per unit effort at age for the 1959 and 1968 year classes.

Often there are differences in growth rate within a population, between different year classes (Fig. 3.8a). Between year fluctuations within populations may be partly the result of population density (Fig. 3.8b) or summer (main feeding period) temperature differences (Fig. 3.9). Temperature may be the only significant controlling factor if food is not limiting. In Lake Tjeukemeer first year variability in body size could be almost entirely explained by temperature for planktivorous perch

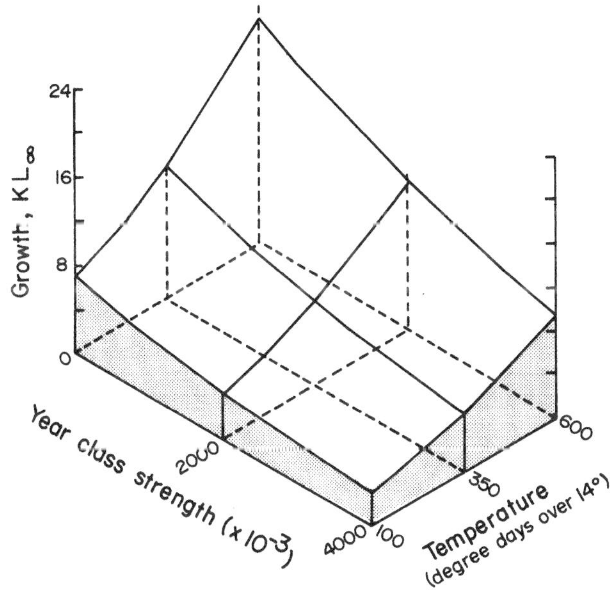

Fig. 3.9 A model of Windermere perch growth (KL_∞ from the von Bertalanffy growth equation) in relation to temperature (degree days >14°C) and density (year class-strength at age two years). (After Craig & Kipling 1983.)

which were not food limited (Mooij *et al.* 1994). The model used could also account for variation in growth of the piscivorous zander by the greater availability of smelt *Osmerus eperlanus* (Linnaeus) in warmer years. Le Cren (1958) found that two-thirds of the variation in the yearly growth of perch in Windermere could be accounted for by temperature (above 14°C) in the growing period. Other factors, which control the growth of Windermere perch, include the negative effect of density (Fig. 3.9) in which the availability of food may become limiting on growth. Craig (1980a) also suggested that the type of food available to the fish was important and perch achieved faster growth from a diet of fish than from invertebrates. Boisclair and Leggett (1989 a, b, c) suggest that prey type, prey availability and the fish community largely determine variation in yellow perch growth which supports the findings of Mills *et al.* (1989) who indicated the importance of the abundance, size structure and species composition of zooplankton in regulating first year growth of yellow perch in Oneida Lake. Post and McQueen (1994) calculated that 93–98% of the variability in growth of yellow perch in their first year was explained by the availability of benthic and/or planktonic prey (temperature data did not strengthen the relationship) and Fox (1991) found juvenile walleye growth to be regulated by the density and size of prey. The availability of prey may be a factor of density-dependent intra-age class competition. In Lake Huron growth of yellow perch was correlated with year-class density, not water temperature. Forney (1966) found that the early summer growth of young-of-the-year Oneida Lake walleye (range in size from 36 to 64 mm on 1 July from 1958 to 1962) was significantly correlated with water temperature. Later in the summer, variations in growth (range in size from 118 to 163 mm on 11 October from 1956 to 1964) were related to the availability of forage fish. Walleye fingerlings grew faster on a diet of fish than on a diet of invertebrates.

In the Baltic Sea differences in growth, negatively correlated with population density, were found between perch caught at 10 sites (Hansson 1985). In Lake Balaton, Hungary, zander grew at different rates in the three basins (Fig. 3.10). The initial growth was higher in the south-west basin but was reduced in older fish which were more abundant in this region. Genetic differences as well as competition within a group (year class or cohort) may lead to differences in growth rate between individuals and these factors may be critical in the first year of life.

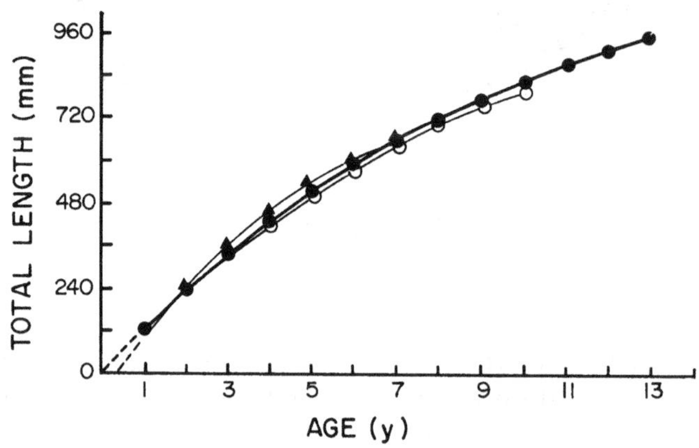

Fig. 3.10 Mean total length at age of Lake Balaton zander from the northwest (•), central (○) and southwest basins (▲). (After Biró 1985.)

Productivity of a water system and the number of fish in that system competing for the available food will influence growth. Bimodality in length has been described for zander in their first year, one group of fast growing littoral piscivores and the other group of pelagic zooplanktivores (Frankiewicz *et al.* 1996; Van Densen *et al.* 1996). Van Densen *et al.* (1996) found this bimodality in zander growth in eutrophic lakes in which macro-invertebrates were at low densities and suggest that there is strong selection pressure in recruitment in these types of lakes. Boisclair and Rasmussen (1996) used a bioenergetics model (see Chapter 5) to show that yellow perch growth decreased as the total fish (whole community) numerical abundance increased and also decreased as the percentage of small prey in the invertebrate community increased. The changes in the invertebrate community may have been brought about by eutrophication and thus perch growth is probably negatively affected by eutrophication. A similar situation was identified in Lake Erie (Hayward & Margraf 1987). However, Hartmann and Knoepfler (1986) suggest that perch growth in Lake Constance was a poor indicator of the trophic state of the lake as temperature was the main controlling factor.

Many *Perca* and *Stizostedion* populations exhibit sexual dimorphism in which females grow faster and reach a greater ultimate size than the males (Fig. 3.11). Artificially reared female yellow perch were observed to start growing faster than the males at 110 mm length (Schott *et al.* 1978).

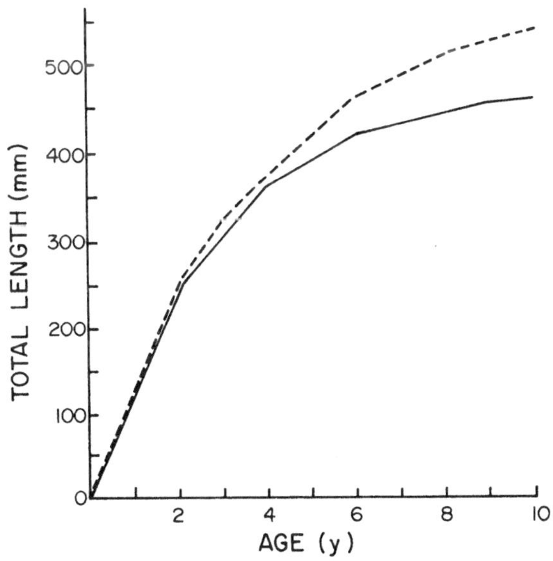

Fig. 3.11 Mean total length at age of Lake Gobebic male (solid line) and female (broken line) walleye. (After Colby *et al.* 1979.)

There is no evidence in *Perca* species that populations are stunted because of genetic factors (Alm 1946; Hoestlandt 1979; Heath & Roff 1987). The growth of percids from stunted populations do not usually differ from unstunted populations when grown under similar environmental conditions as the latter. Persson (1987a) suggests that perch are more likely to become stunted in lakes where the pelagic area is much greater than the littoral. This, coupled with high summer temperatures, favours small perch. Lott *et al.* 1996 established that 'high' quality populations of yellow perch resulted from the availability of large macro-invertebrates compared to 'low' quality populations feeding on zooplankton. Fish were not a substantial component in the diet in which yellow perch reached a large size (>30 cm).

Growth is an anabolic process and only occurs when catabolic processes have been satisfied (see Chapter 5). Growth may be an increase in somatic tissue or energy may be directed towards the development of gonads. In some situations energy stored in somatic tissue is later transferred on a seasonal basis to gonad tissue (Craig 1977a; Tanasichuk & Mackay 1989; Hayes & Taylor 1994; Griffiths & Kirkwood 1995; Bryan *et al.* 1996). Age of maturity will play an important role in continued somatic growth as development of the gonad takes up some of the available energy. Craig (1977a) found that the amount of stored energy decreased (or inversely the amount of water increased) as perch grew older. Thus, the drain of surplus energy to reproductive processes will slow down somatic growth and possibly lead to eventual senescence and death. Populations which are slow growing tend to be late maturers. Female perch and yellow perch mature between two and four years and males in the same water body usually mature one year earlier. Walleye show more disparity between ages of maturity in different localities and like growth, differences can be related to temperature (Colby & Nepszy 1981). Maturity of females ranges from two years in the southern states of the USA to ten years in the Northwest Territories, Canada. Zander mature at between two and five years.

3.5 Mortality

The main proximate causes of fish mortality are probably predation, parasitism, reproductive demands and senescence. The life of a fish population takes it through a succession of different stages in which the mortality rates of the population vary. At the juvenile stage when mortality is usually the result of predation or competition for food, faster growing fish will have an advantage over slower growers (Craig 1985; Post & Prankevicius 1987). Size-dependent mortality can also occur over winter in the first year of life due to lack of energy reserves (Post & Evans 1989a). Later in life faster growing fish will often have a higher natural mortality rate than slow growing fish. Parasites have been implicated in fish mortality but are often overlooked in natural populations. Szalai and Dick (1991) suggested that parasite induced mortality, especially infection with *Raphidascaris acus* (Bloch), could account for most of the unexplained mortality of yellow perch in Dauphin Lake, Manitoba. In this lake the total annual natural mortality was estimated as 61% of which 19% in females and 45% in males could be explained by pike, *Esox lucius*, predation.

Forney (1967) calculated annual mortality rates of adult walleye in Oneida Lake, USA, by marking the fish when spawning and recapturing marked fish during the following summer and autumn. The annual mortality rate, A, is the sum of rate of death from angling (rate of exploitation, u) and from natural causes (the expectation of natural death, v) where $A = u + v$. For the years 1957–8, 1958–9 and 1959–60 values for u from angling were 0.14, 0.10 and 0.47 and values for v were 0.06, 0.01 and 0.07 respectively. From these values the total instantaneous mortality rate, Z, could be calculated and also the instantaneous fishing mortality, F, and instantaneous natural mortality, M, rates where $Z = F + M$. For the three years given above for Oneida Lake, $M = 0.07$, 0.01 and 0.10 respectively. Forney (1967) considered that adult natural mortality probably remained fairly constant with time and $M = 0.06$ was a best estimate of the population's natural mortality rate. Fluctuations from this were caused by errors in population statistic estimates. However, studies on perch in

Windermere, England, have shown that M can be very variable in a population and can also vary considerably between year classes and between sexes (Craig *et al.* 1979; Craig & Kipling 1983). Before a major fishery for perch was started in 1941 the natural mortality rate of the adult population was probably stable at about $M = 0.41$. After the fishery ceased M continued approximately at this level until about 1969 when between 1969 and 1975 the rate increased and varied from 0.65 to 1.28. This was a time of rapid growth in year classes in the population. Values of Z ($Z = M$ since F was negligible) were calculated for each year class of mature perch from 1955 to 1972. Values for males ranged from 0.25 to 1.11 (Craig *et al.* 1979). The mortality rates for males for the 1959 and 1968 year classes are illustrated in Figure 3.8b. These mortality rates were estimated from catch curves and relate to fish three years or older since younger fish were not fully recruited by the fishing gear. Adult mortality rate remains remarkably constant throughout nearly all the life span of each year class. The natural morality is usually greater for males than females. Lang (1987) estimated mortality of female perch by counting the number of egg strands laid by a strong year class in successive years. As has been found in other studies, the highest mortality was found in the larval and juvenile phases (99% from egg to two years).

Perch and yellow perch longevity can range from six to about 21 years, walleye from three to 19 years and zander from six to 17 years. Mortality and thus maximum age is related to temperature. In the northern limits of distribution, slower growing percids have a longer life expectancy than those living further south (Fig. 3.7).

Craig (1980a; 1982) has shown that the parameters M and K (the latter from the von Bertalanffy, 1938, growth equation) are linearly related for year classes of perch in Windermere (Fig. 3.12). The reasons for the relationship are not yet fully clear. It has already been suggested that percids, and fish in general, are very plastic in their response to environmental cues (especially temperature). Genetic variations have

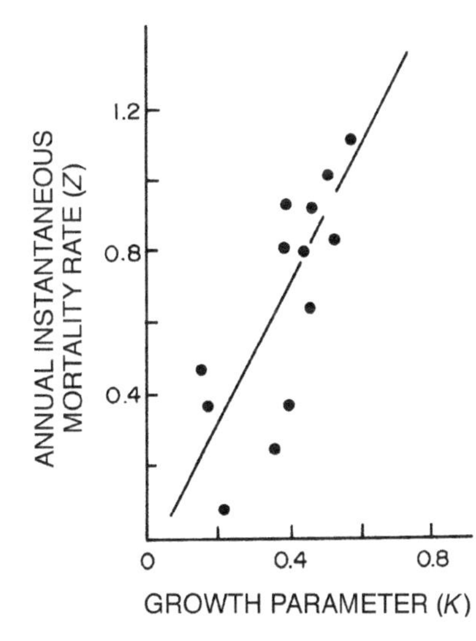

Fig. 3.12 The relationship between the annual instantaneous mortality rate (Z) and the growth parameter K (from the von Bertalanffy growth equation) for Windermere perch. The points represent different year-classes. (After Craig 1982.)

always been difficult to elucidate. Alm (1946) found that stunted perch improved their growth rate when transplanted to a more favourable environment and fertilised eggs from the stunted population grew equally well compared to a normal population. Both M and K are correlated with temperature (Pauly 1980). Craig (1985) suggests that in perch (and fishes in general) physiological processes which are genetically determined, but are controlled by a biological clock, control longevity. Thus, speeding up of these physiological processes may induce the release of immune responses and resultant early death. External factors influence these internal systems.

Chapter 4
Reproduction and Development

4.1 Introduction

The development of the gonad requires energy in excess of that needed for maintenance or catabolism (Ferroni *et al.* 1992; Chapter 3). The mechanisms that allocate energy to growth, maintenance and reproduction are not fully understood.. The cost of egg production is high. For example the energy lost in eggs by perch can be > 86% of the energy stored by the soma in a year (Craig 1977a). The energy lost by the testes in milt is much less, ≤ 10%. Kazanskiy (1985) suggests that there is an optimum ratio between energy loss by an individual (zander) from the soma and the development of gametes which maximises the total biomass of the population. The consequences or costs of reproduction may be reduced growth and survival. Studies on yellow perch have shown that sex steroids have a profound effect on growth (Malison *et al.* 1986). It appears that attainment of a specific body size (80–100 mm) and maturation of specific hormone receptors may be necessary for a growth response to these hormones. Similar conditions appear to control spermatogenesis and vitellogenesis. Oestradiol-17β injections stimulated yellow perch growth by affecting appetite but not metabolism. Androgens retarded growth probably by altering metabolism. When gonadectomies were performed on yellow perch, it was found that males had reduced blood levels of testosterone and growth improved. Castrated female fish had reduced blood levels of oestradiol-17β but growth was poorer.

4.2 Gonads

The sex of *Perca* and *Stizostedion* species is not easy to identify from external appearances except at spawning time when the gravid female is more rounded and the male expels milt on handling. The primary sex cells are present at the time of hatching and they migrate into the region of the prerenal ducts. The female is the homogenetic sex (Malison *et al.* 1986). Sex differentiation begins in the perch at a length of 10–12 mm (Mezhnin 1978; Zelenkov 1982). Oogenesis in hatchery reared yellow perch proceeded at 35 mm and vitellogenesis and spermatogenesis at 85 mm (Malison *et al.* 1986). The single ovary of the female is formed by fusion of two rudiments and it has no external oviduct (Chapter 2). A summary of the stages of the developing ovary is given in Table 4.1. The ovary of the gravid perch contains hyaline oocytes which are filled with yolk (Fig. 4.1). The cells of postovulatory follicles of perch undergo disintegration and autophagocytosis. Cell debris is reabsorbed and the size of the follicle is reduced. Other cells of the follicular epithelium become synthetically active cells and take on a secretory function (Lang 1981).

The paired ovaries of walleye are small and not well developed in fish <75 mm. They are heavily pigmented by melanophores (Eschmeyer 1950). As the fish increases in size the pigment becomes more scattered over the ovaries until in adults it is confined to a few areas in the anterior dorsal region. Large, immature females have transparent ovaries which are cylindrical in shape although the anterior end is rounded. Transverse blood vessels become more conspicuous as the fish matures. The dorsal longitudinal blood vessels lie on the surface of the ovaries. As in perch, the ovaries of walleye develop through an annual cycle. In Lake Gogebic, USA, the ovary mass to total body mass increased from 0.7% in August to 4.7% in October and to 16.3% in May of the following year when spawning took place (Eschmeyer 1950). The maturation stages of *Stizostedion* are very similar to those of *Perca* species (Table 4.1).

A summary of gross anatomical changes of the perch and yellow perch testes is given in Table 4.2. The spermatozoon is a primitive-type found in other teleosts (Koenig *et al.* 1978). It is asymmetrical as the flagellum inserts medio-laterally on the nucleus. It lacks an acrosome, has an ovoid head and a small mid-piece with one mitochondrion (Lahnsteiner *et al.* 1995). The number of sperm per unit volume varied from 1.14×10^{10} to $3.02 \times 10^{10}\,\mathrm{ml}^{-1}$ in the sample of 10 yellow perch studied by Koenig *et al.* (1978). Ciereszko and Dabrowski (1993) counted $4.16 \pm 0.35\ 10^{10}\,\mathrm{ml}^{-1}$. This compared to 3.7×10^{10} to $12.74 \times 10^{10}\,\mathrm{ml}^{-1}$ (mean 7.62 sperm $\times\ 10^{10}\,\mathrm{ml}^{-1}$) for 42 perch studied by Piironen and Hyvaerinen (1983). In distilled water at 10°C, $85 \pm 27\%$ of spermatozoa were found to be motile mostly swimming by linear motion at $122.4 \pm 21.9\,\mu\mathrm{m\,s}^{-1}$ (Lahnsteiner *et al.* 1995). Perch gonads can develop normally in acidified water (pH 4.0–5.0) (Talikina *et al.* 1996).

Up to a fish length of about 75 mm, walleye testes are thread-like with little or no pigment and are smaller in cross-section than ovaries from fish of a similar size. Testes of larger immature males are elongate and uniform in diameter along their length. The transverse blood vessels are not very apparent in the testes although they are conspicuous in the ovaries. Testes in a resting phase are greyish-white in colour. In Lake Gogebic, walleye testes had this appearance during July and August but by mid-October they were large, soft and milky-white. They reached a similar size and mass to the ovaries of the female at this time (Eschmeyer 1950). The testes became firmer through the winter months until the spawning season when milt was produced. The spent testes were similar in appearance to resting gonads. Testes to body mass varied from 0.2% in July and August to 4.3% in October. It remained at about 4% until spawning then declined to 0.4%, in late June.

Table 4.1 External and histological appearances of each stage of development of the perch ovary. (After Treasurer & Holliday 1981.)

Maturity stage	External	Histological
Virgin	Small thread-like organ lying underneath the bladder; no oocytes visible.	Two primordial ovaries separated by an epithelium layer; connective tissue matrix containing primary germ cells (3.3–6.7 μm), oogonia (13.3–20.0 μm), and primary oocytes (13.3–60.0 μm); ovary wall 4.2–10.4 μm thick.
Maturing virgin	Small pear-shaped pink organ; oocytes not yet visible.	Internal organisation advanced; ovary traversed by ovigerous lamellae; primary oocytes have increased to 20.0–116.6 μm diameter; ovary wall is 13.3–30.0 μm.
Developing (early)	Ovary is opaque and reddish with blood capillaries; occupies about half of body cavity; eggs visible to the eye are whitish and granular.	Developing oocytes 162–346 μm; follicle comprises theca externa and granulosa; chorion comprising tunica propria, a loose network, and the lower striated zona radiata; formation of yolk vesicles; ovarian wall is 40–97 μm thick.
Developing (late)	Ovary reddish and opaque eggs clearly discernible; ovary occupies about two-thirds of body cavity.	Distinct size separation between developing and resting oocytes, the latter forming distinct groupings; oocytes up to 773 μm diameter, cytoplasm almost filled with yolk vesicles; chorion much wider.
Gravid	Ovary fills ventral cavity; eggs completely round and almost appear translucent a few days prior to spawning.	Hyaline oocytes present (these usually collapse in histological sectioning); at best the yolk appears as a homogeneous mass filling the interior of the oocyte.
Spawning	Roe runs with slight pressure; most eggs translucent with a few opaque eggs left in the ovary.	Histologically similar to the gravid stage; intense vasodilation; oviduct wall is thinner.
Spent	Ovary empty, flaccid and red, occupying about two-thirds of the body cavity; ovary wall thick and tough; a few residual oocytes may be visible.	Ovigerous lamellae disorganised; follicles contracted and folded; tunica wall is folded with an increase in diameter; occasional residual oocytes undergoing reabsorption.
Resting	Ovary grey-red; firmer than spent stage; length about half that of the body cavity; occasional residual oocytes may be visible.	Varying rates of recovery but normally organisation of lamellae is advanced; follicular remains arranged on outer edge of ovigerous lamellae; increase in number of small oocytes; advanced oocytes developing in granulosa layer.

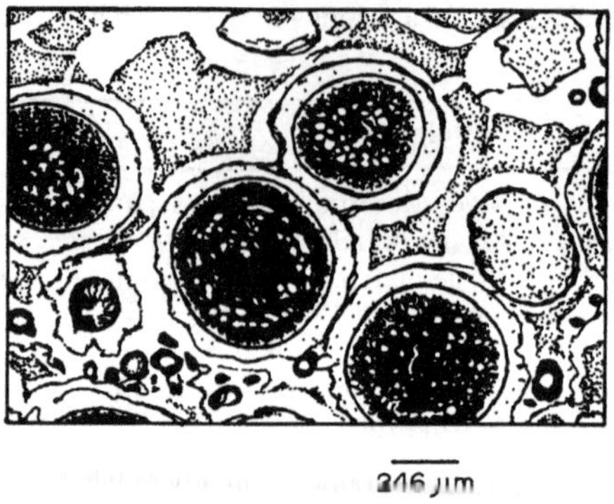

Fig. 4.1 The gravid ovary of perch with hyaline oocytes.

Perch, walleye and zander all lay a large number of eggs but, for a given size, zander is more fecund than walleye (Fig. 4.2). Within a species, for a given length, the absolute and relative fecundity can vary widely between fish from different areas although this is not necessarily a function of latitude. Thorpe (1977) for example could find no significant relationship between relative fecundity and latitude in different perch populations. Fecundity can vary between different populations in the same body of water as is illustrated by two perch populations in the Rybinsk Reservoir, Russia, and by walleye populations in the two main basins of Lake Erie (Fig. 4.3). The number of eggs laid by a female is influenced, as is growth and age of maturity, by food supply or energy intake of the fish (Baccante & Reid 1988). Johnston (1997) found that walleye egg dry mass from Lake Manitoba fish was positively related to both female length and age. Hatching success of the walleye eggs was also positively related to the age of the female. Differences in

Table 4.2 The external appearance, at each stage of development, of the perch testes.

Maturity Stage	External
Virgin	Very small paired organs found close under the vertebral column; transparent.
Maturing virgin	Translucent; length about half of the ventral cavity.
Developing (early)	Opaque and reddish with blood capillaries; occupying about half of the ventral cavity.
Developing (late)	Reddish-white in colour; occupying about two-thirds of the ventral cavity: no milt is produced with pressure.
Gravid	The testes occupy the length of the ventral cavity; they are white and produce milt with pressure.
Running	As for gravid but milt is produced with only slight pressure.
Spent	The testes start to shrink; they are reddish in colour and some milt is still apparent.
Resting	Translucent; length less than half of the ventral cavity.

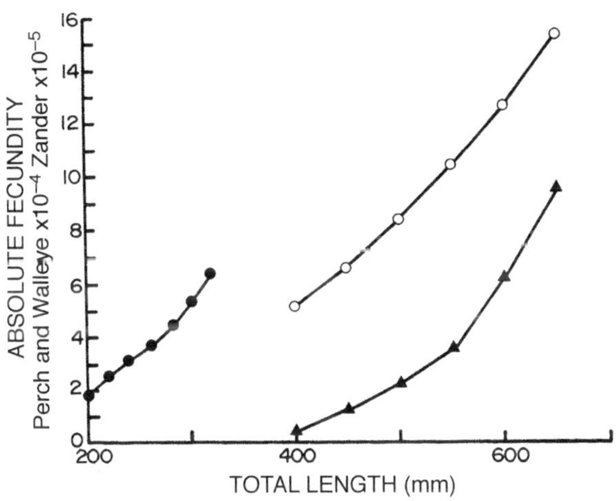

Fig. 4.2 Absolute fecundity at total length for Windermere perch (●), Dauphin Lake walleye (○) and Lake Ijssel zander (▲). Note different scales for absolute fecundity between species.

fecundity can occur in the same population with time as differences in food availability occur. Kuznetsov (1982; 1986) noted that Volga pikeperch and zander produced more eggs with higher fat content following a year of good feeding conditions compared to a year of poor food supply. Deficiency of food and poor growth in a preceding summer can have a marked effect on reproduction in the following spring. Poor conditions delay the reabsorbing processes after spawning and delay oocyte development so that the females may miss the next spawning period. Insufficient energy stored in the soma during the summer for translocation to the ovaries as they develop in the winter may cause the death of females as illustrated for yellow perch by Newsome and Leduc (1975).

There is still speculation on the effect of age on egg production. Evidence suggests that as perch females grow older the number of eggs per unit mass (relative fecundity) declines but the size of the egg increases (Craig 1974b; Volodin 1980). Absolute fecundity and age follow a power law relationship until the age when senescence causes egg numbers to decline.

4.3 Environmental cues and hormonal changes

Teleost reproduction is influenced by a whole series of interacting environmental, endocrine and social factors (Dabrowski *et al.* 1996). Key environmental elements are photoperiod and water temperature. In females, endocrine factors include hypothalamic neurohormones and neurotransmitters, gonadotrophins, sex steroids, prostaglandins and growth factors. Some of the regulatory substances such as prostaglandins and free or conjugated steroids can act as pheronomes in teleost fishes

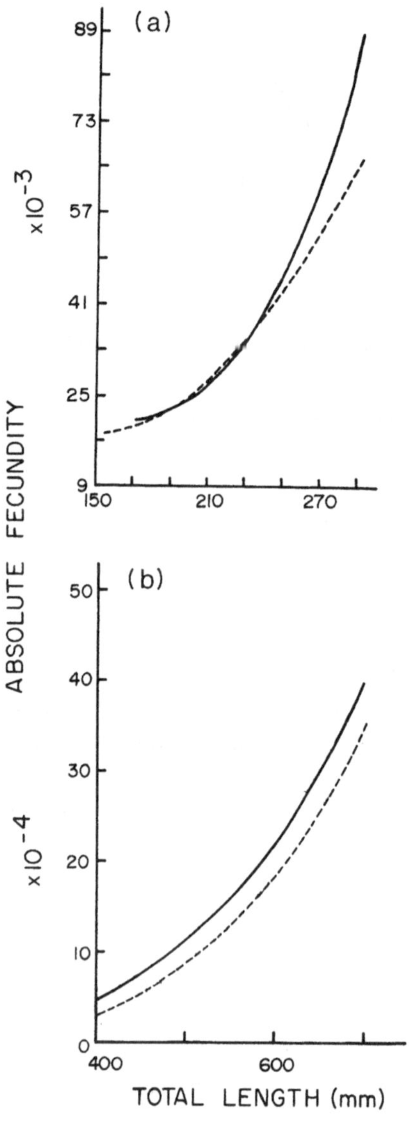

Fig. 4.3 Absolute fecundity at total length of: (a) Rybinsk Reservoir perch and (b) Lake Erie walleye. Solid and broken lines represent different regions of the water bodies. (After (a) Volodin 1980 and (b) Wolfert 1969.)

(Dabrowski *et al.* 1996). The stringency of conditions in male teleost reproduction is less rigid than that of the females. The principal endocrine factors involved in the regulation of reproduction are pituitary gonadotrophins and gonadal steroids.

The reproductive cycle can be divided into two main phases, gametogenesis and spawning. During gametogenesis highly specialised gametes (oocytes or spermatozoa) are formed from simple germ cells (oogonia or spermatogonia). The duration of gametogenesis varies according to species and temperature. Spawning involves a sequence of events which is complex and more variable in duration than gametogenesis. This leads to liberation of gametes, meiosis resumption, oocyte maturation, ovulation and oviposition in females and spermiation and sperm release in males. Endocrine changes during the reproductive cycle include variations in the levels of pituitary and plasma gonadotrophin (GTH) and plasma sex steroids (Figs 4.4 & 4.5). There is strong evidence for a gonadotrophin-releasing hormone (GnRH) in the teleost brain and probably a gonadotrophin release-inhibiting factor (GRIF) (Stacey 1984). GTH release appears to be controlled by the interaction of these two hormones. There are possibly two types of GTH, one with a low carbohydrate content, which stimulates vitellogenin uptake in the oocytes, and one with a high carbohydrate content which stimulates steroid oogenesis, oocyte maturation, ovulation and spermiation. GTH increases in the blood serum during the preovulatory stage. Oestradiol-17β increases rapidly in the serum of walleye from low values in October to peak levels in November, coinciding with the period of the most rapid ovarian growth (Fig. 4.5) (Malison *et al.* 1994a). The effect of GTH and steroids varies between percid species. GTH produced by the ovarian follicles just prior to ovulation, induces final maturation of the oocyte indirectly in yellow perch by stimulating the production of steroids. In this species and probably in walleye, the most potent steroid for this function is 17α,20β-dihydroxy-4-pregnen-3-one (Goetz & Theofan 1979). Steroids produced by yellow perch are also responsible for inducing germinal vesicle breakdown (Theofan & Goetz 1983). Germinal vesicle

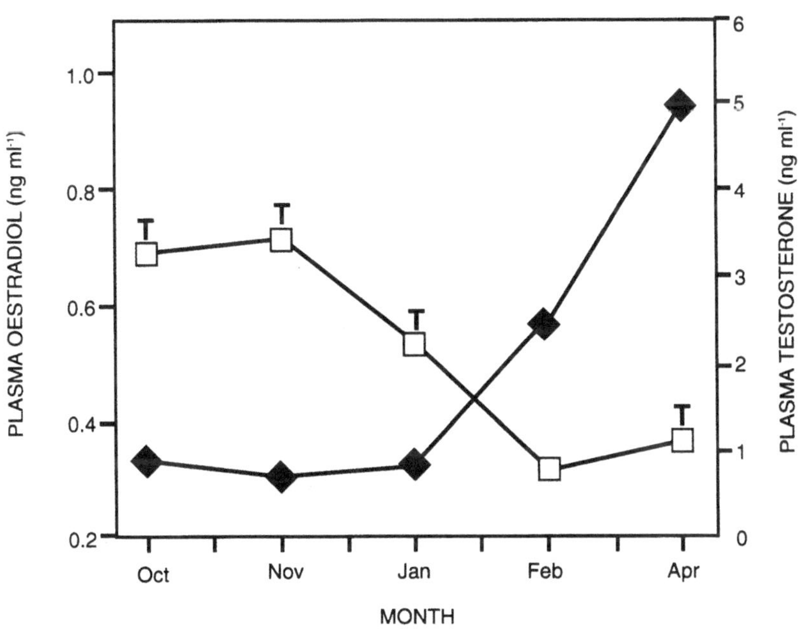

Fig. 4.4 Monthly means (± SE) concentrations of oestradiol (□) and testosterone (■) in female yellow perch. (After Dabrowski *et al.* 1996, with kind permission from Blackwell Science.)

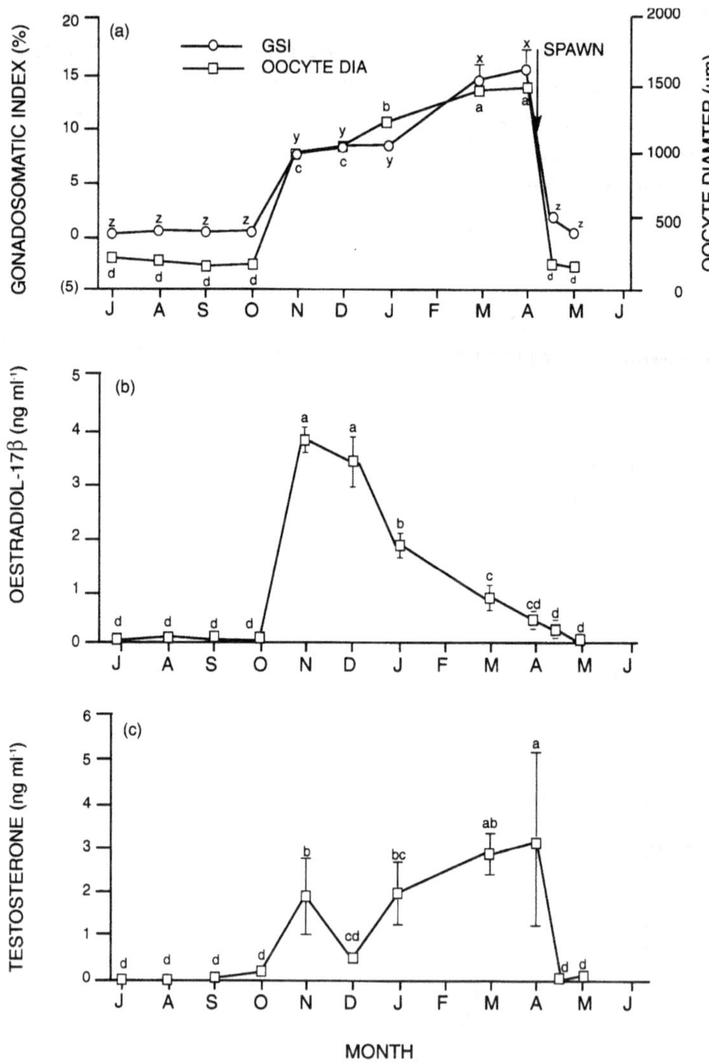

Fig. 4.5 Monthly means (± SE) of: (a) gonadosomatic index (GSI) and oocyte diameter, (b) oestradiol-17β and (c) testosterone in female walleye. Means with the same letter (*a*, *b*, *c* or *d* for oocyte diameter and *x*, *y* or *z* for GSI) are not significantly different ($p < 0.05$). (After Malison *et al.* 1994a, with kind permission from Kluwer Academic Publishers.)

breakdown in walleye can be directly influenced by GTH. *In vitro* effects of various GTH preparations on yellow perch and walleye germinal vesicles caused breakdown in walleye but no significant effects in yellow perch (Goetz & Bergman 1978). Lessman (1978) demonstrated the direct effect of GTH on ovulation in walleye. Preparations of carp and human GTH induced ovulation in prespawning walleye but cortisone and progesterone were ineffective. In males, just before spermiation, there is an increase in androgens (testosterone and 11-ketotestosterone).

Perca and *Stizostedion* spawn once a year in the spring. This annual rhythm is controlled by exogenous factors which are difficult to quantify. Indeed little is known about how these factors control spermiation and ovulation in teleosts as a whole. Some possible environmental conditions which influence this reproductive cycle include temperature, photoperiod, substrate conditions, water levels and velocity and social interactions (including the possible production of pheromones). Temperature probably plays the major role and the following examples emphasise its importance. Experiments on temperature effects demonstrated that optimal conditions for yellow perch maturation occurred when fish were held at a water temperature of 6°C or lower for 185 days from 30 October. The number of maturing females declined as the holding temperature increased (Hokanson 1977a). In the southern USA yellow perch is at the limit of its range. Warm winter temperatures (8–10°C) may impair reproductive success (Clugston *et al.* 1978). When walleye were introduced into California reservoirs where the winter temperature ranged from 10–12.5°C, they failed to reproduce (Miller 1967). The timing of spawning is also influenced by temperature, both by absolute temperature and by temperature change with time. *Perca* species in the Northern Hemisphere spawn from February to July depending on latitude and introduced populations in the Southern hemisphere spawn from August to October. Walleye in the southern limits of their range spawn early in the year. For example, they migrate up the Pearl River, Mississippi, USA, to spawn in late January or early February (Cook 1959), whereas those in northern Canada spawn in late June (Scott & Crossman 1973). Zander spawn from February to early July (Deelder & Willemsen 1964) depending on latitude and water temperature. Two populations of zander in Turkey, living at the same latitude, spawn at different times due to the temperature disparities between two lakes. Zander spawn in the warmer lake in late February or March and in May or June in the colder lake (Aksiray 1961). An accelerating temperature appears to be the major factor inducing spawning (Hokanson 1977a; Thorpe 1977) and photoperiod may only play a minor role (Kayes & Calbert 1979). An interruption of this temperature acceleration by a cold period may cause oocytes to be reabsorbed and spawning is missed. This has been noted in perch at Windermere, England, where many perch failed to spawn and females were found to be reabsorbing their eggs after a severe winter in 1962–63 (Le Cren 1965; Kipling 1976). However, some races of perch are exposed to ice conditions annually and perch have been noted to spawn under the ice at 4°C in the Aral Sea (Filatov & Duplakov 1926). An interrupted spawning season caused by cold weather induced walleye females to reabsorb their eggs in Heming Lake, Canada (Derback 1947). In the Don River, Russia, a cold spring delays the migration of zander and many of those entering the river in May are in the process of reabsorbing their eggs (Golovanenko *et al.* 1970). In Lake Geneva a rise in water temperature to 14°C stimulates perch spawning. However, if the temperature drops to 10°C the normal spawning period of one month is extended (Gillet & Dubois 1995). Perch spawn in the lake in May and water temperature appears to have only a minor influence on the commencement of spawning

(Gillet *et al.* 1995). The main effect is on spawning intensity. The choice of freshwater or seawater habitats for perch was more strongly influenced by suitability of temperature rather than salinity in the Northern Quark, Gulf of Bothnia (Karås & Hudd 1993).

4.4 Spawning behaviour

Usually spawning migrations of *Perca* species are short-ranged. In a lake system this involves movement from deep water, where the fish have over-wintered, to shallow water spawning areas. Other movements may be from lakes into rivers or from brackish areas into rivers. Aalto and Newsome (1990) and Jennings *et al.* (1996) provided evidence that yellow perch and walleye respectively return to their natal location to spawn. In the latter case two stocks of walleye broodstock from both river (gravel bars) and lake (rocky substrate) were artificially spawned producing 'genetically tagged' fry. They were introduced into a system containing both types of spawning habitat. At maturity (three to four years) walleye from river spawning parents moved to the river and those from lake spawning origins moved to the lakeshore. Male *Perca* species arrive on the spawning grounds earlier than the females. Male yellow perch in south-eastern Lake Michigan, USA, concentrate in water depths of 6–12 m in early May. Females stay in deep water (> 18 m) until late May when they join the males and spawning takes place (Dorr 1982). The duration of spawning varies. In south-eastern Lake Michigan yellow perch spawning lasts for about two weeks compared to Slapton Ley, England where spawning of perch occurs over about an eight week period (Craig 1974b).

Perch and yellow perch will spawn over a wide variety of substrates including boulders and gravel, aquatic macrophytes, roots of trees, dead branches and other materials normally in water depths from 0.5–8 m. Spawning can take place throughout the day and night and the fish lose their normal diel behaviour patterns during the spawning season (Craig 1977b).

A single female *Perca* species is accompanied by a number of males, usually four or five but sometimes two or three. The female swims in a non-directional manner through, over and around the spawning substrate. She often pauses between these movements. Some movements are circular (anticlockwise and clockwise) others are spiral and upward. As a rule one male follows the female in these movements swimming alongside and slightly beneath her, the other males remaining stationary. The active male often touches the vent of the female with its snout. If the female swims to another site the group of males follow her. Sometimes the male following the movements of the female may break away and be replaced by another individual. In vegetation this type of behaviour can go on for about 30 min with frequency of movements and swimming speed increasing until ovulation (Treasurer 1981). The egg strand is released as the female swims in a spiral clockwise movement flexing herself into a U-shape. All the eggs are released in this egg strand which becomes twisted around and attached to the spawning substrate. Treasurer (1981) noted that at least two males, one following and another swimming in the opposite direction released milt. Egg and sperm release takes about five seconds (Fabricius 1956; Hergenrader 1969; Treasurer 1981). The sperm of perch is only viable for a few minutes after release (Lindroth 1947) probably less than two minutes (Piironen & Hyvaerinen 1983). There is little if any protection of the eggs. After spawning females and males show signs of loss of equilibrium which may take up to one hour to regain.

Walleye will start spawning migrations before the ice has melted from a lake, particularly into an inflowing stream which has a higher water temperature than the lake. Spawning migrations can be blocked if the streams are colder than the lake water (Forney 1967). Walleye have been observed to spawn in lakes under ice (Eschmeyer 1950). In Southern Indian Lake, Canada, walleye commenced spawning migrations into streams when the lake was still covered in ice but the streams had reached a temperature of 5°C (Bodaly 1980). When the temperature dropped below this value the upstream movements ceased. The size of stream discharges did not have a direct effect on these upstream migrations. Movements only take place at night and in Southern Indian Lake occurred over 10–18 days in May. The time of spawning migrations depends on latitude but in general occurs from April to May. In some extreme situations, on the southern limits of walleye distribution, spawning occurs earlier. For example, in the Canton Reservoir, USA, the peak of spawning occurs in mid March (Grinstead 1971). Walleye do appear to return to the same general vicinity to spawn each year. Bodaly (1980) found that 20% of the spawning population returned to the same stream whereas 62% exhibited homing to the same general vicinity. Although more studies are needed, there is evidence that distinct stocks of walleye use different areas of a lake to spawn. In Southern Indian Lake stocks of walleye with identifiable phenotypic differences spawned in streams 75 km apart (Bodaly 1980). As in *Perca* species, male walleye arrive at the spawning areas before the females and will remain there for a few days after the females have left. Males always predominate in numbers and at peak of spawning can reach a ratio of male:female = 52:1 (Bodaly 1980).

Walleye will spawn in streams or lakes in depths ranging from 0.1–4.6 m (Colby *et al.* 1979). Spawning areas include river and stream mouths, shallow shoreline areas and gravel reefs of lakes and tributary streams. Walleye tend to spawn in quiet water rather than in riffles (Corbett & Powles 1986). Eggs are laid over a substrate of rock, rubble or gravel although if these are not available spawning may occur over sand or silt. Survival is greater on sand-gravel-rock substrate than on mud-detritus. Submerged twigs or branches or vegetation in flooded marshy areas may sometimes be used. In the upper Mississippi walleye spawn over a substrate of sand, gravel and rocks in water depths of 0.6–6.1 m and a current velocity of 42.7–115.8 $cm\,s^{-1}$ (Pitlo 1989). Suitable habitat is often reduced by human activities. Auer and Auer (1990) found that only 3% of the Fox River, Wisconsin, USA, was optimal walleye spawning habitat. Low oxygen and high levels of hydrogen sulphide and ammonium nitrogen made at least 75% of the habitat unsuitable (Pallas 1771). In western Lake Erie there is an abundance of zebra mussels, *Dreissena polymorphae*, on walleye spawning grounds (Fitzsimmons *et al.* 1995). However, no apparent affect was noted on walleye egg deposition or egg viability and recruitment.

Spawning of walleye, unlike *Perca* species, is normally nocturnal and shows a diel pattern (Ellis & Giles 1965). Activity increases in the evening when illumination falls below about 0.172 lux. Courtship involves approaches by males or females to fish of the opposite sex either from behind or laterally and pushing the other fish. Sometimes a fish will drift back into another one pushing it backwards. The dorsal fins are alternately raised and lowered during these activities and in shallow water the fins can appear above the surface. Courtship may last one to two minutes before the release of eggs and milt. There is no pairing or territorial behaviour and courtship is promiscuous. This courtship activity increases with time and one or more females start to make forward and upward movements which culminate in a sudden rush to the surface by the females followed by one or more males. At the surface this group swims around vigorously releasing eggs and milt. This rush to the surface, release of sexual products and return to the bottom may last only for about five seconds. An individual female may spawn in this way up to three times in one minute. Ellis and Giles (1965) observed sizes of spawning groups ranged from one female with one or two males to two females with six males but the average is about seven fish (Eschmeyer 1950).

Walleye exhibit no parental care. The eggs have a sticky outer envelope, composed of muco-polysaccharides (Patzner & Glechner 1996), and adhere to rocks and gravel. The Volga pikeperch which spawns in the spring (May) releases its eggs in one batch (Danilenko 1992).

Zander spawn in April and May with extremes from February to July normally at a temperature of about 12°C but spawning can occur at 8°C. The faster the water warms the earlier the time of spawning. Specific nursery areas are chosen and spawning migrations occur within lakes, from lakes into rivers and from brackish areas to rivers. Zander in the Sea of Azov spawn in the Kubon River, in the Kubon Estuary, in the river flood plain and in lower sections of the river's tributaries. Over a period of 50 years study the spawning behaviour of these fish has changed and this is thought to be due to increased salinity in the Sea of Azov (Troitskiy & Tsunikova 1976). More adult zander now approach the Kubon coast in the autumn. The main spawning run takes place in late March or early April over 20–27 days. Fifty years ago spawning took place about one month later and could last for 43 days. A protracted period of spawning may increase the survival potential of the young as it increases the chances that some larvae will hatch at an optimum period.

In rivers, zander spawn in depths from 0.5–1.0 m usually in water velocities of 0.1–0.2 $m\,s^{-1}$ (Deelder & Willemsen 1964). This water velocity helps to remove silt from the eggs. In lakes, spawning sites are selected in depths from 0.5–17 m (far deeper than the 4.6 m maximum depth of walleye). The substrate selected is sand or stones in which male fish build nests. The male, while nest building, exposes plant roots on which the eggs are later deposited and where they stick. Female and male zander are monogamous as the female lays all eggs simultaneously and the male stays to protect the eggs and young fry.

Spawning of zander occurs at dawn. The female remains stationary over a nest while the male circles rapidly around her at a distance of about one metre. The male then assumes a vertical position. This behaviour continues for 20–25 min until both fish swim around very energetically, the water becomes very turbid and during this time eggs and milt are released. The female then moves away and the male remains to aerate the eggs with his pectoral fins (Botsjarnikowa 1952).

The sea pikeperch in the Caspian Sea has rather a similar reproductive biology to zander. Spawning occurs in the sea at depths from 3–12 m in nests built by the males in a sandy substrate (Guseva 1974). Spawning takes place from mid April to mid May at a temperature range of 10–17°C. All the eggs of one female are laid in one nest and the male guards these eggs.

4.5 Development of the embryo

Perca species produce a unique transparent gelatinous accordion-folded egg strand which can be up to 2.1 m long and 51–102 mm wide (Scott & Crossman 1973). Gillet *et al.* (1995) found the width of the egg strand was proportional to the size of the spawning female. The egg strand formed in the ovary is slightly heavier than water and becomes entangled in debris, submerged vegetation or attached to the lake or river bottom. Eggs are individually connected to the mucilaginous sheath which withstands mechanical shock and provides protection from predators and disease. Water can circulate and aerate the eggs between the outside and the central canal through spaces between the eggs. Milt also has to penetrate through these gaps into the centre as the spermatozoa enter the eggs through the micropyles which face inward (Thorpe 1977).

The egg of perch is clear, pale amber in colour and has a thick membrane which consists of an outer adhesive layer, a wide middle area composed of fine radially arranged fibres and an inner layer called the zona radiata. Before hardening, the diameters of the fertilised egg, yolk and the single oil droplet are 1.0–2.0 mm, 1.3 mm and 0.6 mm respectively (Thorpe 1977). Within three minutes of fertilisation the membrane swells and loses its adhesive properties and the egg reaches a diameter of 1.9–2.8 mm. The eggs of walleye and zander are spherical, translucent and pale yellow in colour (Deelder & Willemsen 1964; Colby *et al.* 1979). The mean egg diameter of walleye is 1.5–2.0 mm and the egg contains an oil droplet about 0.8 mm in diameter at the vegetative pole. The egg of zander is slightly smaller, 1.0–1.5 mm, and contains an oil globule of 0.4 mm. Both species lay adhesive eggs which clump together and to the substrate. The eggs swell after fertilisation due to uptake of water.

Embryonic development depends on water temperature. Mansueti (1964) studied the development of yellow perch eggs at an incubation temperature of 8.5–12°C (Fig. 4.6). The time intervals given here are based on Mansueti's studies. However, the incubation time to hatching of 25–27 days is longer than normal, 8–10 days for yellow perch recorded by Scott and Crossman (1973). The oil droplet in the fertilised egg shifts to one side after 14 min pulling yolk material with it. Frequently an indentation in the yolk appears in the area of the oil droplet. Within 30 min the transparent colourless blastodermal tissue grows over the oil droplet which continues to bulge from the yolk. Cleavage of the egg occurs within five hours into a two-cell stage. The late blastula has a peripheral germ ring and a blastocoel at about 21 h. The gastrula stage is reached in 29 h and the germ ring has grown one-third of the way around the yolk by this time. The embryo develops slightly off centre and up to six days no chromatophores are found. By the sixth day the tip of the tail becomes freed from the yolk. Between days 6–11 body melanophores appear and auditory vesicles, pectoral buds and caudal finfolds form. From days 14–16, pigment appears in the eyes, the pectorals are well developed and the myotomes are nearly complete. Melanophores cover most of the yolk and the vent is visible at the junction of the tail

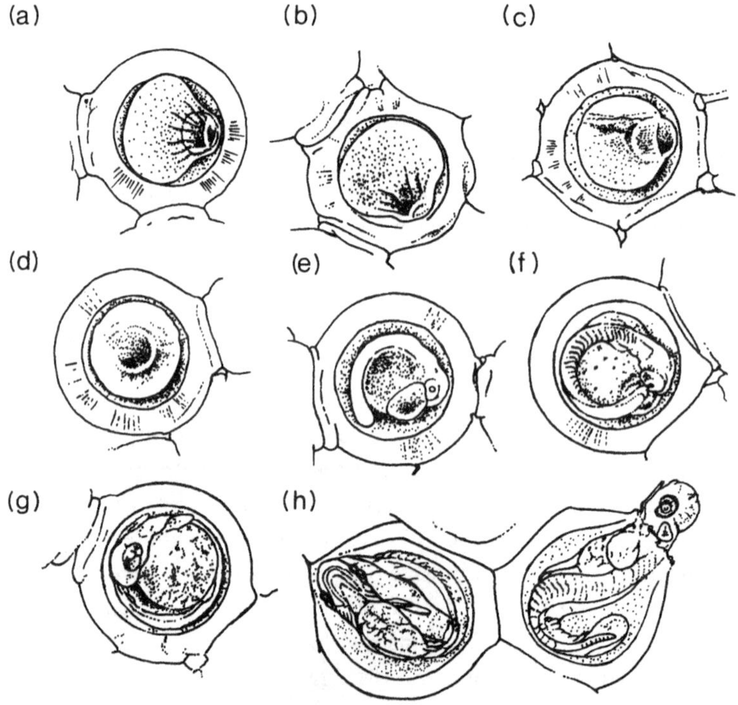

Fig. 4.6 Embryonic development of yellow perch: (a) prior to blastula formation (14 min), (b) blastula (22 min), (c) early gastrula (21 h 25 min), (d) gastrula (29 h 9 min), (e) tail-free (144 h), (f) formation of myotomes and pectoral buds (11 days), (g) pigmentation of the eyes and development of the pectoral fins (16 days) and (h) hatching (27 days). Egg diameter is *c.* 2 mm. (After Mansueti 1964.)

and the yolk. Primitive gill structures and the mouth are present at this stage. The structure of the heart and auditory vesicles can be plainly seen and the tail is extended in a curve over the head. At about 24 days the egg case begins to soften and hatching occurs from day 25–27. Emerging fry range in length from 4.1–6.6 mm (Thorpe 1977).

Embryonic development in *Stizostedion* species is similar to *Perca* species. A detailed account of walleye development under laboratory conditions where the eggs were held at 15°C has been given by McElman and Balon (1979). Cleavage starts after 2.5 h and ends just after 48 h (Fig. 4.7a-d). Cleavage of the blastodisc brings about reduction in blastomere size but an increase in their number. At just under 14 h the periblast is discernible on the margins of the blastodisc. The blastodisc assumes a dome shape and becomes compressed. Cells become individually indistinguishable and the translucent cell mass becomes dark and opaque. At nearly 29 h the embryonic shield and germ ring appear in the thicker areas of the blastoderm and by about 31 h the blastoderm envelopes about 50% of the surface area of the yolk (Fig. 4.7e).

The embryonic phase lasts from 46 h to 9 days when hatching occurs. It is separated from the cleavage phase by the formation of primordial optic vesicles. The first step is the differentiation of head and trunk by 2 days 22 h. At this time the germ ring closes around the top of the oil globule forming a trunk–tail mound (Fig.

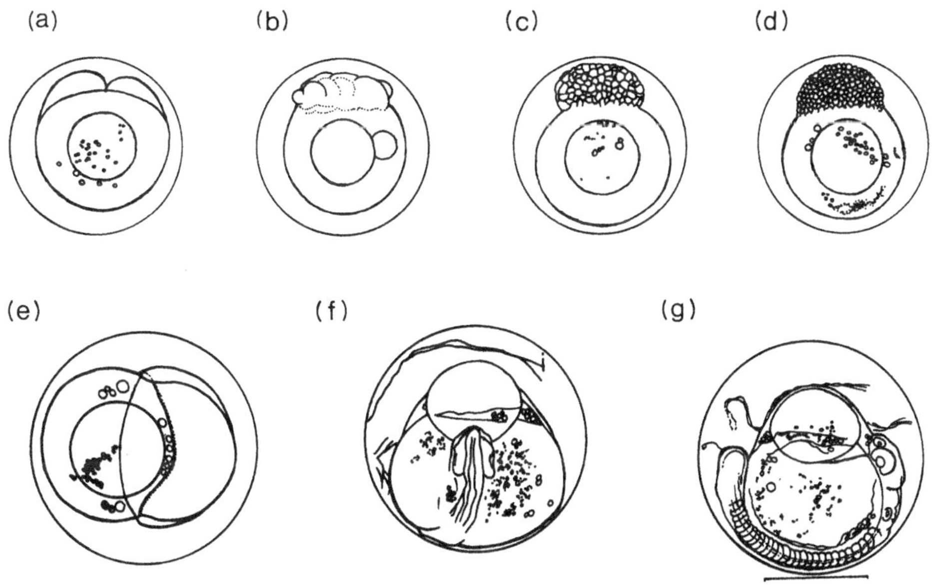

Fig. 4.7 Early embryonic development of walleye: (a) two blastomeres (3 h 47 min), (b) 8 to 16 blastomeres (6 h 41 min), (c) late cleavage – vertical view (9 h 20 min), (d) late cleavage (13 h), (e) mid epiboly vertical view (1 day 7 h 10 min), (f) head and trunk development (2 days 22 h) and (g) left side view of embryo (3 days 50 min). (After McElman & Balon 1979.)

4.7f). The body undergoes a 90° anterior translocation during this step. By 2 days 1.5 h the trunk–tail has moved off the top of the large oil globule, six somite pairs appear in the mid-body region and a notochord becomes visible. Small oil globules accumulate around and below the posterior half of the body especially in the trunk–tail region. At 2 days 6 h the head is elevated to 0.25 mm above the yolk blastoderm and has a maximum width of 0.55 mm at the optic vesicles. Somites total 9–11. By 2 days 10 h the head is 0.30 mm above the yolk blastoderm and three slight constrictions of the hindbrain are apparent. Optic stalks develop between the brain and each optic vesicle. Optic cups form from these vesicles by just over 2 days 16 h, but due to expansion of the fore- and midbrain regions, they are not connected to the brain. The trunk–tail is now below the lower margin of the oil globule and the head is above. By 2 days 17.25 h, the embryo re-establishes its bilateral symmetry with respect to the yolk and large oil globule. It has a total length of 3.20 mm and 17–19 pairs of somites. The small oil globules have disappeared from the trunk–tail area and the yolk starts to diminish from the lower half of the large oil globule. The trunk–tail region becomes elevated. Development up to nearly 3 days 7 h (Fig. 4.7g) includes formation of spherical lenses and olfactory and auditory vesicles. The trunk–tail becomes elevated to 0.25 mm. The dorsal median fin fold is apparent anterior to the tip of the trunk–tail band and posterior to the 29–31 pairs of somites most of which are slightly V-shaped. The cerebellum is discernible due to lateral expansion and the hindbrain becomes split by a midline cleft.

From 3 days 7 h to 4 days 10 h melanophores form, contraction of the primordial skeletal and cardiac musculature (heart rate contractions are about 87 min^{-1}) occurs, otoliths appear and there is a rapid increase in length up to 4 mm (Fig. 4.8a). Somites total 45 pairs. During the next stage up to 5 days 10 h, somite segmentation is completed (52 pairs), eye pigments develop and a simple blood flow occurs in a sub-intestinal vitelline circulatory system. The embryo reaches a total length of 5.5 mm. It becomes more mobile and a thickening of tissues in the ventral finfold posterior to the yolk indicates the position of a primordial rectal region. Hatching cells appear on all exposed surfaces of the head to just behind the auditory vesicles. Heart contraction rate increases up to 7 days 4 h of development (the highest mean value observed was 164 min^{-1}, McElman & Balon 1979). New blood vessels increase circulation (Fig. 4.8b), blood circulates to head, liver and myomeres (somites become differentiated into muscle fibres), and haemoglobin is synthesised. The urinary bladder develops, adjacent and posterior to the lower half of the rectum. Melanophores are concentrated in the lateral and ventral yolk surfaces and along the ventral surface of the caudal fin. They eventually form a continuous series on the ventrum of the post-anal myomeres. The head continues to lift away from the large oil globule and mouth formation becomes noticeable, as do the otoliths. Eggs appear black and opaque. The final stage to time of hatching at about nine days (Fig. 4.8c) includes an increase in liver size. Heart rate contractions decrease to around 106 min^{-1} and vitelline blood circulation reaches a maximum. Cartilage forms in

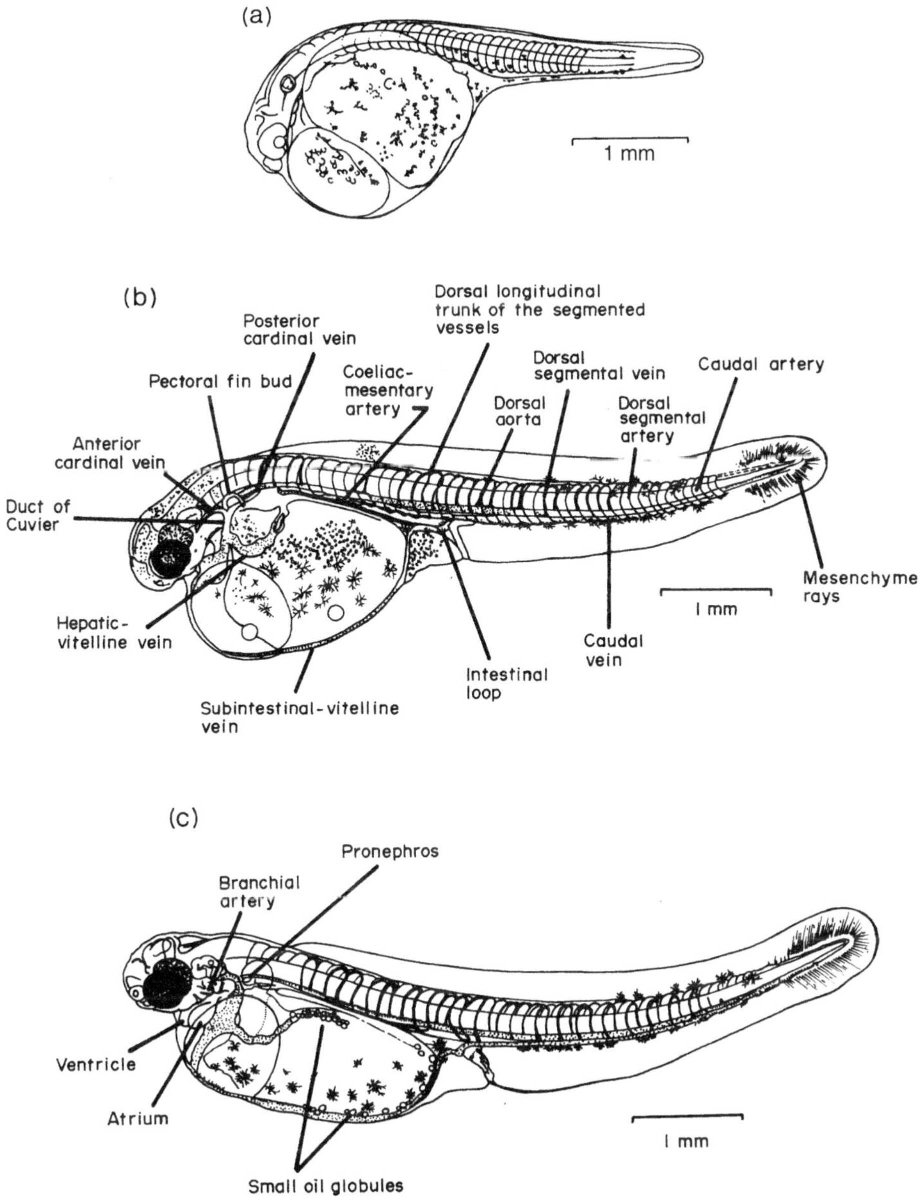

Fig. 4.8 Late embryonic development of walleye: (a) embryo at 4 days 5 h 15 min (right side, after removal from the egg membranes), (b) embryo at 6 days 18 h 10 min (left side, after removal from the egg membranes) and (c) embryo at 8 days 1 h 35 min, immediately after hatching (left side). (After McElman & Balon 1979.)

the head. The embryo shows movements within the egg membranes and there is a straightening of head and trunk.

As in perch and walleye, zander hatch with well developed eyes but with little pigment in the eyes or on the yolk sac. Sauger also hatch with only a few chromatophores. Walleye hatch at a length of 6.0–8.6 mm and zander hatch at a length 4.0–5.0 mm.

Guma'a (1978b) incubated perch eggs at ten different temperatures (range from 6–22°C) and established that development time decreased with increasing temperature (Fig. 4.9). It is the same for yellow perch (Hokanson & Kleiner 1974). The relationship between speed of embryonic development and incubation temperature is exponential (Fig. 4.10). However, the degree-days required by perch for 10, 50 and 90% of the eggs to hatch is a constant regardless of the incubation temperature. These are (with their 95% confidence limits): 91.4 (83.3–102.0) degree-days above 4.6°C for 10% hatch; 97.1 (90.9–104.2) degree-days above 4.9°C for 50% hatch; 101.0 (94.3–108.7) degree-days above 5.0°C for 90% hatch. The largest and healthiest perch and zander larvae hatch at 14°C (Kokurewicz 1969; Guma'a 1978b). The highest viability of yellow perch and walleye embryos are produced by parents acclimatised at 6–9°C and incubation of the embryos at 9–15°C.

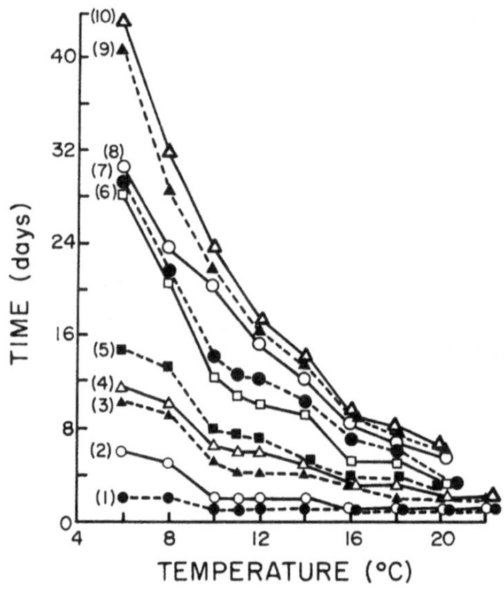

Fig. 4.9 The effect of temperature on the embryonic development of perch: (1) complete blastula, (2) complete gastrula, (3) early embryo, (4) free tail, (5) embryo moving but eyes still unpigmented, (6) eye pigment developed, (7) hatching started, (8) 10% hatched, (9) 50% hatched and (10) 90% hatched. Points represent the means of six observations. (After Guma'a 1978b.)

The viability of spawned eggs ranges from 3.4–100% in walleye (Colby *et al.* 1979). In a batch of perch eggs, Reyes *et al.* (1992) recorded that 91.6% embryos were viable, 1.1% were infertile, 5.1% were dead and 2.2% had abnormalities. A similar study by Treasurer (1983) found that later in development up to 10% of the embryos were deformed. Zeh *et al.* (1989) estimated perch survival from egg to eye pigmentation as 92–99%. A number of environmental factors affect the mortality of fertile eggs. Higher temperature causes higher mortality especially at late stages of

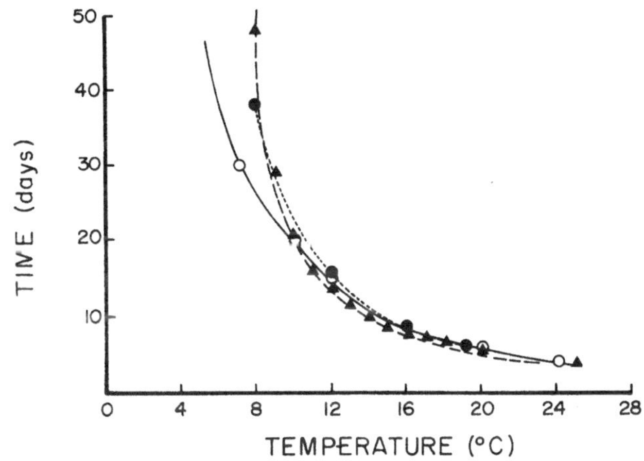

Fig. 4.10 The effect of temperature on the mean incubation time of perch (●), walleye (○) and zander (▲). (After Guma'a 1978b, Colby *et al.* 1979 and Deelder & Willemsen 1964.)

development (Hurley 1972; Guma'a 1978b). In walleye 3–5% of mortality occurred from fertilisation to eyed stage, compared to 23–44% from the eyed stage to hatching (Hurley 1972). Oxygen concentration is important but levels down to 3.4 mg 1^{-1} show no significant effects on developing embryos. Below this value larvae are smaller at hatching (Siefert & Spoor 1974). In streams and rivers, currents remove debris which might cover the eggs and reduce oxygen concentration. Death of embryos may occur when eggs spawned in flooded marshy areas become exposed to air following a drop in water level. Eggs laid in lakes may be destroyed by high winds which cause the eggs to be washed ashore. Clady and Hutchinson (1975) estimated that 0.8% of the total eggs laid by yellow perch in one area of Oneida Lake, USA, perished for this reason.

Mortality of eggs increases with a lowering in pH. Perch eggs develop normally above pH 5.5 (Runn *et al.* 1977). At pH 5.0 mortality is still *c.* 10% but incubation time increases (Rask 1984). The mortality at pH 4.0 is 59% and development time is increased by 29%. At pH 3.5 no eggs hatch. Walleye eggs are more sensitive to low pH than perch. Mortality of 90.5% from fertilisation to eyed stage of walleye occurs at pH 5.4 (Hulsman *et al.* 1983) although mortality is comparatively low (25.5–33.5%) at pH 6.0.

In an examination of perch egg strands, Diamond and Wakefield (1986) found two species of flatworms and three species of caddisfly larvae (19 taxa were recorded) were preying on the eggs. Invertebrates were not considered to be important predators on perch eggs. Newsome and Tompkins (1985) found no predation on yellow perch egg masses by eel, *Anguilla rostrata* (Leseuer), white suckers, *Catostomus commersoni* (Lacépède), brown bullheads, *Ictalurus nebulosus* (Leseuer), yellow perch, and three- and nine-spined sticklebacks (*Gasterosteus aculeatus* (Linnaeus); *Pungitius pungitius* (Linnaeus)). They suggested that an unpalatable rather than a toxic substance is present to deter predation. Several examples have been given of predation on walleye eggs including by spottail shiners, *Notropis hudsonius* (Clinton) (Corbett & Powles 1986) and by white perch, *Morone americana* (Gmelin), (Schaeffer & Margraf 1987). In the latter case, predation decreased as the white perch themselves spawned.

4.6 Development of the larva

At the start of exogenous feeding most of the yolk has been absorbed. There is no stomach for 4–5 weeks after hatching and the pancreas is probably not functional during early exogenous feeding (Richter *et al.* 1991). The larva from the time of hatching until this stage is referred to as a prolarva. As yolk absorption continues, the mouth becomes fully developed (Fig. 4.11). Larvae of *Perca* and *Stizostedion* species develop large mouths, well-developed jaws, teeth and eyes. At the end of the prolarva stage the gut is a simple straight tube and the larva has teeth, buccal cavity, oesophagus, stomach, liver and gall bladder. Blood vessels develop in the gills and gut as the importance of gill respiration increases and with the onset of feeding. At 7.0 mm the yellow perch operculum can be distinguished behind the pre-operculum and four gill arches can be identified (Mansueti 1964). At this length the larva begins active exogenous feeding although food is still gained from the yolk sac. At 7.2 mm the head is more elongate. There is more extensive development of chromatophores particularly over the head, auditory vesicles and developing air bladder. In the postlarva phase pigmentation increases on the head and body. The vertical bands on yellow perch appear at about 20 mm length. In this species all fins are differentiated at a fish length of 14 mm but not all are fully formed with spines and rays until fish are 21–27 mm. The scales appear on the caudal peduncle at a fish length of about

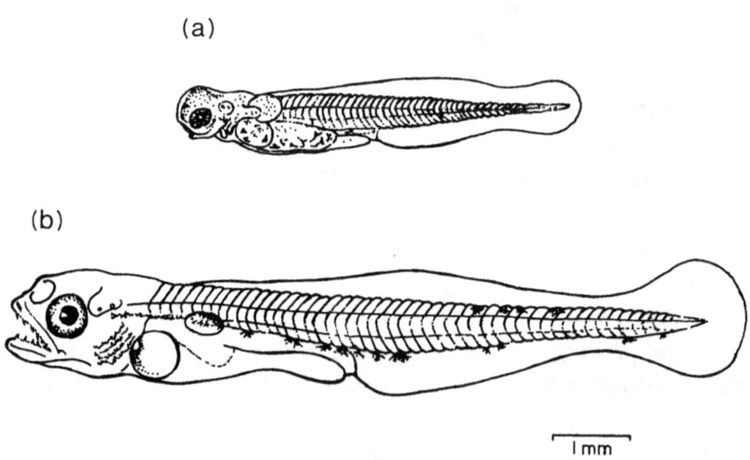

Fig. 4.11 The prolarvae of: (a) yellow perch and (b) walleye. (After Mansueti 1964 and Li & Mathias 1982.)

20 mm and scalation proceeds along the lateral line (Pycha & Smith 1955). Perch are fully scaled at 36–37 mm. One or two scales are present in the peduncle region of walleye at 24–30 mm (Priegel 1964; Glenn & Mathias 1985). Scale development proceeds in a similar fashion to perch and is complete at 45 mm. Adult walleye colouration is completed at 35 mm (Nelson 1968).

Urho (1996) has described methods of distinguishing between perch, zander and ruffe larvae and juveniles. At the early yolk sac stage, the number of myomeres from yolk to anus is 2–3 for ruffe, 4–6 for perch and 7–9 for zander. In older larvae the ruffe has 13–16 pre-anal myomeres and both perch and zander have at least 17. The number of post-anal myomeres at this stage is 22–24 in ruffe, 23–26 in perch and 27–31 in zander.

Li and Mathias (1982) divided the postlarva phase of walleye into two stages. Stage I postlarva has an oil globule but no yolk. The gut is very similar to that of the prolarva although a sphincter divides the gut into intestine and rectum. Some exogenous feeding occurs, combined with the endogenous energy source from the oil globule. Stage II of the larval development commences when the oil globule is fully absorbed. The stomach wall thickens and expands and the first loop appears in the intestine (Fig. 4.12). The gill filaments increase in size and the gill rakers begin to differentiate. At the end of Stage II the larva is 16–19 mm TL. During the juvenile stage the gut develops further with expansion of the stomach and the formation of pyloric caecae. Three caecae are formed at 19 mm. Gill rakers continue to differentiate and take on a filtering function.

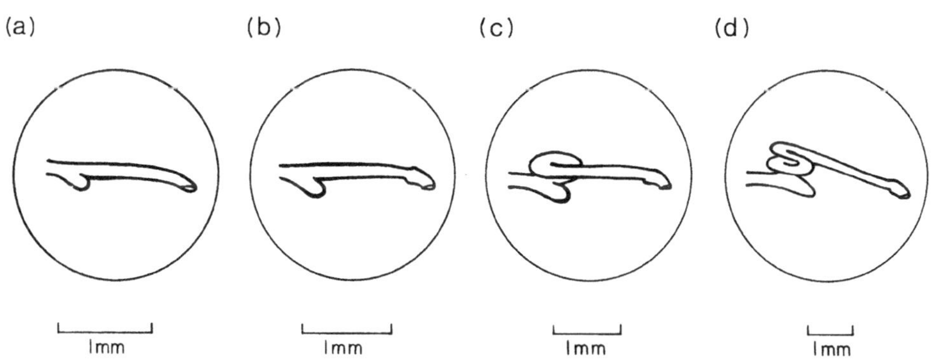

Fig. 4.12 Gut development of walleye: (a) prolarva, (b) postlarva–I, (c) postlarva–II and (d) juvenile. (After Li & Mathias 1982.)

Newly hatched *Perca* species larvae tolerate a temperature range from 3–28°C but are inactive below 5.3°C (Hokanson 1977a). Optimum survival occurs when the temperature is raised 1°C day^{-1} from initial incubation temperature to reach 16–18°C at the prolarvae stage. For *Stizostedion* species this optimum temperature lies between 14–23°C. During the postlarval feeding stage the optimum temperature is 21°C for *Perca* species rising to a physiological optimum of 25°C for juveniles. No postlarvae survive at temperatures $\leq 10°C$. These optima are probably similar for *Stizostedion* species although zander juveniles and adults have a higher physiological optimum temperature of about 27°C. Smith and Koenst (1975) found that no walleye or sauger postlarvae survived at temperatures $\leq 15°C$ or less compared to 8% at 21°C. Critical levels of oxygen concentrations for the larvae are those <50% (4.8 mg l^{-1} at 17°C) (Siefert & Spoor 1974). Growth and survival of larvae and juveniles will be strongly influenced by other factors such as food availability and predators. Li and Mathias (1982) identify the critical period for walleye larvae to occur at the commencement of exogenous feeding when food density and availability need to be at an optimum. Both *Perca* and *Stizostedion* species larvae are cannibalistic on their siblings (Smyly 1952; Li & Mathias 1982). Cannibalism by adults also takes place in both species when larvae are >18 mm. Other predators include fishes, in particular pike, and invertebrates. A cyclopoid, *Diacyclops thomasi* Forbes has been found to prey heavily on yellow perch larvae (Hartig & Jude 1984). The alewife, *Alosa pseudoharengus* (Wilson), preys on yellow perch larvae in Lake Ontario and could affect recruitment of the latter in the lake (Mason & Brandt 1996).

Under laboratory conditions, Bein and Ribi (1994) tested the effect of population density and salinity on survival and initial swim bladder inflation of larval perch. Survival was highest at low (4 larvae l^{-1}) compared to high (16 larvae l^{-1}) densities and at salinities of 0.6 ppt and 1.2 ppt compared to freshwater. Initial swim bladder inflation was not affected by density or salinity. From lakes in central Europe, Egloff (1996) found that 0.1–7.9% of larval perch failed to inflate their swim bladders.

4.7 Behaviour of larvae and fry

The embryos of *Perca* and *Stizostedion* species show some movement to the surface and they have been observed to do so with the egg membrane still attached to the head (Belyi 1972; Reighard 1890). The larvae eventually escape from this membrane by energetic movements. The prolarvae swim upward at an angle of about 30 degrees to the vertical and then sink passively head down at an angle of about 80 degrees. The oil globule gives some buoyancy at this stage since the swim bladder is not fully formed and inflated (Belyi 1972; Guseva 1974; Ross *et al.* 1977; Li & Mathias 1982). Usually the movements upwards are directed towards the surface film. Belyi (1972) noted that zander larvae spent about seven seconds in active movement followed by sinking for 8.6 s. The active movement occurred for 46% of a 24 h period over the first four days from hatching. In a fluvial system the larvae are carried passively by currents usually in the middle layer of the flowing water, speed of transport varying with stream velocity and occurring during periods of decreasing light intensity (Corbett & Powles 1986). At the postlarvae stage the fish become stronger and swim almost continuously at 30 degrees to the horizontal, near the surface. They find areas of little flow and cease to move downstream. Many populations of walleye and zander have larvae which migrate passively from a fluvial to a lacustrine habitat. The latter environment will, as a rule, have a more suitable food supply. Larvae of *Perca* and *Stizostedion* species are positively phototactic. One- to eight-week-old walleye larvae (9–32 mm in length) were attracted to the highest light intensity (7800 lux) used under experimental conditions (Bulkowski & Meade 1983). Fish > 8 weeks (32–40 mm) were attracted to the lowest light intensity (2–4 lux). Fingerlings of zander migrate downstream over 24 h in turbid rivers but at dusk or in the night in clear water.

Most studies of lake *Perca* species have shown that after hatching the larvae become limnetic (Chirkova 1955; Guma'a 1978c; Coles 1981; Whiteside *et al.* 1985). They are distributed over the lake by currents controlled by wind action. They remain pelagic until full development of fins which varies according to water temperature. This may be June in Bala Lake, North Wales (Coles 1981) or mid July in Windermere, England (Guma'a 1978c). The fingerlings then move inshore to the littoral zone, in depths usually < 3 m, and form shoals. At the end of the summer (end of August in Bala Lake and end of September in Windermere) the perch move to deeper water to overwinter. In the Volga Delta, perch larvae are negatively buoyant to aid them in remaining inshore and preventing their movement downstream (Pavlov *et al.* 1996).

Walleye prolarvae in lakes are transported by lake currents. Surface currents generated by wind action transport older larvae. In Oneida Lake, USA, Houde (1969a) found that larval yellow perch and walleye were distributed according to prevailing winds. Most of the yellow perch were in two metres depth but the walleye were in two metres depths when the winds were light and in three metres depth when winds were stronger, $> 5\,m\,s^{-1}$. Walleye are able to distribute themselves more actively by this behaviour at an earlier stage, within one to two weeks after hatching, than yellow perch (Colby *et al.* 1979). In Lake Gogebic, USA, walleye fry move inshore at a length of 25 mm in late June or early July (Eschmeyer 1950). This compares to a length of about 35 mm attained by Oneida Lake walleye before they migrate to inshore areas (Forney 1976).

Chapter 5
Food and Feeding

5.1 Introduction

As well as obtaining energy from its food, mainly from lipids and proteins, and sometimes lesser amounts from carbohydrates, a fish must take in essential vitamins and mineral salts (Ketola 1978; Milliken 1982). This whole process is very complex and involves many interactions between the external environment and the internal system of the fish. This chapter describes the type of foods eaten, prey selection, food consumption, digestion, respiration and allocation of energy intake of percids. Percids are common members of temperate freshwater communities and often play a major part in food web dynamics. Many bodies of water have become eutrophic producing nuisance algal blooms. The role and use of percids in biomanipulation is described.

5.2 Food items in the diet

Percids tend to be opportunistic feeders utilising available prey although in young fish prey size may be limiting. The larvae of *Perca* and *Stizostedion* species commence exogenous feeding on algae, protozoans, immature copepods and cladocerans, and sometimes rotifers (Smyly 1952; Belyi 1972; Kukuradze 1974; Mathias & Li 1982; Whiteside *et al.* 1985). Faurot and White (1994) found overlap in the diets of larval walleye and yellow perch in Lake Roosevelt, USA. The most abundant prey in the stomachs of walleye was calanoid copepods compared to calanoid and cyclopoid copepods in yellow perch stomachs. Guma'a (1978c) identified algal cells and ciliates in the stomachs of larval perch from Windermere, England (Fig. 5.1). As the fish increase in length they consume larger members of the plankton. In two Scottish lochs the first prey of larval perch was *Keratella* sp. (Bory de St. Vincent), *Volvox* sp. (Linnaeus), copepod nauplii and Stage 1 copepodites of *Cyclops strenus abyssorum* Sars (Treasurer 1990). Larger copepodites and adults of *C. s. abyssorim* and *Diaptomus gracilis* (Sars 1863) were consumed as the larvae grew. Cladocerans were eaten by larvae > 11 mm. In a study of the food of perch fry in Windermere, England, Smyly (1952) observed that zooplankton were the dominant prey from May to September. Some benthic organisms were eaten

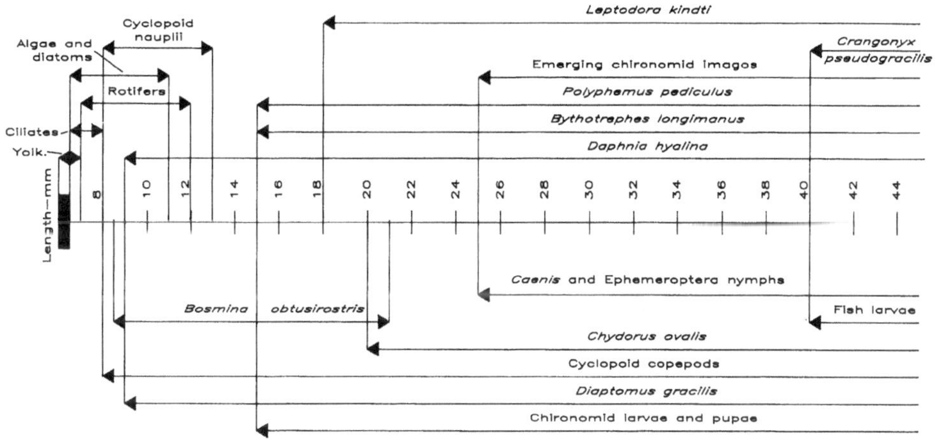

Fig. 5.1 Changes in the diet as a function of body length of young-of-year Windermere perch. (After Guma'a 1978c.)

after the fry had reached the age of one month and cannibalism by larger fry on smaller siblings took place only in mid July. In ponds, young walleye feeding progressed from zooplankton to *Chaoborus* sp. Lichtenstein and Ephemeroptera nymphs, then to *Gammarus lacustris* Sars, and finally to fishes (Mathias & Li 1982). In many lakes young walleye are piscivorous by midsummer (Jackson *et al.* 1992). Swenson and Smith (1976) found that by August in Lake of the Woods, USA, walleye fry fed predominantly on young-of-the-year yellow perch. Zander in their first year show similar trends to walleye. Early in the summer the fry feed on large crustaceans such as mysids and on the young of other fishes (for example *Clupeonella* sp. Kessler and roach, *Rutilus rutilus* (Linnaeus)). Van Densen (1985a) established that piscivory of first year zander was influenced by the availability of suitable prey fish. If there was an abundance of slender bodied larval smelt, piscivory started at *c.* 20 mm (mid June). Cannibalism and predation on larval perch was less because of their larger circumference. Larval bream, *Abramis brama* (Linnaeus), and roach were only preyed on if their growth was retarded. Survival of fish-eating zander was higher than the smaller, planktivorous feeding individuals.

Perca eat a wide range of organisms at all stages of their life (Fig. 5.2) and they are cannibalistic when young-of-the-year are abundant. In Slapton Ley, England, perch became piscivorous when they were >150 mm in length (Craig 1974a). In Windermere, England, perch over a large size range (32–311 g) fed on organisms ranging from zooplankton to perch fry, but there was little relationship between size of fish and the type of food organism eaten (Craig 1978). Differences in prey can usually be explained more by seasonal vulnerability than by size of predator (Fig. 5.3). Lott *et al.* (1996) found that the fastest growth of yellow perch was attributed to a diet of macro-invertebrates, in particular chironomids, amphipods and corixids. *Hexagenia* sp. (Walsh) is important in the diet of yellow perch in Lake Erie (Shaeffer *et al.*

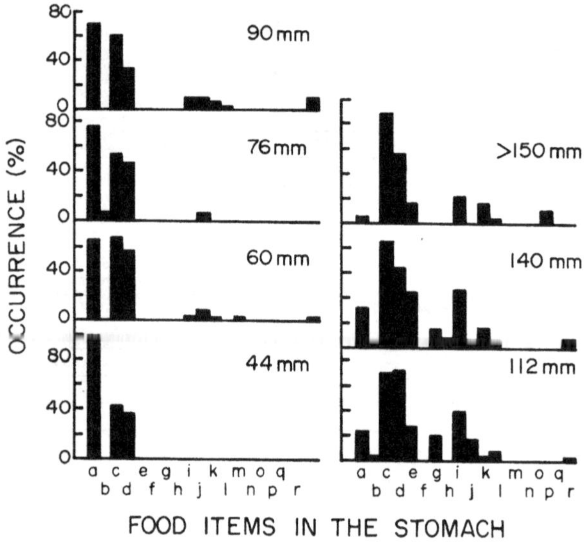

Fig. 5.2 Changes in the diet as a function of body length of juvenile and adult Slapton Ley perch: (a) plankton, (b) *Chaoborus*, (c) chironomid larvae, (d) chironomid pupae, (e) *Asellus*, (f) *Sialis*, (g) tricopteran larvae, (h) *Hydroporus*, (i) emphemeropteran nymphs, (j) *Corixa*, (k) others, (l) insect remains, (m) leeches, (n) tubificids, (o) perch, (p) rudd and roach, (q) fish remains and (r) empty. (After Craig 1974a.)

1994). In small humic lakes in Finland, *Asellus aquaticus* (Linnaeus) is a major food item in the food of perch (Rask & Hiisivuori 1985).

Juvenile and adult walleye consume a wide range of fish species depending on what is available. In many lakes invertebrates are also eaten by all size classes, especially in late spring and early summer. These include mayfly nymphs, chironomids, amphipods and leeches (Colby *et al.* 1979). In the central and northern areas of walleye distribution, yellow perch are the most common prey. Walleye food items in western Lake Erie changed seasonally but remained fairly consistent from year to year (Knight *et al.* 1984). Walleye aged one year or older were almost entirely piscivorous. In the spring they fed on yearling emerald shiners, *Notropis atherinoides* Rafinesque, and spottail shiners, *N. hudsonius*, but changed to young-of-the-year gizzard shad, *Dorosoma cepedianum* (Lesueur), and alewife, *Alosa pseudoharengus*, in late July. In the autumn both shiners and clupeids were eaten. Spiny-rayed fish such as other percids and troutperch, *Percopsis omiscomaycus* (Walbaum), were less important than shiners and clupeids in the diet. Invertebrates, mainly chironomids, were only found in the stomachs of one-year-old walleye, chiefly in the spring. From this study Knight *et al.* (1984) concluded that yellow perch were more opportunistic feeders than walleye in western Lake Erie, a view supported by others for these species generally in North America (Ney 1978).

Zander up to two years of age may continue to eat mysids and gammarids if these are abundant and forage fish are scarce. Zander eat mainly shoaling fish including

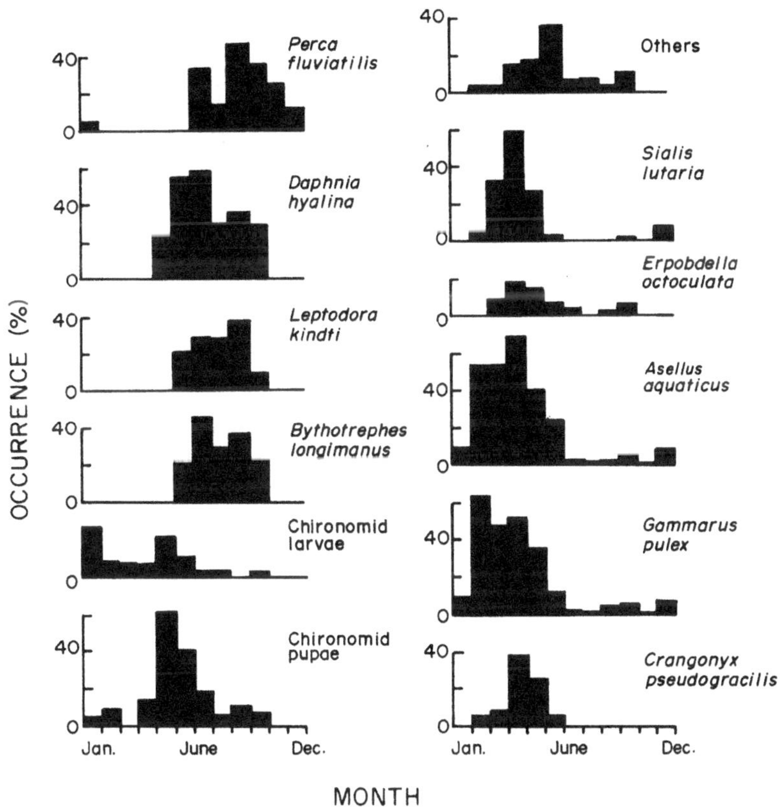

Fig. 5.3 Monthly changes in the diet of adult Windermere perch. (After Craig 1978.)

their own young (Biró 1973). Typical foods include roach, ruffe and perch. In the northern part of the zander range, smelt, *Osmerus eperlanus*, and vendace, *Coregonus albula* (Linnaeus), are frequently eaten while in the south gobiid fishes are preferred (Fig. 5.4). Like other percids, zander exhibit extensive plasticity in prey selection. When smelt were low in abundance in the Baltic Sea the zander switched to the ruffe (Samokhvalova 1982). In Himmerfjärdin Bay, in the northern Baltic Sea, zander feed on marine prey including crustaceans, gobies (*Pomatoschistus* sp. (Malm)) and herring (*Clupea harengus* Linnaeus). The size of the prey fish is *c.* 30% of the length of the zander (Hansson *et al.* 1997). In the Danube Delta, zander consumed white bream, *Blicca bjoerkna* (Linnaeus), crucian carp, *Carassius carassius* (Linnaeus) and rudd, *Scardinius erythrophthalmus* (Linnaeus) (Kukuradze 1974). They fed away from the delta from May to August, when the brackish water was diluted with freshwater, and prey included anchovy, *Engraulis encrasicolus* (Linnaeus), sprat, *Sprattus sprattus* (Linnaeus), silverside, *Atherina presbyter* Valenciennes, and scad, *Trachurus trachurus* (Linnaeus).

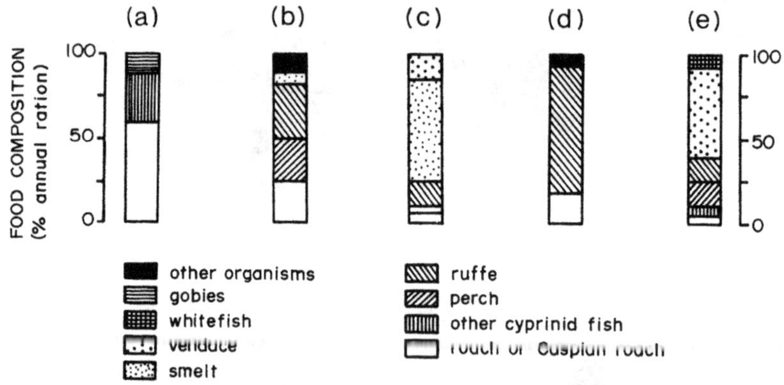

Fig. 5.4 Changes in the diet of adult zander as a function of the water body they occupy. (a) Volga Delta, (b) Rybinsk Reservoir, (c) Pskov-Chudskoy Lake (with high smelt abundance), (d) Pskov-Chudskoy Lake with low smelt abundance and (e) Syam Lake. (After Popova & Sytina 1977.)

5.3 Food selection

Light and size and availability (density) of prey are important in food selection but a more difficult factor to measure is the acceptability of the prey. Young yellow perch eat only the longer-bodied daphnids at low light intensities. As light intensity increases they feed on smaller and a wider array of prey (Mills *et al.* 1986). The size of the mouth gape in young percids determines the size of the prey eaten (Guma'a 1978c; Treasurer 1990; Graham & Sprules 1992). Guma'a (1978c) related perch mouth gape-height to the maximum carapace width of *Daphnia hyalina* Leydig found in the stomach of the perch from Windermere, England (Fig. 5.5). Before

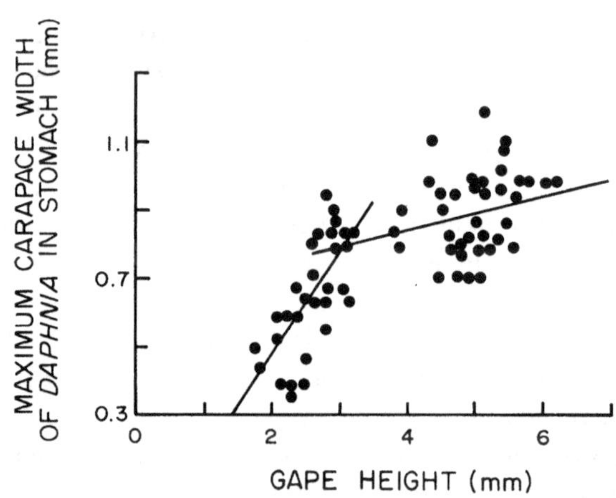

Fig. 5.5 Maximum carapace width of *Daphnia hyalina* and the gape height of young-of-the-year Windermere perch. (After Guma'a 1978c.)

August, fry could not take *D. hyalina* with a carapace width >0.8 mm, but from August onwards, rapid growth in the gape of perch enabled even the largest *D. hyalina* in the population to be consumed (Fig. 5.6). The perch selected the larger prey; the mean length of the prey in the perch stomachs was greater than the mean in the surrounding water. Before mid July yellow perch were eating small (*c.* 1.3 mm length) *Daphnia pulicaria* Brandl in West Blue Lake, Canada. An increase in gape size allowed the yellow perch fry to eat all sizes of the cladocerans present after this time (Wong & Ward 1972) although these fry did not always eat the largest available cladocerans. Young yellow perch at 30–35 mm in Oneida Lake, USA, shifted their preference for prey from *Diaptomus minutus* Lilljeborg to *Daphnia pulex* Leydig (Mills *et al.* 1984). They were capable of consuming the largest *D. pulex* but did not preferentially select them. Confer *et al.* (1990) suggest that young fish would select prey that would provide the maximum ingestion rate at low prey densities but would select for prey that were digested more rapidly and provided the greater growth efficiency at higher prey densities. The young fish may switch from familiar to novel prey if there is a nutritional advantage (Mills *et al.* 1987). Walleye larvae (9.8–18.0 mm) selected medium-sized prey (0.3–0.6 mm width), mostly *Ceriodaphnia*

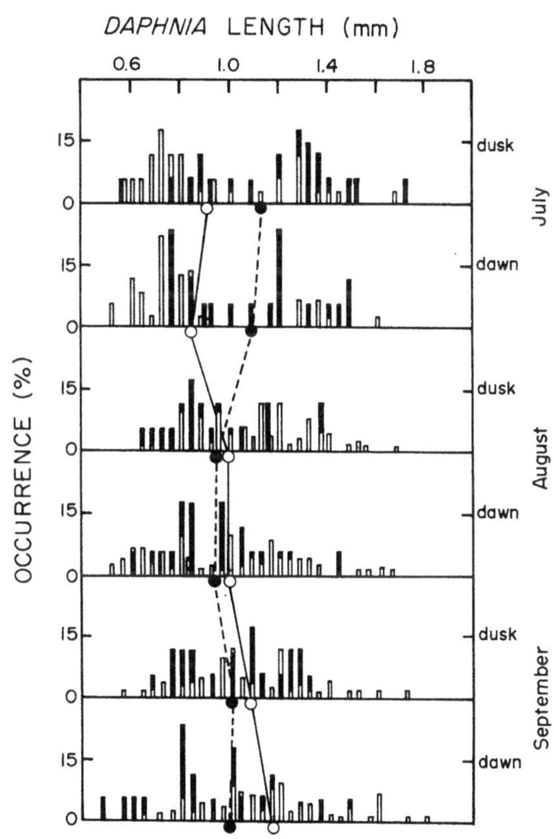

Fig. 5.6 The body length of *Daphnia hyalina* eaten by young-of-the-year Windermere perch (□) and that of the plankton (■). The solid and broken lines represent the mean lengths of *D. hyalina* in perch stomachs (○) and in the plankton (●) respectively. (After Guma'a 1978c.)

quadrangula (O.F. Muller 1785) in the laboratory (Johnston & Mathias 1994). They were negatively elective for prey <0.3 mm, mostly cyclopoid copepods, and neutrally elective for prey ≤0.6 mm, mostly *Daphnia* sp. Muller, Johnston and Mathias (1994) suggest that feeding success of walleye larvae may be limited by both the abundance and size composition of the zooplankton community. In adult, piscivorous fish diets may be a reflection of the abundance of prey fishes (Campbell 1992; Hartman & Margraf 1992).

Perch fry may show a preference for particular prey. Chabot and Maly (1986) found that within a particular size class of prey, yellow perch had a definite species preference. Furnass (1979) demonstrated that as long as the preferred prey *Eudiaptomus gracilis* Sars were visible, the perch would continue to strike at this species even though another prey species, *Daphnia hyalina* var. *galatea* Sars, of similar size was more abundant. Movement of the prey is an important factor in the feeding of both *Perca* and *Stizostedion* species. Under experimental conditions both perch and walleye strike at glass containers holding live food organisms (Furnass 1979; Rottiers & Lemm 1985). Both species also respond to certain odours in the water. For example, walleye reacts positively to washings from live *Daphnia* sp. and *Artemia* sp. Leach (Rottiers & Lemm 1985).

Significant differences can occur in feeding behaviour between individuals within a population (Ansari & Qadri 1989). Perch and yellow perch are shoaling predators although some become ambush predators living a solitary life and having a home range. They are active in the daytime. In the early morning the stomachs of demersal young-of-the-year yellow perch in Oneida Lake, USA, were empty (Noble 1972). The fish then fed on zooplankton and the stomachs quickly filled. The stomachs remained full until about 1600 h when the volume of stomach contents declined. Feeding followed a bell-shaped curve in which the stomach was fullest

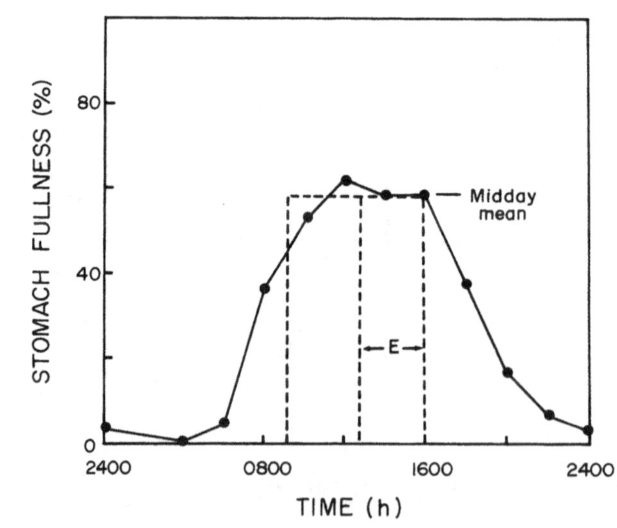

Fig. 5.7 The mean daily feeding period of young-of-the-year Oneida Lake yellow perch. E = the stomach evacuation time. (After Noble 1972.)

from 1200 to 1600 h (Fig. 5.7). For most of the year, peaks of activity of adult *Perca* species occur at sunrise and sunset (Craig 1977b; Eriksson 1978; Helfman 1979; Hubert & Sandheinrich 1983) although in Baptiste Lake, Canada, yellow perch stomach fullness increased throughout the day, peaking in the late evening (Jansen & Mackay 1992). There are some indications that maximum feeding coincides with these peaks although there is so much variability that this is not easy to establish (Fig. 5.8). At any time during the day the stomach contents of individuals within a sample of Windermere perch could range from being empty to full (Craig 1978). A number of regression equations were determined for each state of stomach fullness apart from those which were empty (using a scale from 1 to 4, where 4 was full). These equations related the total dry mass of the stomach contents to the total wet mass of the fish. By determining the proportion of stomach fullness for individuals in the sample population, it was possible to determine the mean mass of stomach contents for a given length of fish at three-hour intervals (Fig. 5.9). Although the perch population fed throughout the day, maximum intake occurred in the late afternoon or evening in most of the months of the year. A small peak occurred in the

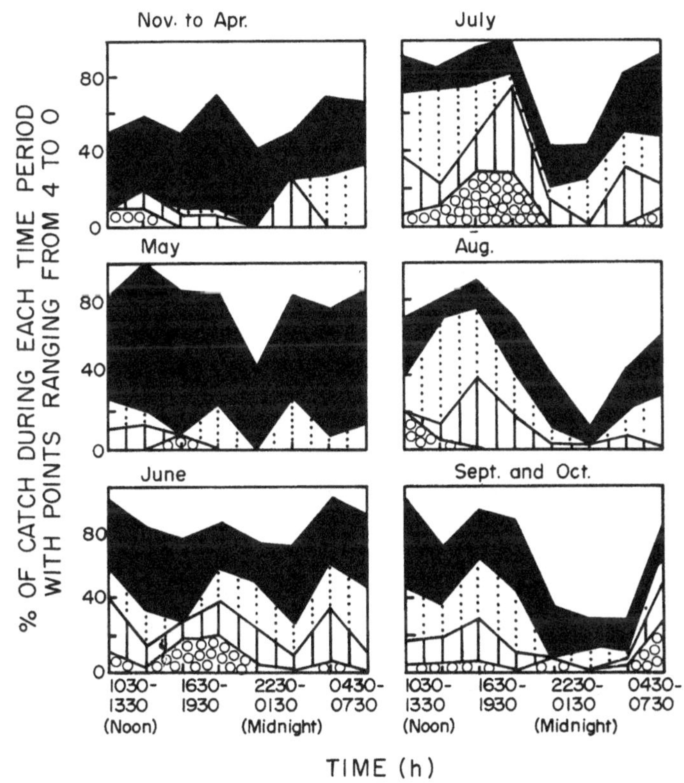

Fig. 5.8 Diel patterns of stomach fullness of adult Windermere perch measured monthly. Coefficients of stomach fullness: [circles] 4 (full), [vertical lines] 3, [dots] 2, [black] 1 and [white] 0 (empty). (After Craig 1978.)

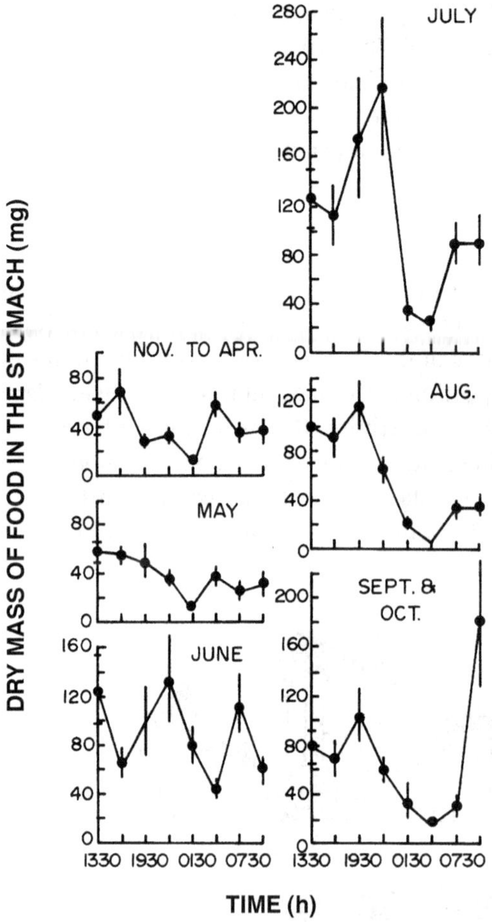

Fig. 5.9 Diel patterns of dry mass of food (mean ± 95% CL) in the stomachs of adult Windermere perch measured monthly. (After Craig 1978.)

early morning. Similar patterns of morning and evening feeding have been observed in different lakes for both perch and yellow perch (Manteifel *et al.* 1965; Ward & Robinson 1974; Thorpe 1977). During the night, both species remain relatively inactive on the lake bottom in shallow water (Hasler & Villemonte 1953; Scott 1955; Helfman 1979). At dawn they move offshore and feed in the open water on plankton and fish and some will feed on the benthos. During the winter perch and yellow perch normally stay in deep water although they continue to capture some benthic prey (Fig. 5.3) and may make some inshore movements to capture these. *Perca* species in early stages are strongly positively phototactic. This response enables the fry to feed continually during the day on zooplankton. Older perch become more sensitive to high light intensity and confine their feeding activities to dawn and dusk (Fig. 5.9).

The shoaling of *Perca* species may aid them in prey capture. If a prey fish species can swim as fast as the predator, in an initial attack the prey will often escape, but the prey has a good chance of swimming into another member of the shoal. This type of behaviour among a group of perch has been observed in aquaria by Deelder (1951) who used roach as prey.

Walleye normally feed from the evening until the early morning. Maximum feeding is apparently associated with optimum light intensity and occurs towards the early morning (Colby *et al.* 1979). In turbid lakes, walleye will continue to feed through the day (Ryder 1977). Feeding occurs in daytime during the winter when light is reduced. Ryder (1977) studied the response of walleye to light (surface illumination) by recording the number of fish caught by angling. He found an inverse relationship between catch per unit effort and surface illumination (Fig. 5.10). Walleye may actively select optimum conditions of light. Scherer (1976) demonstrated that walleye positioned themselves in aquaria inversely to light intensity. At 2 lux they were near the top and at 200 lux they were at the bottom of the tank. Underwater diver observations show that during the day walleye inhabit turbid regions of the lake but in clear water areas they remain near the bottom and find some physical shelter so that the ambient light level is reduced (Ryder 1977). The development of the tapetum lucidium (Chapter 2) in *Stizostedion* species has enabled this group to utilise dim light conditions. Young-of-the-year walleye like perch and yellow perch are positively phototactic enabling them to feed during the day on zooplankton. This positive reaction to light lasts from hatching to the postlarva phase and then changes gradually with age to a negative response. This is an adaptation to their change in feeding habits. Walleye primarily hunt by vision but hearing, taste and smell probably also play a part in prey detection. Walleye normally hunt near the bottom and may enter shallow water to feed. They feed over macrophyte beds or rocky areas. Walleye shoal and form large schools even when they become piscivorous and this behaviour continues as adults. Zander also shoal as young fish when they feed on plankton but as they become piscivorous they are

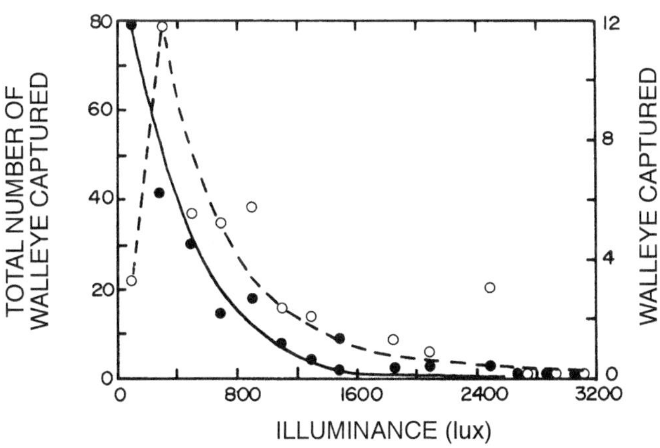

Fig. 5.10 Total number of Shebandowan Lake walleye caught by angling at dusk (•) and catch per hour (○) in relation to surface illumination. (After Ryder 1977.)

solitary (Woynarovich 1962). In the Sulejow Reservoir, Poland (Frankiewicz *et al.* 1997) the main prey of planktivorous juvenile zander is *Leptodora kindtii* (Focke) and copepods. *Leptodora kindtii* predominates in the food of the zander close to the surface while copepods are selected in deep water. This shift with depth from large particle selection to filtering reduces intra-specific competition and avoids predation by adult conspecifics found in the surface layers. In energy terms, *L. kindtii* is important for young zander (Van Densen 1985b).

Zander hunt prey in the open water, on the bottom and amongst vegetation. Like walleye they feed at low light intensities, mainly during twilight. The tapetum lucidium of zander is well developed although the tapetal area is not as extensive as in walleye. Thus, zander are less well adapted to water of high turbidity than walleye but they can feed more easily during the daytime at higher levels of illumination (Jovanovic 1970). In Lake Ladoga, Russia, which has clear water, zander ingested most prey in the early morning (0500–0600 h) from June to August (Fedorova & Drozzhina 1982). In September two peaks of feeding activity occurred, 2100–2200 h and 0500–0600 h. There was virtually no feeding during the day due to strong light intensity and very little during the middle of the night when there was insufficient light.

After chasing their prey, walleye and zander seize them sideways. The prey fish is turned around so that it can be swallowed headfirst by walleye (Colby *et al.* 1979) and tail first by zander (Neuhaus 1934; Steffens 1960). Cannibalistic walleye fry also tend to try to swallow their siblings tail first (Mathias & Li 1982).

In teleosts, the mechanics of feeding based on hydrodynamic principles have been extensively studied (Osse 1969; Elshoud-Oldenhave 1979; Van Leeuwen & Müller 1984a; 1984b; Müller & Osse 1984). When a perch senses the presence of prey it erects its dorsal spines and gently swims towards the prey, and when close the prey is captured. As the prey is approached the buccal and opercular cavities of the perch are greatly enlarged and the mouth is opened. This produces a considerable localised reduction in water pressure which lessens the speed of the prey and increases the speed of the predator. This 'suction' causes water surrounding the prey to be drawn into the enlarged mouth. The sequence of head expansion is as follows Osse (1969): lowering of the lower jaw is combined with dorsal movements of the opercula. There is caudal movement of the shoulder girdle, dorsal movement of the neurocranium, abduction of the suspensoria and protrusion of the upper jaw. This opening movement takes about 0.04 to 0.15 s. For adult perch (250–300 mm) at 0.04 s the volume of water sucked into the mouth is 500 cm^3 at a velocity of 200 $cm\ s^{-1}$. If the prey is separated from the predator by a distance equal to or less than the latter's head length the predator will draw it in. The head returns to normal shape more slowly than during pre-feeding expansion. The mouth is closed during this phase, although the upper jaw still protrudes and there are ventral movements of the opercula. Adduction of the suspensoria occurs and the opercula start to adduct. The branchiostegal rays move dorsally and the neurocranium moves ventrally into its normal position. Adduction of the gill covers and of the suspensoria continues, the hyoid symphysis moves rostrally and the angle between the hyoid bars diminishes.

The symphysis of the pectoral girdle moves dorsally and rostrally. Eventually, the adduction of the gill covers and suspensoria is complete. During these later stages the pectoral fins are brought forward, probably to stabilise the fish's position. If the prey is small, this cycle is sufficient to bring the prey to the oesophagus. If the prey is large the cycle may be repeated several times but the amplitude is reduced in each subsequent cycle. The prey is prevented from escaping by teeth in the mouth and on the branchial arches and pharyngobranchial plates (Table 2.2).

The zander uses both suction and grasping to capture its prey (Elshoud-Oldenhave 1979). As it approaches the prey it accelerates and when the prey is within reach it opens its mouth and seizes the prey. If the food particle is small it is sucked in. A larger particle is drawn towards the buccal cavity and is normally grasped crosswise. It is turned loose by the zander and drawn in by reduced pressure in the mouth and buccal cavity caused by rapid expansion of the head. The expansion of the head involves 24 muscles of the head including the branchial basket, and the anterior body muscles. The main muscular action in head expansion involves the epaxial and hypaxial body muscles (Fig. 2.4a) which bring about movements of the neurocranium, pectoral girdle and hyoid arch (Elshoud-Oldenhave 1979). The neurocranium can be moved in a number of directions so that the suction can be aimed toward the prey. Head muscles regulate head expansion and contraction. The successful capture of prey depends on predator velocity and the timing of head enlargement.

5.4 Food consumption

The amount of food eaten by percids is affected by a number of factors including temperature, prey density and physiological state of the fish. Several methods have been used to calculate daily ration for percids in various bodies of water. The appetite of a perch is dependent on the quantity of prey already in its gut. This is determined by the rate of gastric evacuation; the time taken for a given mass of food to pass through the fish's stomach. This rate has been measured from laboratory experiments and for many fish species including perch (Griffiths 1976; Thorpe 1977; Persson 1981) and walleye (Johnston & Mathias 1996) the rate is exponential (Fig. 5.11). The formula to describe the amount of matter in a fish's stomach is given by:

$$Y_X = Y_0 \exp(-RX)$$

where X is time (h), Y_0 is the mass of a meal (mg), Y_X is mass of food present in the stomach (mg) at time X and R is the instantaneous rate of gastric evacuation.

The rate (R) is dependent on temperature (T, °C) and the relationship between the R and T is usually positively exponential (Fig. 5.12):

$$R = a \exp(b)T$$

where a and b are constants.

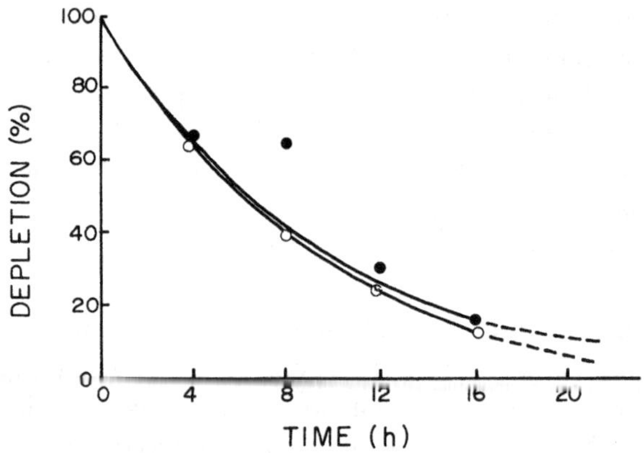

Fig. 5.11 The gastric evacuation of *Gobiomorphus cotidanus* from perch stomachs at 17°C. A comparison is made between force (•) and voluntarily (○) fed fish. (After Griffiths 1976.)

Different types of invertebrate prey do not appear to affect the rate of gastric evacuation in perch. This was tested by Persson (1979) using *Gammarus pulex* (Linnaeus), *Chaoborus* sp., chironomids and zooplankton as prey. In a later experiment, Persson (1981) determined that the rate of gastric evacuation of prey such as fish larvae was slower than that of the invertebrates. Generally, meal size, meal frequency and predator size do not affect the rates of gastric evacuation. However, Persson (1981) found the first phase of digestion was slower for fish eaten when perch were almost satiated. The ration at which this occurred was lower for smaller perch. Swenson and Smith (1973) applied linear models to their walleye data from gastric evacuation experiments. Meal size affected the rate. At small meal sizes ($<10\,\mathrm{mg\,g^{-1}}$ wet mass of predator) the rate appeared to fit the exponential model. At greater meal sizes ($>10\,\mathrm{mg\,g^{-1}}$ fish wet mass) a linear model applied. When

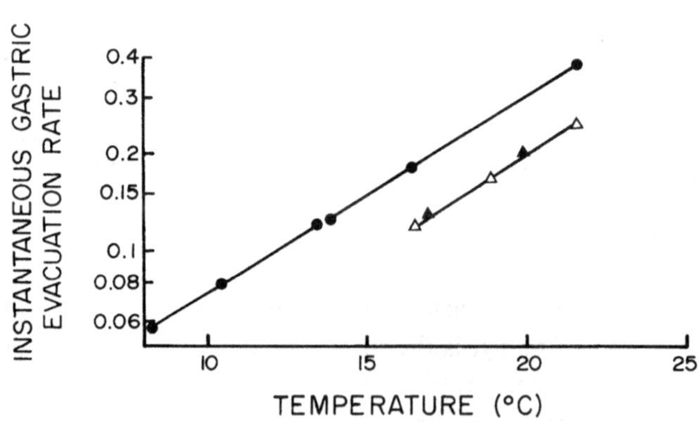

Fig. 5.12 The relationship between instantaneous gastric evacuation rate and temperature for perch fed *Gammarus pulex* (•) and fish larvae (▲). Points are single values except △ which are geometric means. (After Persson 1981.)

walleye were nearly satiated, the first part of digestion took longer than when the stomachs were only partly filled.

If the rate of gastric evacuation is exponential and the rate of food consumption is constant, then the amount of food consumed in *t* hours (ΣC_t) is given by:

$$C_t = \frac{[S_t - S_0 \exp(-R_t)]R_t}{1 - \exp(-R_t)} \qquad \text{Equation 5.1}$$

where S_0 and S_t are the stomach contents at the start and end of sampling respectively (Elliott & Persson 1978; Craig 1979). Craig (1978) made modifications to this equation. Samples were taken at fixed time intervals (three hours) over 24 h so that S_0, S_t in Equation 5.1 are estimated by S_0 and S_1 for the first sampling interval, S_1, S_2 for the second interval and up to S_{m-1} and S_m for the last sampling interval. If the food in the stomach is the same at the beginning and end of the 24 h period so that $S_0 = S_m$, daily food consumption (ΣC_t) is then given by:

$$\sum C_t = \frac{R_t}{1 - \exp(-R_t)} \cdot \sum_{i=1}^{i=m} S_i \ (1 - \exp - R_t)$$

Therefore:

$$\sum C_t = Rt_m \ \bar{S}$$

where:

$$\bar{S} = \frac{\sum S_i}{m}$$

and in 24 h:

$$\sum C_t = 24\bar{S}R \qquad \text{Equation 5.2}$$

Craig (1978) used Equation 5.2 to calculate the daily consumption of perch for each month in Windermere, England (Table 5.1).

Prey density affects food consumption. In a laboratory study on larval perch and zander, Kudrinskaya (1970) found a reduction of zooplankton prey from densities of 3.0–6.0 mg^{-1} to 0.1–0.5 mg l^{-1} at 19.3°C caused 70% of the larvae to cease feeding and predator food consumption was reduced by a factor of 3 for perch and 2.6 for zander. A reduction in temperature to 16.0°C increased non-feeding fish to 80–90% and also reduced food consumption at all levels of prey density. Swenson and Smith (1976) found that food consumption of adult walleye increased with density of prey up to about a density of 400 mg m^{-3}. At this density it remained fairly constant at 3% of body mass.

The daily consumption of juvenile fish is higher than that of adults. In Oneida

Table 5.1 Energy equations for 150 g male perch from Lake Windermere. Q_c = energy of consumed food calculated from the components which are Q_w* = energy lost from egestion and excretion, Q_r = energy of resting metabolism, Q_a = energy expended in activity, Q_g soma = energy changes in the soma and Q_g gonad = energy changes in the gonad. Also given are energy values for consumption based on field measurements Q_{cst} and calculated by Equation 5.2 (see text). All values are in joules and calculated on a daily (24 h) basis. (Modified from Craig 1978.)

Month	Q_{cst}	Q_c	Q_w	Q_r	Q_a	Q_g soma	Q_g gonad
November to April	5 785	5 590	1118	2384	2384	−201	−95
May	4 581	8 694	1739	3716	3716	247	−724
June	11 456	9 350	1870	5441	5441	−3063	−339
July	32 501	24 845	4969	6794	6794	6623	−335
August	15 141	19 046	3809	8334	8334	−1456	25
September to October	10 404	14 439	2888	5547	5447	−466	923

Lake, USA, yellow perch fry consumed 14–43% of their body mass per day (Mills & Forney 1981) during June and July when they fed almost exclusively on *Daphnia pulex*. In the Mozhaisky and Uchinsky Reservoirs, former USSR, perch fry fed on a daily ration of 12–16% body mass in June when *Daphnia* were the chief prey item (Spanovskaya & Grygorash 1977). Many of the studies on both adult perch and yellow perch have shown that these two species exhibit similar seasonal trends in daily food intake. In the study already described for Windermere (Craig 1978), daily food intake per body mass rose from 0.7% in May to 5.2% in July and declined to 2.4% in August and 1.6% in September and October. The perch of Loch Leven, Scotland, consumed 6.5% in June and July declining to 3.2% in September (Thorpe 1977). Three year old perch from Sövdeborgssjön, Sweden, had a maximum food intake of 3% body mass in July declining to 0.1% in November (Persson 1983). In two small lakes further north in Sweden, Nyberg (1979) found the highest percentages, 4–6%, from May to August falling to 1% in September. Yellow perch in Lake Memphremagog on the Canada and USA border, consumed 6.1% of their body mass in July which then declined to 3.7% in August, 1.6% in September and 2.4% in October (Nakashima & Leggett 1978).

Food consumption of walleye is usually greatest in the autumn. In Lake of the Woods, USA, walleye consumed 1% of their body mass in June, 2% in July and 3% in August and September. Although the water temperatures were the same in June and September, more food was available in the latter month (Swenson & Smith 1973).

Zander in the Volga Delta had the highest food intake in April (Orlova 1976), amounting to 2.6% body mass. Consumption then declined to about 0.5% during the summer.

5.5 Digestion

The prey of percids tends to be swallowed whole. Acid gastric fluids are secreted in the stomach and in perch the pH of stomach fluids ranges from 3.0 to 5.0 (Fish 1960).

The acidity increases after food intake. Several proteases are present in the gastric fluids, the most important being pepsin. This is secreted by the gastric gland cells in the form of pepsinogen which is activated in an acid environment (Fänge & Grove 1979).

Pancreatic secretions in the intestine contain enzymes involved in the digestion of proteins, carbohydrates, fat, nucleotides and probably bicarbonates to neutralise acids entering the intestine (Fänge & Grove 1979). Many of the intestinal enzymes are most active at a neutral or slightly alkaline pH. Proteases from the pancreas include trypsin, chymotrypsin, elastase and carboxypeptidase. Amylase has been isolated from the diffuse pancreas of perch (Fish 1960). Chitinase is probably produced in perch species because these fishes feed extensively on insect larvae and crustaceans. This chitinase will come mainly from the pancreas although some may be produced by bacteria in the intestine. The pancreas is probably also the main source of lipases.

Secretions into the intestines from the gall bladder are similar to those in mammals. Bile contains bile salts, cholesterol, phospholipids, bile pigments, glycoproteins, organic anions and inorganic ions (Fänge & Grove 1979). The secretion is weakly alkaline and has a high sodium and a low chloride concentration.

The brush border cells of the intestinal epithelium secrete some enzymes although most enzyme activity is brought about by secretions from the pancreas. Enzymes thought to be produced by the intestinal mucosa include aminopeptidases, di- and tripeptidases, acid and alkaline nucleosidases, polynucleotidases, lecithinase, lipase and other esterases and various carbohydrate digesting enzymes (Fänge & Grove 1979). Activity of enzymes in the intestinal lumen is low especially in the middle and posterior regions and most breakdown occurs at the intestinal mucosa brush border. α amylase, which works at an optimum pH of 6.7 (Fish 1960), is less active and probably less abundant in perch and zander intestine than more omnivorous species. Kuzmina (1984a) found the enzyme to be more active in the intestinal mucosa (54–59%) than in the gut lumen of zander but there was no significant difference in activity between the two areas in perch. Significantly more maltase activity (66–77%) was found in extracts of mucosa than in the luminal fluid of perch and saccharase was found to be active only in extracts from the intestinal mucosa of perch.

Phosphatase activity from the intestinal mucosa of zander had two pH optima (5.0 and 9.0–10.0) at 0°C and 20°C (Kuzmina 1984b). When the seasonal activity of zander alkaline phosphates was examined, Kuzmina (1989) found the maximum enzymatic activity in the spring and summer while autumn and winter activity was significantly lower.

Amino acids are absorbed against a concentration gradient and some proteins and peptides are absorbed by pinocytosis in the middle section of the intestine. Fats are absorbed by epithelial cells in the anterior part of the intestine including in perch, mucosal cells of the pyloric caeca (Jansson & Olsson 1960).

Absorption efficiency, the proportion of ingested matter that is absorbed by the gut, declines with increasing food consumption. Absorption efficiency of walleye larvae, held at 20°C and fed zooplanktonic crustacea, increased with gut retention time and declined with increase in walleye dry mass (Johnston & Mathias 1996). Maximum digestibility was 79% of organic matter from food retained in the gut for 6 h.

5.6 Energetics

The basic equation, which summarises the fate of the energy intake, is as follows:

$$C = R + \delta B/B\delta t + F + U \qquad \text{Equation 5.3}$$

where, in energy units, C = food consumption, R = metabolism, $\delta B/B\delta t$ = specific growth rate, F = egested loss and U = excreted loss. Metabolism, R may be divided up into a number of components: $R = R_s + R_a + R_d$, where R_s = the energy used for standard metabolism in a resting non-feeding fish, R_a = the energy used in activity, such as swimming and R_d = the energy used in the digestion, movement and deposition of food materials including specific dynamic action (Beamish & MacMahon 1988). Specific dynamic action is the energy released from the hydrolysis of the peptide bonds between amino acids and other conversion processes.

Kitchell *et al.* (1977) derived an energetics model for yellow perch and walleye, a computer-based version of which was produced by Hewett & Johnson (1992). They used data derived by Solomon and Brafield (1972) for perch and Kelso (1972) and Swenson and Smith (1973) for walleye as a basis and applied other data, including those from other species (Brett 1971; Brett *et al.* 1969; Elliott 1976a; 1976b) to fill in gaps necessary for their model. Temperatures used in the simulations were taken from Hokanson (1977a). These are given in Table 5.2. The specific dynamic action was considered to remain constant at 15% of the food consumed. Bartell *et al.* (1986) used statistical analysis of Monte Carlo simulations to rank the importance of the different parameters of the energetics equation. They showed that it was essential to estimate maximum consumption accurately and the allometric parameters for consumption and respiration. Excretion and egestion parameters contributed little to the predictable errors. The model could then be simplified to a temperature-dependent feeding model without loss of ability to predict growth.

Faeces contain small quantities of secretions from the fish and some nitrogenous waste products as well as undigested food. Solomon and Brafield (1972) kept perch at a fixed temperature (14°C) and found that the percentage loss in faeces increased with increase in ration. Elliott (1976b) demonstrated that percentage loss decreases with increase in temperature in the brown trout *Salmo trutta* Linnaeus. Faecal loss was also affected by fish size but when faecal loss was expressed as a proportion of fish size, it remained fairly constant for different sized fish. Simulating this in the

Table 5.2 Optimum and maximum temperatures (°C) for consumption and respiration for young-of-the-year, juvenile and adult yellow perch. Derived from Hokanson (1977a) and utilised in the bioenergetics model of Kitchell *et al.* (1977).

Temperature regime	Fish	T (°C)
Optimum temperature for consumption	Young-of-the-year	29
	Juveniles and adults	23
Maximum temperature for consumption	Young-of-the-year	32
	Juveniles and adults	28
Optimum temperature for standard respiration	Young-of-the-year	32
	Juveniles and adults	28
Maximum temperature for standard respiration	Young-of-the-year	35
	Juveniles and adults	33

model for percids produced a curve, which is shown in Figure 5.13. The main excretory product of perch is ammonia lost through the gills (Godemar 1994). Other excretory products of teleosts include urea, uric acid, creatinine, amines and amino acids (Elliott 1979) but in energetic terms losses from these products are minimal. Elliott (1979) found that losses from excretory products increased with fish size, water temperature and ration size. Cai and Summerfelt (1992) measured an increase in the ammonia excreted by juvenile walleye from 20°C (17 mg kg^{-1} h^{-1}) to 25°C (19 mg kg^{-1} h^{-1}). With increase in size, the mass specific rate significantly decreased (Fig 5.14). However, like egestion, the proportion of the daily ration lost in excretion probably remains fairly constant (Fig. 5.13). The net loss from both egestion and excretion ranges from 26–32%.

The model of Kitchell *et al.* (1977) shows the highest maintenance values for yellow perch of all sizes at 27°C (Fig. 5.15). Winberg (1956) made the assumption that energy used by fish in activity was roughly equal to that used in resting

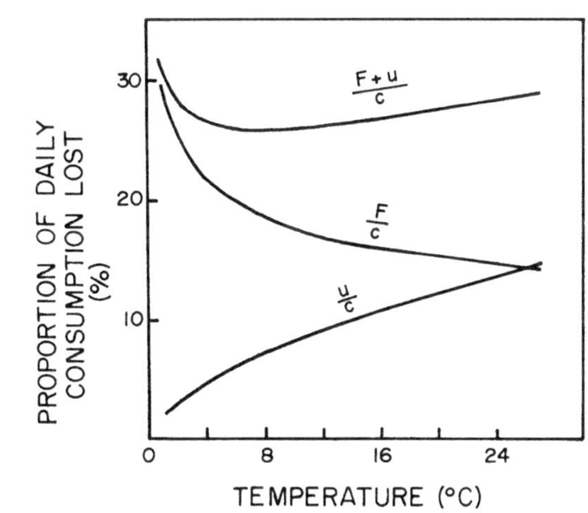

Fig. 5.13 The proportion of consumed food egested (*F*/*C*) and excreted (*U*/*C*) by yellow perch and their sum ((*F* + *U*)/*C*) in relation to temperature. (After Kitchell *et al.* 1977.)

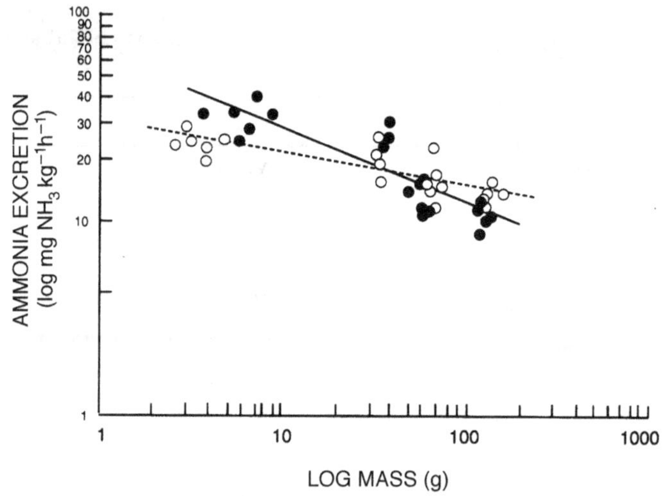

Fig. 5.14 The relationship between ammonia excretion and mean body mass of walleye at 20°C (○) and 25°C (●). (After Cai & Summerfelt 1992, with kind permission from Elsevier Science Publishers.)

metabolism ($R_a = R_s$). Boisclair and Leggett (1989d) estimated that for an actively foraging fish such as yellow perch, activity costs ranged from 0–40% of the budget. They discovered, that as Kerr (1982) suggested, activity costs increased as food consumption increased. However, they found no significant relationship between predicted (models of Kitchell *et al.* (1977) and Kerr 1982)) and observed food consumption values. The activity component of the energy budget can make a major contribution to inter-population variation in yellow perch and other actively foraging fish species growth (Boisclair & Leggett 1989d). In studies on juvenile walleye, Fox (1991) found that the Kitchell *et al.* (1977) model overestimated consumption. Replacement of a 'reasonable' standard metabolic rate gave estimates

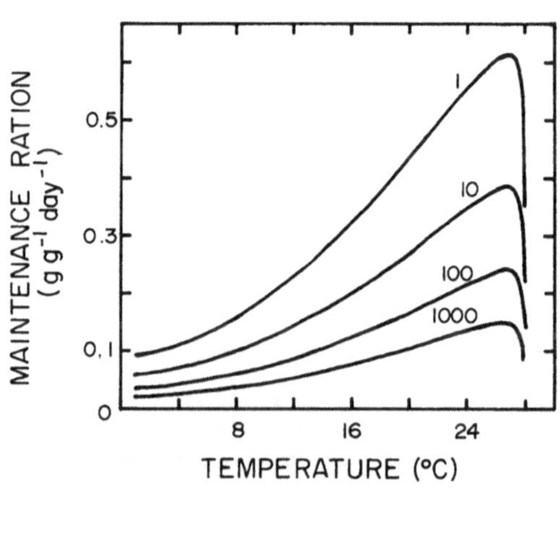

Fig. 5.15 The relationship between estimated maintenance rations and temperature for 1, 10, 100 and 1000 g yellow perch. (After Kitchell *et al.* 1977.)

that were closer to *in situ* measurements (Table 5. 3). Activity should be modelled as a variable component, the variation being a function of prey biomass (Madon & Culver 1993). Consumption and growth of larval and juvenile fish are more sensitive to prey availability and temperature. The active metabolism of young-of-the-year yellow perch can be over four times the standard metabolism of adult fish (Post 1990).

Table 5.3 Comparison of *in situ* juvenile walleye consumption estimates (% body mass day^{-1}) with predicted outputs from the model of Kitchell *et al.* (1977) using original (*K*) and the respiration submodels of Mills and Forney (1981) for active feeding metabolism of young-of-the-year yellow perch (M&F) and Minton and McLean (1982) for standard metabolism of sauger multiplied by 1–3 to estimate active metabolism (M&M). (After Fox 1991, with kind permission of the Canadian Department of Fisheries and Oceans.)

Pond	Age (days)	*In situ* estimate	*K*	M&F	M&M	2M&M	3M&M
2B	17	39.2	41.1	41.7	33.9	36.1	38.2
5A	17	25.7	42.8	42.2	35.7	37.6	39.6
5B	38	4.9	8.7	7.1	3.7	5.2	6.7
5D	38	4.6	11.2	8.1	5.7	7.1	8.6
Mean model: estimate:			1.73	1.48	1.06	1.25	1.44

The bioenergetics model has also been used to simulate specific growth rates related to water temperature and ration size (Fig. 5.16 for a 10 g yellow perch). Growth rates decline with increasing temperatures at all ration levels. Maintenance rations (g g^{-1} day^{-1}) range from 1% at 8°C to 4% at 27°C. The figure also shows the effect of temperature and ration size on gross and net efficiency. Using the highest rations the gross efficiency exceeds 45% in the temperature range 4–20°C. In a field situation, young-of-the-year yellow perch in Oneida Lake, USA, had a gross conversion efficiency of 27% (Mills & Forney 1981). The simulated model illustrates that net efficiency exceeds 64% in the yellow perch at low ration levels and temperatures below 10°C. The Kitchell *et al.* (1977) model appears to fit the Lake Erie observed data for yellow perch (Fig. 5.17). Karås and Thoresson (1992) modified the parameters of the model to fit it to young-of-the-year perch. Predictions from this modified model usually matched data from field observations more closely (Fig. 5.18).

The energy that is available for growth can be used to increase the energy content of the soma or for gonad development in mature fish (Chapters 3 and 4). As perch in Windermere, England, get older, a greater proportion of available energy is directed towards gonad development. Most studies on energetics have been performed on immature fish because the growth component of the energy budget is easy to determine as a result of high somatic growth rates. However, Craig (1978) calculated the energy budgets for large mature perch and details are given in Table 5.1 for male Windermere perch. Consumption values are based on field data already

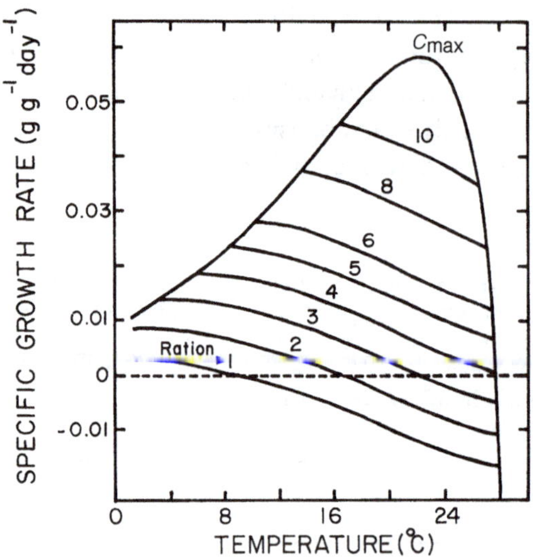

Fig. 5.16 Changes in specific growth rate of 10 g yellow perch as a function of temperature and ration level (expressed as percentage body mass, ranging from one per cent to maximum, C_{max}). (After Kitchell *et al.* 1977.)

described earlier in the chapter (C_{st}) and summation of the components of the budget (C; Equation 5.3). There was no significant difference ($p > 0.05$) between these two values and they are reasonably similar. This is reassuring because of the assumptions and approximations which had to be made in their derivation. In male and female perch, energy was stored in the soma in July. Energy was transferred to the testes in the late summer and autumn and to the ovary during the winter. Spawning behaviour may have accounted for the large amount of energy lost in June from the male soma (it was lost through Q_a; Table 5.1). The energy value of the tissues does not remain constant but varies with energy intake and requirements of the fish. Perch fed on fishes have a higher calorific value per unit mass than those fed

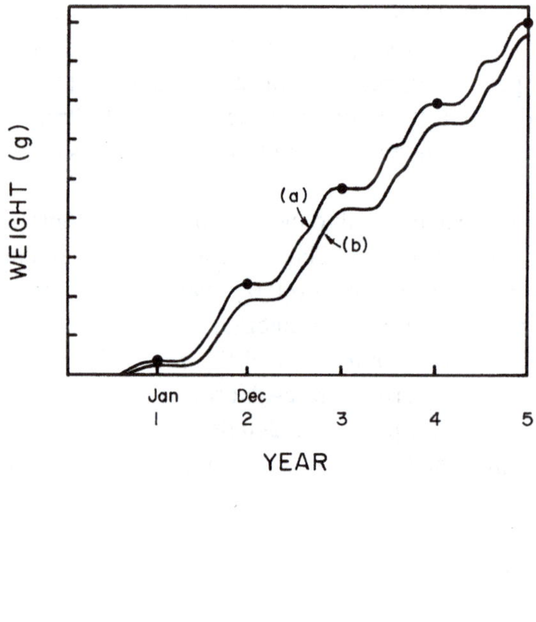

Fig. 5.17 Simulated growth of yellow perch from (a) western and (b) eastern Lake Erie based on the bioenergetics model of Kitchell *et al.* (1977). Field values (•) are given.

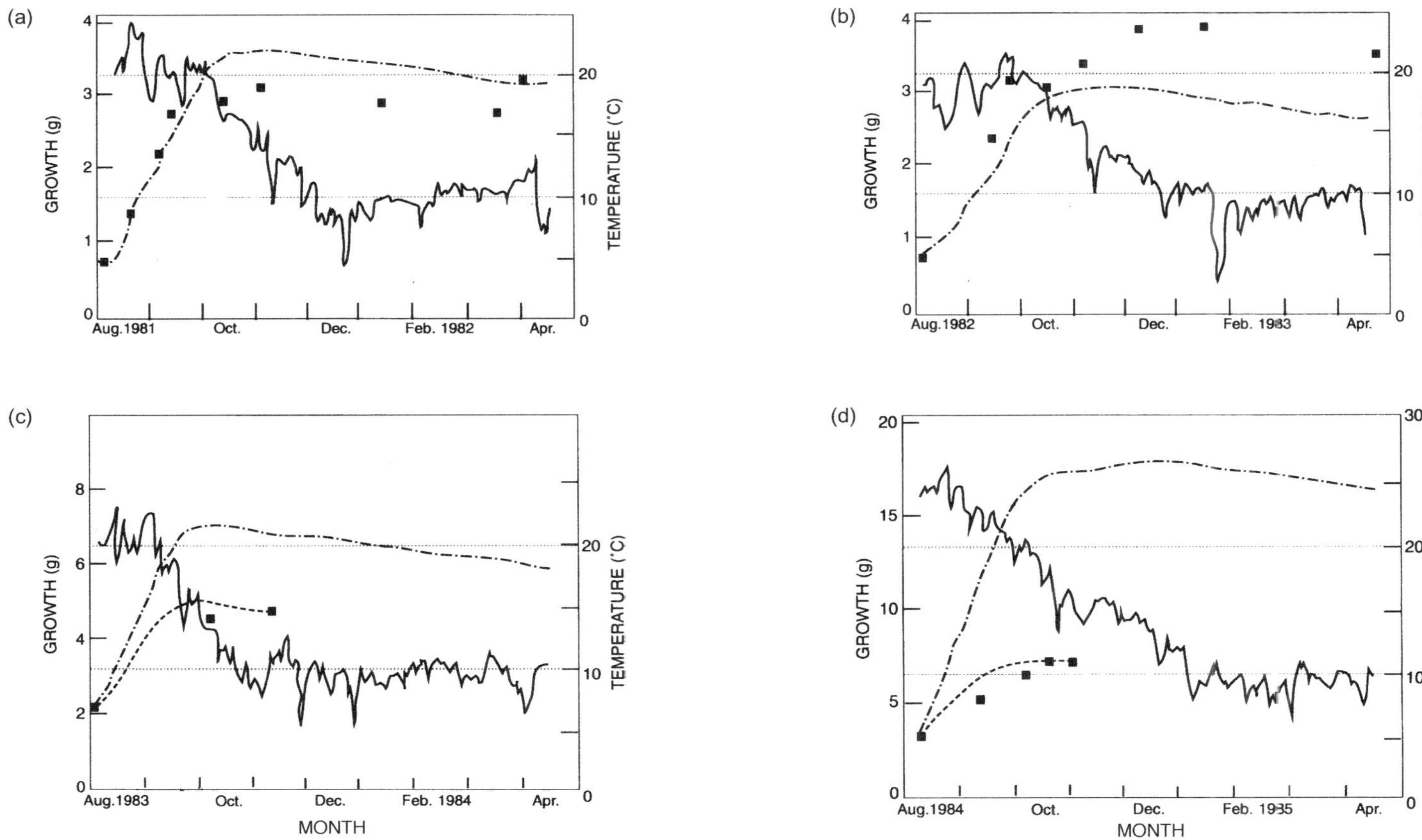

Fig. 5.18 Simulated (under the assumption that consumption is at a maximum) (dash and dot) and actual (broken line) growth of young-of-the-year perch, 1981–1984, from the Swedish Baltic coast based on a bioenergetics model of Karås and Thoresson. Mean mass from field data (■) and temperature (solid line) are given. (After Karås & Thoresson 1992, with kind permission of Academic Press.)

on invertebrates (Jezierska 1974). The energy content of perch tissue in Windermere increases during the period of intensive feeding in July, declines in August and increases again during the autumn before it declines once again in the winter (Fig. 5.19).

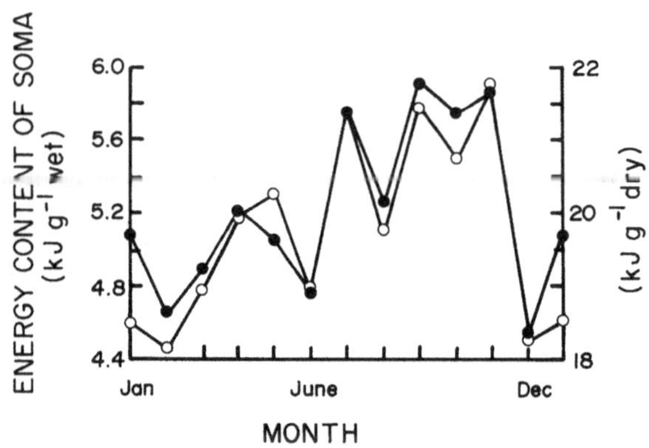

Fig. 5.19 The monthly energy content of wet (•) and dry (○) somatic tissue of male Windermere perch. (After Craig 1977a.)

5.7 Respiration

The demand for oxygen or oxygen consumption by fish is dependent on temperature and fish mass (Fig. 5.20). Energy requirements, and thus oxygen, also increases with activity. Lowered oxygen concentrations can restrict activity (Fig. 5.21). Oxygen is taken up from the surrounding water through the gills. The flow of water over the gill surface is rhythmic or continuous depending on activity and brought about by movements of the walls and floor of the oral cavity, the gill covers and the branchiostegal apparatus. A hydrostatic pressure difference exists between the oral cavity and the opercular region. Occasionally, the flow is reversed by a 'coughing' reflex, which cleans the gills of foreign matter. The skin takes up very little oxygen and in perch it is insufficient to meet cutaneous demands let alone supply other organs (Nonnotte 1981). Oxygen is absorbed through the surface of the epithelium lining the gill lamellae into the blood capillaries over a diffusion gradient. The exchange of gases is limited by the surface area of the gill, A, the permeability of the gill surface to gases, K, the distance from the water to the blood, d, and the mean difference in oxygen partial pressure between water and blood, ΔPg. Thus oxygen uptake (OU):

$$OU = \frac{KA(\Delta Pg)}{d}$$

(Jones & Randall 1978; Randall & Daxboeck 1984).

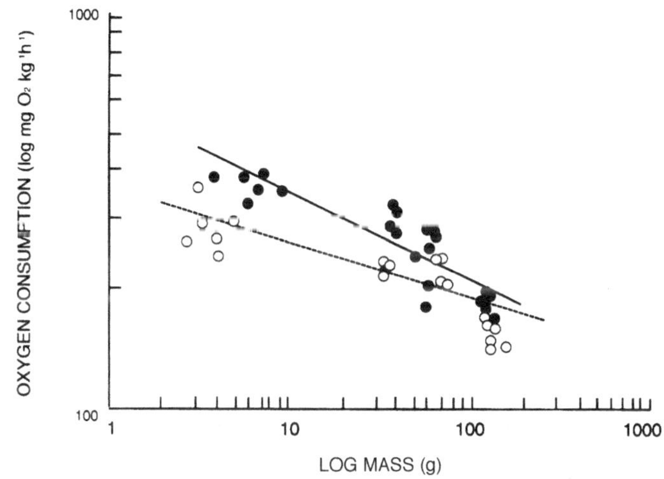

Fig. 5.20 The relationship between oxygen consumption and mean body mass of walleye at 20°C (○) and 25°C (●). (After Cai & Summerfelt 1992, with kind permission from Elsevier Science Publishers.) Note log scale.

There is extensive variability between species in their response to oxygen concentrations in the external medium. At one extreme, fish in freshwater sometimes encounter supersaturation of oxygen in the water and this can lead to their death. Gas supersaturation in Lake Waubesca, USA, in 1940 was due to extensive algal blooms (Woodbury 1942). Concentrations of 30–32 mg l^{-1} were recorded and large numbers of walleye died in the lake. This was caused by gas emboli which blocked circulation of blood through the gill capillaries and led to death through respiratory failure. A detailed study on the response of zander to reduction in oxygen concentration in the water was made by Dolinin (1974). He conducted two experiments, one for fish at rest and one in which he induced fish activity. He carried out his tests

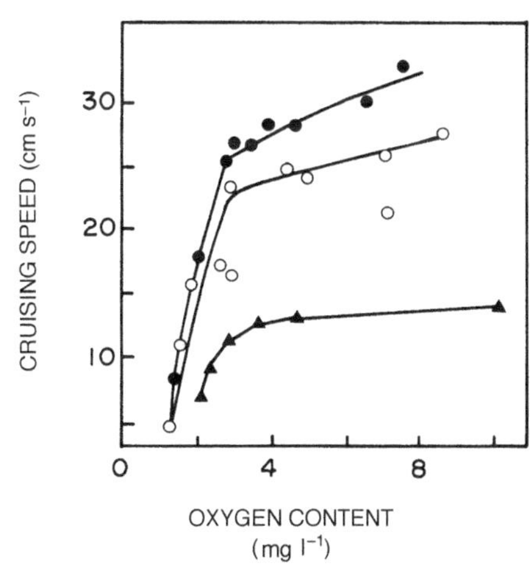

Fig. 5.21 Cruising speed of yellow perch in relation to temperature 25°C (●), 15°C (○) and 10°C (▲) and oxygen content. (After Fry 1957.)

at four temperatures 5, 10, 15 and 20°C covering a range normally encountered by the fish. Dolinin (1974) defined three important levels of oxygen concentration. When oxygen concentration is reduced in the water the rate of gas exchange by the fish remains unaltered until a 'critical' concentration is reached when the amount of oxygen taken up by the fish declines. This critical level varies with temperature and from 5–20°C increases by a factor of 1.7. At 5°C the critical concentration is 4.18 $\pm$ 0.29 mg l^{-1} ($\pm$ SE) rising to 6.93 $\pm$ 0.81 mg l^{-1} at 20°C. The point where the rate of decline starts to increase is called the 'threshold' concentration. This concentration remains fairly constant with temperature and has a value of 4.23 $\pm$ 0.22 mg l^{-1}. At the 'limiting' concentration of 1.41 $\pm$ 0.14 mg l^{-1} extraction of oxygen ceases. The amount of water pumped through the gill apparatus increases gradually as the oxygen concentration is reduced and then falls sharply when the threshold value of 4.23 mg l^{-1} is reached. The maximum amount of water pumped through the gills increases with a rise in temperature, as does the coefficient of oxygen uptake and the number of respiratory movements in fish held at the threshold oxygen concentration (Table 5.4). Coefficient of oxygen uptake is equivalent to the transfer factor (TO_2):

$$TO_2 = \frac{O_2 \text{ uptake} \times 100}{PO_2 \text{ water} - PO_2 \text{ blood}}$$

Table 5.4 Ventilation volume, coefficient of oxygen uptake and respiratory movements of zander held in water of temperature 5–20°C. (After Dolinin 1974.)

	Temperature (°C)			
	5	10	15	20
Ventilation volume (ml min^{-1} kg^{-1})	400.0 $\pm$ 20.6	489.1 $\pm$ 18.3	581.6 $\pm$ 42.0	634.0 $\pm$ 18.3
Coefficient of oxygen uptake (%)	29.4 $\pm$ 2.6	39.3 $\pm$ 2.2	44.0 $\pm$ 4.8	54.8 $\pm$ 2.3
Respiratory movements (min^{-1})	25 $\pm$ 2	33 $\pm$ 2	42 $\pm$ 4	50 $\pm$ 3

where PO_2 is the partial pressure of oxygen (Jones & Randall 1978). A reduction in oxygen concentration below the threshold value causes the coefficient of oxygen uptake to fall to zero. There is also a reduction in ventilation volume. In concentrations above the threshold value the oxygen uptake coefficient is inversely related to the ventilation volume (Fig. 5.22). Respiratory movements are linearly related to ventilation volume (Fig. 5.23) and as the temperature increases the intercept of the line relating the two factors increases although the slope remains the same. There appears to be no difference in the relationship for active or non-active fish. In active fish the changes in gas exchange are similar in nature to those of a resting fish and oxygen consumption halts at about a concentration of 1.47 mg l^{-1}. However, the alteration of ventilation volume varies from that of resting fish. The

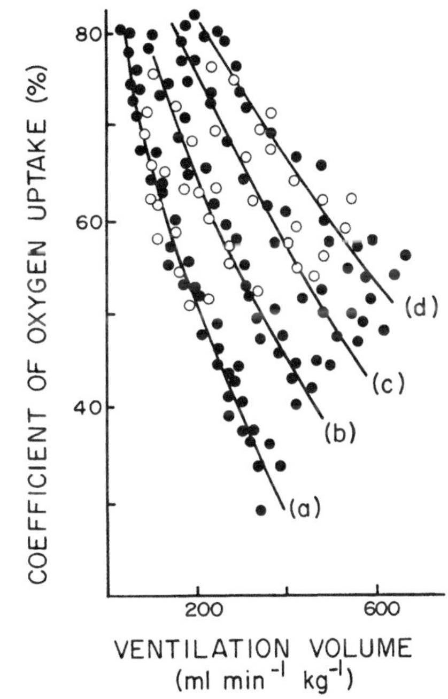

Fig. 5.22 Zander ventilation volume and coefficient of oxygen uptake from the water at four temperatures: (a) 5°C, (b) 10°C, (c) 15°C and (d) 20°C. Results were obtained with no motor activity (●) and induced motor activity (○). (After Dolinin 1974.)

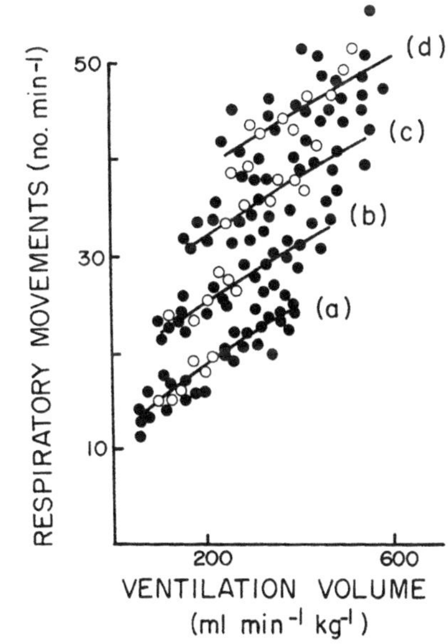

Fig. 5.23 Zander ventilation volume and respiratory movements at four temperatures: (a) 5°C, (b) 10°C, (c) 15°C and (d) 20°C. Results were obtained with no motor activity (●) and induced motor activity (○). (After Dolinin 1974.)

maximum ventilation is not reached at the threshold value as in a resting fish but at the 'critical' value. The ventilation volume remains constant below the critical value until it begins to fall to a similar value to that for a resting fish. The relationship between ventilation volume and oxygen uptake coefficients remains the same in active and resting fish above the threshold value. A reduction in ventilation volume from 400 to 200 ml min^{-1} kg^{-1} results in an increase in the coefficient from 52.3% to 71.5% in active zander held in water containing oxygen above the threshold value. However, below threshold concentrations there is a decline in the coefficient from 38.6% to 17.1%.

Studies of the tolerance of other percids to a reduction in oxygen concentrations are not as complete as the one described for zander. Jones (1964) suggested that the 'limiting' oxygen concentrations for perch were 1.1–1.3 mg l^{-1} at 16°C and 2.25 mg l^{-1} at 20–26°C. Activity is reduced at 7 mg l^{-1} in yellow perch at a summer temperature of 20°C (Fry 1957). Yellow perch can survive in oxygen concentrations down to 0.25 mg l^{-1} at 2.5–4°C, conditions often experienced under the ice in northern lakes of North America. Ventilation of the gills increases as oxygen is reduced from 4 to 1 mg l^{-1}. Maximum ventilation occurs at 0.5 mg l^{-1} at 2.5–4°C (Petrosky & Magnuson 1973). The yellow perch were observed to swim up and nose at the underside of the ice at 0.25 mg l^{-1} oxygen. Walleye held in the laboratory can survive low levels of oxygen (2.0 mg l^{-1}) without showing ill effect. At an oxygen concentration of 1.0–1.5 mg l^{-1} (at 22°C) walleye rise to the surface and at 0.6 mg l^{-1} they lose co-ordination and equilibrium (Scherer 1971). Unlike yellow perch, walleye are not able to obtain oxygen from the surface film and are thus less able to withstand hypoxic conditions sometimes experienced in lakes during the summer (Gee *et al.* 1978). Walleye tend to inhabit waters with oxygen concentration >3 mg l^{-1}. Oxygen depletion causes a reduction in negative phototactic behaviour in walleye. There is no negative response to light at oxygen concentrations of 1–2 mg l^{-1} which still allow the walleye to be mobile. Scherer (1971), using aquaria over which lights were shone, found that walleye remained in their shelters down to an oxygen concentration of 5 mg l^{-1}. Below this level they moved in and out from the shelters but at 1–2 mg l^{-1} they swam to the surface. At optimum oxygen concentrations the walleye remained in their shelters if carbon dioxide tension was kept below 5 mm Hg, whereas if the carbon dioxide level was raised the fish were more active in the light and exhibited upward movements. No light avoidance was shown at 14–18 mm Hg (30–40 mg l^{-1}). Free carbon dioxide in the water suppresses oxygen uptake.

Good quantitative data on the oxygen requirements of eggs and young stages are lacking. At low oxygen concentrations walleye eggs take longer to hatch (Fig. 5.24). Oseid and Smith (1971) found that there was no difference in mortality of eggs held at constant temperature in the range of 2 to 7 mg l^{-1} of oxygen. However, low levels of oxygen reduced the size of larvae at hatching and retarded larval growth. Walleye larvae raised at 2.4 mg l^{-1} oxygen were found to be weaker swimmers than those reared at normal oxygen levels.

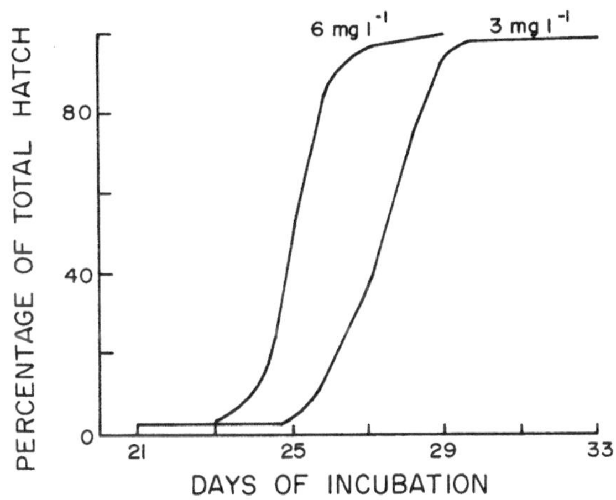

Fig. 5.24 The effect of dissolved oxygen on the hatching rate of zander eggs. Two oxygen concentrations were used, 3 and 6 mg l^{-1}. Temperature was 4–5°C for the first 21 days and 12.3°C thereafter and flow rate of water through the eggs was 300 ml min^{-1}. (After Oseid & Smith 1971.)

Percids in general have a well-developed respiratory apparatus and they live in oxygen rich environments. They are not a very active and mobile species and thus their demands are low. When exposed to low levels of oxygen they can show considerable tolerance. Perch exhibit diel patterns of oxygen consumption with low rates in the dark compared to the light period which can be related to their activity (see Section 5.6 and Chapter 7). In starved, juvenile fish this pattern broke down after 7–8 days (Mehner & Wieser 1994). The breakdown was related to the exhaustion of glycogen reserves in the muscle and liver. A switch in energy sources was associated with a change in swimming activity. The authors found a negative correlation between the scope for spontaneous activity and the specific rate of oxygen consumption in the dark phase. This suggested that the resting rate of metabolism responded more strongly to food deprivation than the rate of spontaneous activity in the light.

5.8 Biomanipulation

As outlined above, in many temperate lakes percids are planktivorous especially in the larval and juvenile phases. Heavy grazing not only reduces the abundance of zooplankton but also the community structure. In enclosure experiments, Post and McQueen (1987) tested the relationship between juvenile yellow perch and zooplankton abundance and found a significantly negative relationship. At high fish densities, the zooplankton community was dominated by small filter-feeding cladocerans, primarily bosminids. At low densities of yellow perch, large filter feeding cladocerans, mainly daphnids, made up most of the zooplankton composition. However, they found no significant relationship between zooplankton and phytoplankton biomasses as a whole, although they did detect lower phytoplankton abundance in the enclosure dominated by daphnids in the summer. Post and McQueen (1987) concluded that the weak relationship at this trophic level suggested that the manipulation (biomanipulation) of fish abundance was ineffective as a management tool in controlling phytoplankton abundance. In another enclosure experiment, Beklioglu & Moss (1995) tested the effects of fish abundance and pH. Increasing pH decreased carbon dioxide and chlorophyll-*a* concentrations and diatom biomass. *Daphnia hyalina* numbers increased, particularly at pH 10, independently of perch abundance (perch reduced their feeding at this high pH) but the daphnid population collapsed at pH 11. An increase in the number of perch did result in an increase in chlorophyll-*a* concentrations and biomass of chlorophytes, although there was no effect on cyanophytes or cryptophytes. In untreated lake water (pH <9) increased numbers of perch led to a decline in *D. hyalina*.

There are a number of case studies which refute the conclusions of Post and McQueen (1987) and suggest that the control of planktivores significantly reduces algae and improves water quality. In a seven year study it was found that in years with the most abundant perch there was a decline in the density of *Daphnia* consumed by the perch and an increase in the biomass of phytoplankton (Zalewski *et al.* 1990a,b). An optimal recruitment curve was developed to maintain water quality and fish stocks. Regulating water levels during the reproductive and early ontogenetic phase controlled recruitment of perch. Lake Mendota, USA, has been subject to a variety of management practices to control nuisance algal blooms. In 1987 walleye and pike fingerlings were introduced to increase predation on the planktivores (Temte *et al.* 1988). However, the main effect was the high summer temperatures resulting in excessive mortality of cisco, *Coregonus artedii* Leseuer, the principal zooplanktivore. The results were a shift in species composition of *Daphnia*, no spring phytoplankton bloom (cryptophytes and solitary and colonial chlorophytes), an earlier onset and increased duration of the clear water phase correlated with *Daphnia* abundance and a 3 m increase in the maximum Secchi disc depth. Yellow perch and white bass, *Morone chrysops* (Rafinesque), are the other dominant predators on the zooplankton although their foraging rate over a lifetime is very different than that of cisco (Johnston & Kitchell 1996). One cisco will consume as much *Daphnia* as 14 yellow perch or eight white bass. About 80% of zooplankton consumption by cisco is by adults while over 50% of yellow perch and white bass predation on the zooplankton is by juveniles. Cisco prey mainly in the spring and autumn affecting the *D. pulicaria* population compared to yellow perch and white bass which consume most in the summer. Sustaining a low biomass of cisco brings about improved water quality of Lake Mendota. Other methods of increasing the biomass of large crustaceans by reducing fish grazing have included catching small fish and introducing or controlling the catches of piscivores such as pike and zander (Benndorf *et al.* 1988; Ronneberger *et al.* 1993; Findlay *et al.* 1994). However, the ability to biomanipulate in this manner to improve water quality will depend on the extent of eutrophication. Its effectiveness has been found to decline with increasing phosphate loading (Benndorf *et al.* 1988).

Chapter 6
Ionic and Water Balance

6.1 Introduction

Percids are essentially freshwater fish although they venture into brackish waters. When in freshwater the blood and body fluids are hypertonic to the external medium. The osmotic and ion concentration gradient favours the influx of water and the loss of inorganic salts. In brackish water exceeding 10 ppt the reverse occurs, the body fluids are hypotonic to the surrounding medium and without control, there is a loss in water and a gain in inorganic salts.

6.2 Water and ion exchange

Ionic and water balance is brought about mainly by action of gills and kidneys (Fig. 6.1). Since the gills are concerned with oxygen uptake from the water to the blood they must have a large surface area. The limitations of the exchange surface (available area and diffusion distance) determines the maximum aerobic metabolic rate. This large area also provides a surface for water and ion transfer. There may be a 'trade-off' between maximum permeability to oxygen and minimum fluxes of water and ions. The branchial gas and ion exchange are not entirely interdependent and the capacity for gas exchange can be increased without there being an obligatory increase in ion 'leakiness' (McDonald *et al.* 1991a). Increase in permeability to ions can be mediated by decreasing the depth of the tight junctions between adjacent pavement cells. These leaky junctions either create the conditions for higher active excretion of acid-base equivalents by the gills or play a direct role in transferring acid-base equivalents by permitting the outward leak on a diffusion gradient. The epithelium of the gill contains a respiratory component, mainly the secondary lamellae, and a non-respiratory epithelium containing chloride cells. The former makes up 96% of the epithelium surface area compared to 4% for the latter (Isaia 1984). Water fluxes take place through the respiratory cells. Over 90% of diffusional water exchange takes place at the gill surface. The non-respiratory epithelium provides a surface for ion transfer. The number of chloride cells in the epithelium increases with rise in salinity of the external environment. These chloride cells contain numerous mitochondria and a well developed endoplasmic

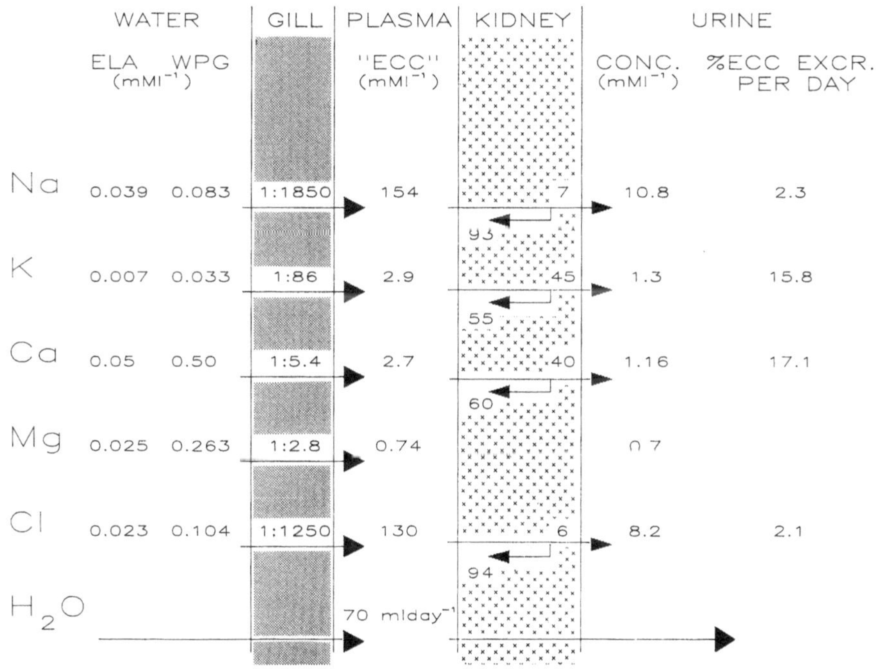

Fig. 6.1 The regulation of water and major ions by gill and kidney action for a 1000 g freshwater fish. Fish were held in the Experimental Lakes Area (ELA) or Winnipeg (Wpg) water. ECC = extra cellular compartment.

reticulum. They are concerned with the regulation of salt uptake either in taking up or secreting ions and use energy in the process. The energy required for osmoregulation is not clearly understood. In some teleosts held in isosmotic solutions oxygen consumption has been shown to be reduced (Rao 1968; Farmer & Beamish 1969). In freshwater the gills absorb Na^+ and Cl^- replacing those lost by renal action. Sodium ions are exchanged with ammonium or hydrogen ions and chloride ions with bicarbonate ions (Fig. 6.2) (Payan *et al.* 1984). Transfer is linked to acid-base equilibrium and elimination of ammonia from deamination of proteins. The ionic transfer across membranes of K^+ and Ca^{2+} involves ATPases. Calcium regulates H^+ uptake and the resulting loss in Na^+. In acid water Na^+ is lost. The kidneys can modify the glomerular filtrate depending on the external medium (Fig. 6.3). The kidney conserves filtered electrolytes. It reabsorbs Na^+ and Cl^- ions from the plasma ultrafiltrate. The urine of freshwater fish is dilute and nearly free of Na^+ and Cl^- (Fig. 6.1) (Hickman & Trump 1969).

In freshwater the primary role of the kidney is to retain electrolytes while excreting the net osmotic gain of water via production of a dilute hypotonic urine. In brackish water urine production is much reduced. The gut may be involved to a lesser extent in water and ion exchange and some salts may be lost in the faeces. Percids are thought to drink little water in contrast to many marine fishes.

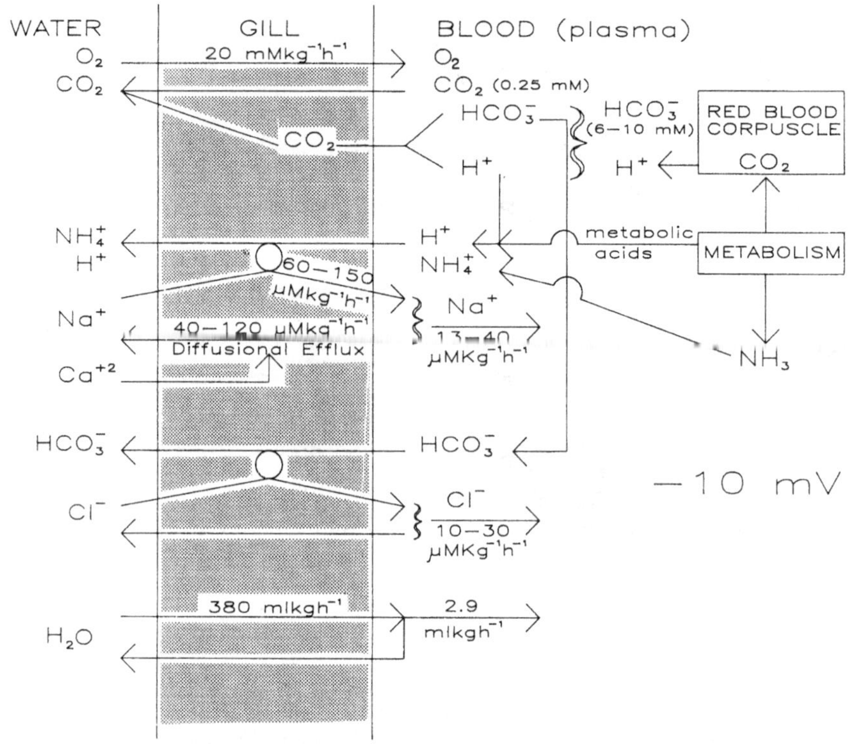

Fig. 6.2 The major rates of water and ion flux across the gill membrane of a freshwater fish and the mechanisms for acid-base regulation of products from body metabolism.

Hormones, principally produced by the pituitary and head kidney, influence water and ion transfer. Hypophysectomy of freshwater fishes reduces the rate of water turnover, which is restored with prolactin injection (Rankin & Bolis 1984). Paralactin, an equivalent to prolactin of mammals, was isolated in cyprinodont fish by Ball (1969). In the head kidney (Chapter 2) interrenal tissue contains corticosteroid-producing cells. Cortisol increases water fluxes and its production is controlled by pituitary adrenocorticotrophic hormone. The head kidney also contains chromaffin tissue consisting of catecholamine-producing cells. The production of adrenaline from this tissue causes an increase in oxygen uptake and a disequilibrium effect on water balance. The result is an increase in gill permeability to water.

Carbon dioxide produced from catabolic processes in the tissues diffuses into the blood and reacts with water to form carbonic acid (Fig. 6. 2). The process is a slow one in the plasma but is speeded up in the red cells by the enzyme carbonic anhydrase (Albers 1970). Carbonic acid dissociates into hydrogen and bicarbonate ions and to a small extent carbonate ions. Hydrogen ions recombine with the plasma proteins and haemoglobin, which represent the buffer agents. Temperature affects disassociation constants of the buffer agents as well as the solubility of carbon

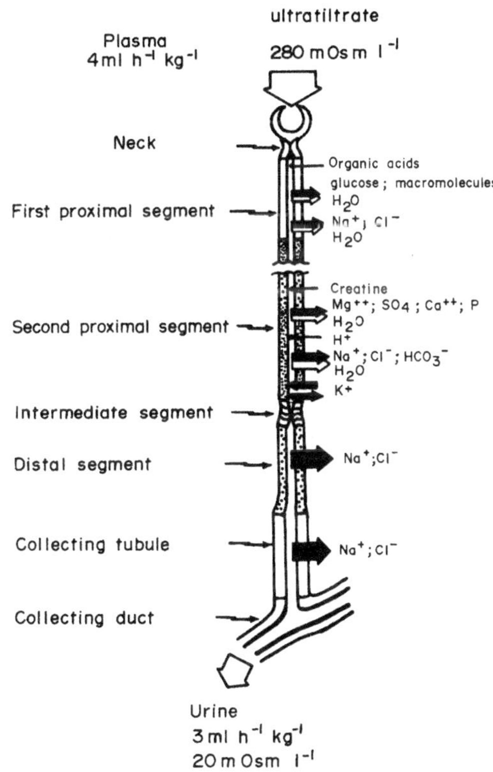

Fig. 6.3 The structure and function of a nephron of a freshwater fish. Active (solid arrows) and passive (open arrows) movements are indicated. The width of the arrows is proportional to the rate of movement of substances across the tubule epithelium. (After Hickman & Trump 1969.)

dioxide. In a steady state any surplus hydrogen ions are excreted at the same rate as they are produced. Temperature increase and stress cause an increase in H^+ and OH^- ion production, which upsets the acid-base balance. The buffering mechanisms of the body and changes in gill ventilation are limited in their ability to deal with this increase. An increase in gill ventilation is restrained by physical limitations in the energy supply for long-term hyperventilation in a water medium (Heisler 1984). Excretion plays the main role in maintaining a steady state and adjusting pH and acid-base balance during a time of stress. Bicarbonate may be directly excreted through the gills. Also the excretion of ammonia causes an influx in Na^+ ions. As a consequence of the high degree of solubility of CO_2 in water and its high permeability coefficient through the gill, fish are unable to regulate acid-base balance by adjusting the CO_2 concentration in the blood. Fishes regulate blood pH, therefore, by controlling plasma bicarbonate levels.

6.3 Salinity

Perch species living in water with a high sodium ion concentration tend to have a higher concentration of Na^+ in their blood serum compared to fishes living in waters with lower concentrations of the ions. The Volga River has a sodium concentration of 1.07 (meq) l^{-1} and the blood serum of perch living in the river contains sodium ions at 150 $\pm$ 8 (meq) l^{-1}. In Lake Baikal (former USSR) where the sodium concentration is 0.18 (meq) l^{-1} the sodium in the blood serum of the resident perch amounts to 114 $\pm$ 6 (meq) l^{-1} (Natochin & Lavrova 1974). Experiments were conducted by Lutz (1972) to test the tolerance of perch to increasing concentrations of salinity. The results of his experiments, in which he subjected the perch to concentrations ranging from freshwater to 15 ppt salinity, half the concentration of sea water, and measured the concentrations of major inorganic ions in the blood plasma and muscle, are given in Table 6.1. Perch survived well in concentrations up to and including 10 ppt (most freshwater fish are isotonic at this concentration) but did not survive in 15 ppt sea water. All ion levels increased in the blood plasma at 15 ppt and there was a large invasion of sodium and chloride ions into the muscle. The whole body lost water although the liver remained fairly resistant to water loss and chloride ion invasion. The water content of the muscle was significantly lower than that of perch held in freshwater. In the Aral Sea, perch inhabit waters with a salinity of 10 ppt (Letichevskii 1946) and in prairie saline lakes in Manitoba, Canada, yellow perch live in salinities of 10.3 ppt (Driver & Garside 1966). Perch and yellow perch are able to live in other areas where the salinity does not exceed about 10 ppt. This includes the Baltic Sea where both perch and zander feed in salinities ranging from freshwater to 8 ppt (Berzins 1949). Yellow perch are found in Chesapeake Bay, USA, at salinities ranging from 5 to 7 ppt (Muncy 1962) and in the Neuse River, USA, in salinities up to 12.2 ppt but preferring areas of 1.8 to 3.5 ppt (Keup & Bayless 1964).

The sudden transfer from freshwater to high saline concentrations can cause damage to the tissues and often death. Zander transferred in this manner are unable

Table 6.1 The concentration ($\mu M l^{-1}$) of various ions in the plasma and muscle of perch placed in freshwater (FW) and 3.6, 10.0 and 15.0 ppt salinities. Also given are the water concentrations (g water g^{-1} dry mass) of the muscle. (After Lutz 1972.)

Tissue	Salinity (ppt)	Concentration ($\mu M l^{-1}$) Na	K	Ca	Mg	Cl	Water (g water g^{-1} dry mass)
Plasma	FW	154.2	3.6	4.4	1.5	120.3	
	3.6	155.6	3.7	4.8	1.5	136.5	
	10.0	155.8	5.3	4.0	2.7	137.9	
	15.0	210.5	6.0	7.3	7.7	195.8	
Muscle	FW	19.9	143.0	2.6	15.2	10.3	4.2
	3.6	19.3	148.8	3.5	16.3	11.5	3.9
	10.0	20.6	162.3	3.4	16.6	12.1	4.0
	15.0	47.8	172.7	3.7	21.9	32.6	3.4*

* Significantly different from FW ($p < 0.05$).

to keep ions in equilibrium and hydration of the cells leads to death. Sodium ions increase in the cells and potassium and calcium ions are lost (Neacsu *et al.* 1981).

The sudden transfer causes an initial reduction in ATPase activity followed by recovery and the intense osmotic claim may cause death. Zander are able to tolerate high salinities if transferred gradually. This results in increased ATPase activity to pump out salts to meet the new osmotic equilibrium (Craciun *et al.* 1982). The fish does appear to tolerate a sudden change back to freshwater and can control and regain ionic equilibrium. Zander tolerate salinities up to about 12 ppt. An increase in salinity from 10.6 to 12.7 ppt in the Sea of Azov has restricted zander to low salinity areas in the Bay of Taganrog and has also caused a decline in growth (Avedikova & D'Yakova 1979). The optimum salinity is probably about 6 ppt. Zander fingerlings have the best growth rates at this salinity and mortality declines in salinities ranging from 3.5 to 11 ppt compared to controls held in freshwater (Zhmurova & Somkina 1976). Optimum salinities for walleye are similar to those of zander and lie in the range 4–8 ppt. They can live in waters of much higher salinity than this, such as saline prairie lakes (15 ppt) into which they have been introduced (Rawson 1946) but where they are apparently unable to reproduce.

6.4 Acid stress

Acid stress in the external medium interferes with permeability and electrolyte transport at the gill surface with the resultant loss of electrolytes, especially sodium. A series of adjustments are made in gill permeability and electrolyte transport and also in acid-base balance. Acid tolerance does not appear to be related to the physical nature of the gills, surface area, lamella thickness and diffusion distance, or with the extent of mucus production at the surface (McDonald *et al.* 1991b). It appears to be correlated with chloride cell density and branchial ion transport activity. In perch McDonald *et al.* (1991b) suggest that the ion permeability may be reduced by an increase in the depth of tight junctions between adjacent gill pavement cells. In soft water, chloride cells proliferate in the gills especially in the epithelium of the secondary lamellae. However, when exposed to acid, the chloride cells of yellow perch were damaged and less abundant in the secondary lamellae (Leino *et al.* 1987). Blood osmolality was reduced at pH 5.0. The authors suggest that chloride cells proliferate in very soft freshwater to help maintain ionic balance but that damage to these cells in acidified soft water diminishes the capacity for ion regulation. Some waters are naturally acidic and contain populations of yellow perch which are hardier to low pH than walleye, which are not normally found in these lakes. Perch and yellow perch occurring naturally in acidic lakes have a higher tolerance to low pH than those living in non-acidic lakes (Rask & Virtanen 1986). This is in part, a genetic characteristic (Rahel 1983). Perch inhabiting naturally acid waters also have gills with lower permeability to hydrogen and sodium ions (McWilliams 1982). Rahel (1983) found that the yellow perch in an acidic Wisconsin brown water lake which had a low salinity (pH 4.6, Na 13 (meq) l^{-1} and K 11 (meq) l^{-1}) had only slightly lower sodium and potassium levels in the blood than those yellow perch from non-acidic, high salinity lakes (pH 7.6, Na 70 (meq) l^{-1} and K 18 (meq) l^{-1}). Perch tend to be more tolerant to acidic conditions than other freshwater species such as minnows (*Phoxinus phoxinus* (Linnaeus)), brown trout, roach and Arctic charr (*Salvelinus alpinus* (Linnaeus)) (Bergquist 1991). In a study of 100 small forest lakes in Finland, Toivonen (1991) divided the lakes into two types. Perch-type lakes had low pH and conductivity compared to roach-type lakes which had higher pH values (>6) and higher conductivity. The latter lakes had more species, although roach were dominant. Roach are more sensitive to acid conditions than perch (Rask *et al.* 1995a,b). Yellow perch are similarly less sensitive to lower pH than other species in lake communities (McCormick *et al.* 1989; Kelso & Johnson 1991; Brezonik *et al.* 1993; Kopp *et al.* 1994). Exposure to acid conditions may cause fish to adapt for hypoxia. Nelson *et al.* (1988) found that haemoglobin concentrations in the blood were highest in yellow perch from a well oxygenated acidic lake (pH 4.5), intermediate in a neutral lake with anoxic zones and lowest in a well-oxygenated neutral lake.

Many lakes have become acidic due to acid precipitation from the burning of fossil fuels and the roasting of sulphide ores. Oxides of nitrogen and sulphur are released into the atmosphere and return as acid snow and rain. Surface waters running into lakes increase the hydrogen ion concentration and reduce the buffering capacity. In the Sudbury area of Ontario, Canada, the pH of many of the lakes is 5.0–5.5. In these lakes the walleye populations are more sensitive to the low pH than the yellow perch. This may be due in part to the sensitivity of eggs and embryos. The snowmelt into the streams where walleye spawn may reduce the pH considerably (Gunn 1982). In southern Finland mortality of perch eggs was nearly 100% in recently acidified lakes (Lappalainen *et al.* 1988). However, female *Perca* species living and laying eggs in an acid environment may lay eggs which produce fry more tolerant to low pH than normal (Rask 1984). In the Lac Cloche Mountain lakes, Canada, young yellow perch grow faster than normal and older fish grow more slowly (Ryan & Harvey 1980). The reversal in growth occurs at about 140 mm (Fig. 6.4). The young yellow perch have a high mortality rate and may grow faster due to more food being available for the survivors. Similar results were found for perch soon after acidification in southern Finland (Rask & Raitaniemi 1988). In a naturally acidic, highly humic forest lake, Lake Karhujaervi, Finland, low pH greatly reduced the reproduction of perch, whereas their growth rate was higher than those in nearby less acidic lakes (Rask *et al.* 1986). *Perca* species recruitment can be affected by delayed spawning in highly acidified lakes (Rask *et al.* 1990) and by the sensitivity of eggs, embryos and fry to these conditions (Vinogradov & Komov 1985; Holtze & Hutchinson 1989). Larval perch can withstand acidic conditions if they are exposed for a short duration (<2 days). After this time acidity alone can be used as a predictor of survival (Hildén & Hirvi 1987).

Liming of lakes may result in the replacement or shift in abundance of acid

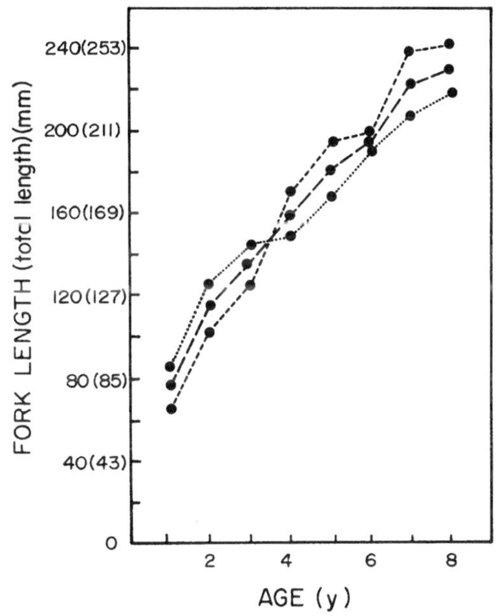

Fig. 6.4 Growth of yellow perch from lakes in the La Cloche Mountain area at pH 4.5 (dotted line), pH 5.5 (large dashed line) and pH 6.5 (small dashed line). (After Ryan & Harvey 1980.)

tolerant species with acid sensitive ones (Raitaniemi & Rask 1990). However, in Lake Fjorda, Norway, populations of fish species other than perch did not recover even after eight years of liming (Saksgaard & Hesthagen 1995). Perch recruitment was normal before and after liming. Yellow perch were unaffected by the addition of calcium carbonate in Bowland Lake near Sudbury, Canada but lake trout (*Salvelinus namaycush* (Walbaum)) increased and competed with the yellow perch (Gunn *et al.* 1990). Aluminium and beryllium may be important toxic agents in acidified waters (Jagoe *et al.* 1993; Vuorinen *et al.* 1992) causing gill abnormalities. Levels of trace elements such as mercury, cadmium and lead tend to be higher in the muscle and liver of yellow perch in acidic lakes (Sprenger *et al.* 1988; Stripp *et al.* 1990). Liming can ameliorate metal accumulation (Andersson *et al.* 1995).

Chapter 7
Swimming, Movements and Migrations

7.1 Introduction

Swimming or locomotion, movements and migrations of percids have evolved to adapt the family to their life in freshwater particularly in feeding, reproduction and ontogenetic development. Locomotion and its sustainability is closely linked with anatomy and morphology including shape, musculature, skeleton and sensory system (Chapter 2) and with respiration and energetics (Chapter 5). Some percid species are lacustrine while others are lotic for their whole life or change during ontogenetic development or during a phase in the reproductive cycle. Some form schools while others are solitary and some change from one to the other during their life cycle. Many undergo marked diel and seasonal movements and migrations.

7.2 Swimming

Compared to many other species of fish *Perca* are inferior swimmers. In experiments on fast-start performance, yellow perch were shown to be poor accelerators. Maximum acceleration rate $\pm$ 2 SE = 23.9 $\pm$ 12.4 $m\,s^{-2}$ compared for example to pike with maximum acceleration 39.4 $\pm$ 8.5 $m\,s^{-2}$ (Webb 1978). *Perca* species may overcome poor swimming ability by schooling, which provides protection for the young fish and social hunting in older fish. The swimming speed of larval yellow perch and walleye is fairly similar and increases with size (Fig. 7.1). Walleye swim more slowly than yellow perch in the size range 7–9 mm since they contain more yolk than the yellow perch. Swimming ability improves from 8 to 9.5 mm as the yolk sac is absorbed and from 9.5 to 14.5 mm they swim equally well.

Temperature plays a major role in swimming speed. Hergenrader and Hasler (1967) measured the swimming speed of yellow perch at different times of the year using sonar. Slowest movements were in the winter at water temperatures of 0–5°C and they increased with a rise in temperature to a maximum in the summer when the water temperature reached 20–25°C. Swimming speeds were found to be linearly

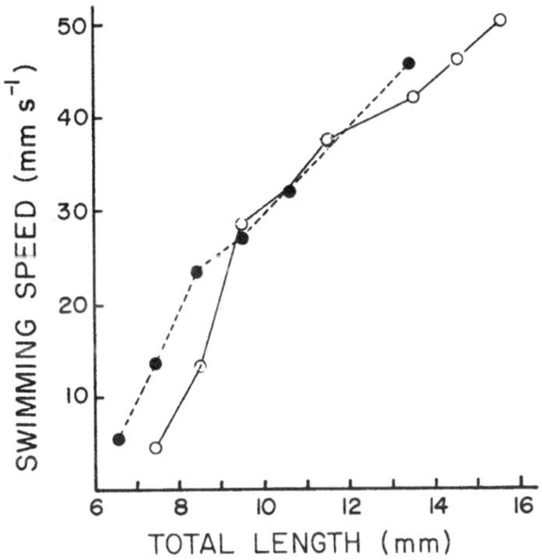

Fig. 7.1 Sustained swimming speed of larval yellow perch (●) and walleye (○) in relation to length. (After Houde 1969b.)

related to water temperature (Fig. 7.2). The greatest variation among individuals was found in the summer. The tracking of individual fish showed that these swam on average at half the speed of the schools. The annual variation in swimming ability of perch was demonstrated by Sandström (1983). He used a rotatory flow technique in which the critical revolutions per minute in the apparatus were the maximum the fish could withstand. This critical value increased in summer and declined during the autumn. Laboratory tests with perch held under constant conditions showed that the response was endogenous and not temperature dependent. It suggests an economic use of stored energy so that, for example during winter months when energy intake is limited, the perch will not use up energy reserves. Thus, swimming ability appears to be endogenous with temperature acting as an external control.

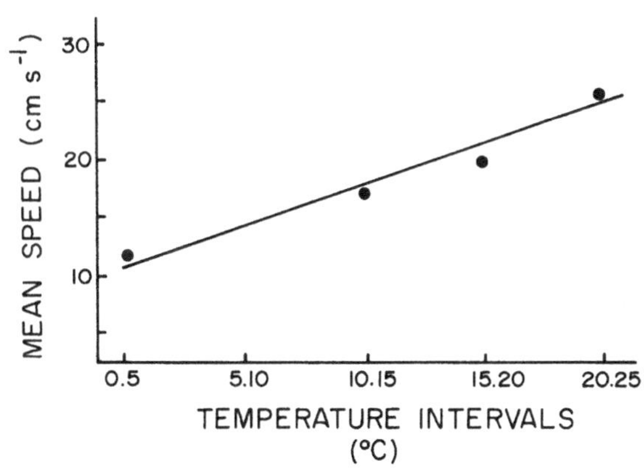

Fig. 7.2 Mean speeds of yellow perch schools in Lake Mendota in relation to temperature. (After Hergenrader & Hasler 1967.)

Walleye have the ability to swim against strong currents (Fig. 7.3). Jones *et al.* (1974) derived a relationship between length (L, cm) of fish and water velocity (V, cm s^{-1}):

$$V = KL^e$$

where the constant $K = 13.07$ and exponent $e = 0.51$ for walleye. The calculated velocity is that which can be maintained for 10 minutes. The relationship is unaffected by acclimatisation to different temperatures or temperature shock of $\pm$ 7°C. There do not appear to be any differences between sexes.

Activity is correlated with temperature, day-length and food consumption (Neuman *et al.* 1996). However metabolic cost of swimming in juvenile walleye (5–7 g) was independent of temperature at each swimming speed tested by Beamish (1990a). Oxygen consumption increased with rise in swimming speeds between 20 and 45 cm s^{-1} from 5 to 15°C but declined from 15 to 23.5°C (Fig. 7.4). Adult fish vary both active and basal oxygen consumption (Chapter 5) seasonally with the maximum at the reproduction period. Photoperiod mediates seasonal changes in basal metabolism associated with gonadogenesis. The mechanism is not fully

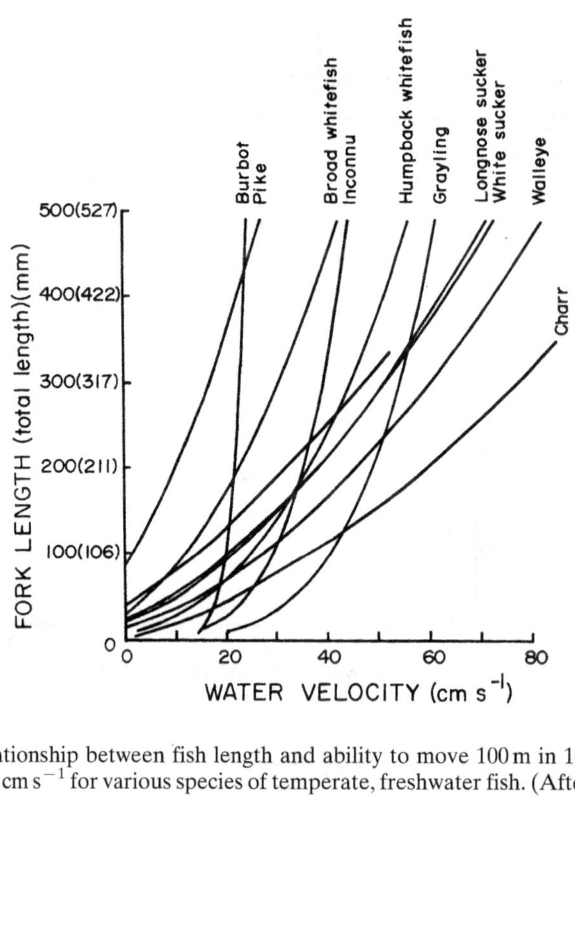

Fig. 7.3 The relationship between fish length and ability to move 100 m in 10 min against water velocities up to 80 cm s^{-1} for various species of temperate, freshwater fish. (After Jones *et al.* 1974.)

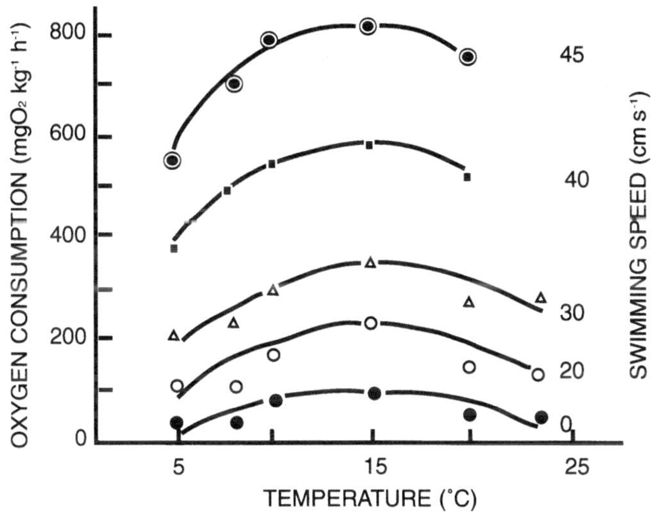

Fig. 7.4 Mean oxygen consumption of walleye in relation to temperature and swimming speed. (After Beamish 1990a, with kind permission from Kluwer Academic Publishers.)

understood but probably the blood oxygen carrying capacity is increased at this time with improved oxygen delivery to the tissues (Beamish 1990b).

The swim bladder plays an important role in swimming. It counteracts the negative buoyancy of the fish. The swim bladder pressure of yellow perch and walleye living in running water is reduced compared to that of fish held in still water (Gee *et al.* 1974). The control of the swim bladder is not well understood. It is probably under secretory cholinergic nervous control. In perch and other physoclists the swim bladder has a postganglionic innervation which exhibits acetylcholine esterase activity (Fänge & Holmgren 1982).

7.3 Movements

Both *Perca* and *Stizostedion* species are schooling fish although adult zander become solitary. Large perch are also often solitary (Bruylants *et al.* 1986; Le Cren 1992). Vision is important in schooling behaviour (Breder 1959) and perch and yellow perch schools break up at twilight and the fishes settle to the bottom (Hasler & Villemonte 1953). From mark and recapture data Craig (1973) found that fish marked at the same place and time could be recaught together at a different location up to 18 months later. Male and female perch were often in separate schools. Helfman (1984) describes school fidelity as the cohesiveness of social aggregations over time as measured by the repeatability with which individuals reoccur as members of a school. He found yellow perch in Cazenovia Lake, USA, to be facultative schoolers with individual differences between fish in schooling tendency (percentage of time in the schools).

Hergenrader and Hasler (1968) established a high negative correlation between yellow perch school size, in terms of space, and light transmission properties of the water. In summer the spaces between individuals was <0.6 m and the distance from top to bottom of the school was 2.5 m. In winter the distance between individuals was >0.6 m and the height of the school was 6.7 m. In the latter period, the yellow perch were not very active and lost some of their tendency to school.

Perch, yellow perch and walleye show distinct patterns of diel activity (Helfman 1986). As a rule perch and yellow perch are more active during the day in contrast to walleye which are more active at night. This can vary among age groups. In some instances juvenile *Perca* species have been found to be active at night. In West Blue Lake, Canada, young yellow perch were active at night and were vulnerable to predation by walleye (Kelso 1976). However, Johnson and Müller (1978) suggest that young perch were active at night, when they migrated down the River Angeran in northern Sweden, to avoid predation by pike and adult perch which were diurnally active (Fig. 7.5). Increased activity at dawn and dusk is a common feature of perch, yellow perch and walleye behaviour (Hasler & Bardach 1949; Lind *et al.* 1973; Kelso 1976; Craig 1977b; Eriksson 1978; McConville & Fossum 1981). The extent of this activity increases with rise in water temperature. As day length shortens and light intensity decreases during the onset of winter, perch and yellow perch become active during the day (Hergenrader & Hasler 1966; Lind *et al.* 1973) including under the ice (Whoriskey & Brown 1994). Perch also become active during the day in the summer on dull days (Svirskii *et al.* 1976) or when the water is turbid or algal blooms (Craig 1977b; Thorpe 1977) reduce light penetration. Hasler and Bardach (1949) noted that in the summer yellow perch moved inshore one hour before sunset and cruised parallel to the shore at the 6 m contour, feeding until after the sun disappeared. Carlander and Cleary (1949) also observed that walleye came into

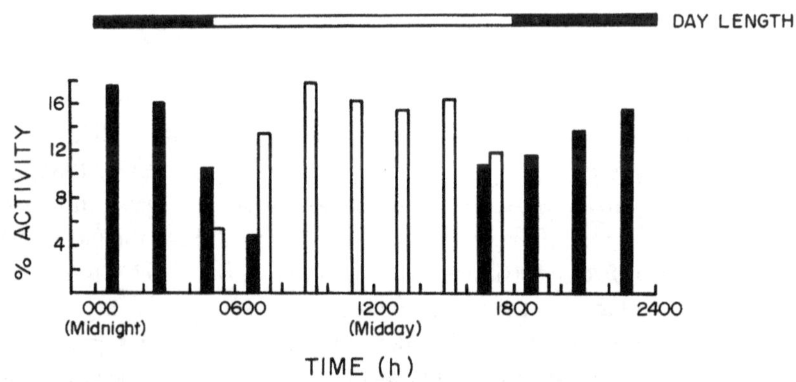

Fig. 7.5 Diel activity of juvenile (■) and adult (□) perch held in experimental tanks. (After Johnson & Müller 1978.)

shallow water to feed at dusk although Holt *et al.* (1977) and Einhouse and Winter (1981) found no such activity. Two types of behaviour have been noted in the daily movements of perch and walleye (McConville & Fossum 1981; Mackay & Craig 1983). Firstly, a variable meandering movement in which the direction changes rapidly and often. This type of movement commonly occurs at sunrise and sunset and may be associated with feeding (Fig. 7.6). Secondly, movements in a relatively straight line suggesting cruising which can occur at any time during the normal activity period of the fish. This latter movement may be a movement from one feeding area to another.

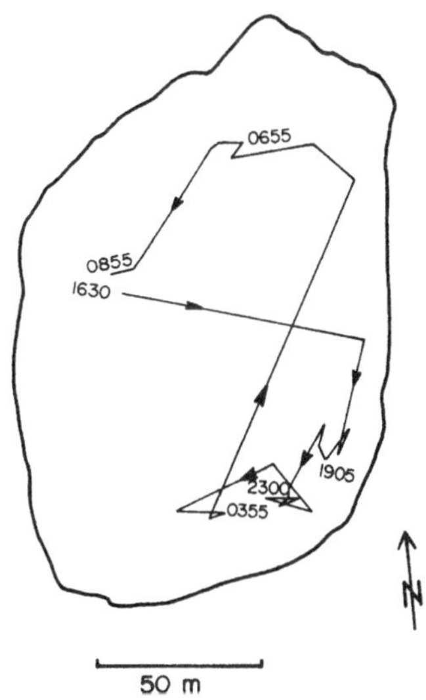

Fig. 7.6 The activity of a perch in Bigland Tarn measured in the summer from 1630 to 0855 h next day with an ultrasonic tag. (After Mackay & Craig 1983.)

The diel movements and distribution of percids not only varies seasonally (Craig 1977b; Goldspink 1990) but also ontogenetically with the age of the fish (Post & McQueen 1988; Chapter 4). In Lake Constance, Germany the summer horizontal distribution of perch was found to be patchy (Imbrock *et al.* 1996). During the daytime the fish were at 3–15 m depth, 2–6 m above the thermocline. Within this layer the age classes were separated with young-of-the-year numbers declining with depth and those two years old and above increasing with depth. At dusk the fish moved to the littoral area and spent the night resting on the bottom. In the winter, when the water was almost homothermal, perch of all ages were located at 35–70 m. During the day they rested on or close to the bottom but at night ascended to about 20 m below the surface. Zander juveniles in the Sulejow Reservoir, Poland, tend to be in deeper water than adults to avoid cannibalism by the latter (Frankiewicz *et al.* 1997).

Percids may move in and out of vegetation on a diel basis in relation to oxygen levels. Suthers and Gee (1986) found that in the Delta Marsh, Manitoba, Canada, juvenile yellow perch avoided severe hypoxia (≤ 1.5 ppm DO) in the cattail (*Typha* spp.) habitat at night but returned to the area during the day, when oxygen levels rose, as it provided protection from predation.

7.4 Migrations

Distances moved by perch over the year are not extensive and most are localised (Smith & Van Oosten 1939; Craig 1974a). In rivers, older, solitary perch may have a home range while this tendency may not be strong in the younger, one to two year old, shoaling fish (Bruylants *et al.* 1986). Perch and yellow perch may return to 'home' areas to spawn (Kipling & Le Cren 1984; Aalto & Newsome 1990; Chapter 4). Perch migrations in the coastal waters of Finland varied in different areas (Boehling & Lehtonen 1985). In the parts where the archipelago was narrow and the area of shallow water was small, migrations were shorter than those areas with extensive shallow waters. Three other factors affected migrations: distribution and volume of warm water, distribution of food resources and the abundance of perch in neighbouring populations. There is also convincing evidence that walleye home to spawning areas. Olson *et al.* (1978) put forward the theory that this is adult learned behaviour and is strengthened by repeated migrations. Since eggs and larvae of walleye are often carried substantial distances in river currents from areas of egg deposition, natal conditioning is unlikely. In Oneida Lake, USA, three distinct walleye populations return to their own specific spawning grounds in successive years (Forney 1963). After spawning, the population gradually disperses from the spawning grounds and there is some mixing between the populations. Widest distribution occurs in the late summer. However, evidence suggests that like perch, the movement of walleye is confined to a limited area except at spawning time. Daily migrations of walleye of the Nipigon Bay region of Lake Superior, North America, even at spawning time are not often greater than 3 km day^{-1} and usually about 0.8 km day^{-1} (Ryder 1968). Apart from individual exceptions (Colby *et al.* 1979) the majority of walleye populations migrate <16 km between spawning and feeding areas. In large areas of water, mixing between different stocks during summer feeding periods may not occur. Tagged walleye in eastern Lake Erie were released on spawning grounds near Barcelona, USA (Wolfert & Van Meter 1978). They moved northeast along the shore to the summer feeding area near the Buffalo-Niagara River inlet. During late summer and the autumn the fish moved back towards the spawning grounds near Barcelona. There were no indications of westward movements, which might have brought them in contact with other stocks.

Zander undertake major migrations for spawning. Other movements are fairly localised. There is also some evidence that this species has a home area for spawning and members of the same population return to this area year after year (Puke 1952). Lehtonen and Toivonen (1988) studied the migrations of zander in the coastal waters of Finland. They found that zander occurred in more or less separate populations each of which showed clear homing to their spawning areas. Annual migrations took place between spawning inlets, feeding areas in the archipelagoes and bays, and the deeper wintering areas. Distances travelled in autumn and winter varied with more extensive migrations occurring in large, shallow, turbid areas than in open coastal waters where the archipelago was narrow.

Chapter 8
Parasites and Diseases

8.1 Introduction

When a parasite has established itself on or within a fish, it can cause the death of the fish, develop an equilibrium state with no evident disease symptoms or be eliminated by the defence system of the fish. *Perca* and *Stizostedion* species are host to a large number of parasites but many of these parasites, as far as is known, do not cause harm to the fish. However, all parasites must deprive the fish of some anabolic energy. In an extreme case, Pitt and Grundman (1957) found that *Ligula intestinalis* reduced the growth of yellow perch by 50% over a five year period. (In this chapter the authorities for parasite species are not given. These can be found in the references cited.)

Host–parasite relationships appear to be evolutionarily adaptive and in some parasites morphological and physiological adaptations are remarkable. Disease outbreaks occur when there is an imbalance between the fish host, the pathogen or disease agent and the environment. Fishes respond significantly to ecological variables, as explained in other chapters, so outbreaks of disease are strongly affected by environmental conditions. The parasitic fauna can be influenced by abiotic and biotic factors including temperature, light, salinity, pH, oxygen concentration, water velocity, the diversity of the local aquatic fauna, the diversity of fish-eating birds, the history and geographic isolation of the fish environment, the size, age and physiological state of the fish host, the size of the host population and the season of the year. Fishes can be intermediate or definitive hosts of parasites. When a fish is an intermediate host it must be eaten before the parasite can be transmitted to the definitive host. The parasite can escape from the definitive host from the body surface (ectoparasite) or through body openings (endoparasite).

Few reports of genetic defects in percids have been made. A skeletal muscle abnormality observed in walleye in some lakes of Minnesota and North Dakota, USA, may be genetic but there certainly is an environmental component in the disease expression since disease prevalence is higher in walleye from more eutrophic lakes (Economon 1978; Holloway & Smith 1982). Economon (1978) suggested that the disease could be likened to muscular dystrophy found in humans. The disease causes pronounced degeneration of the lateral trunk musculature.

Affected tissue is yellowish-brown and coarsely fibrous. Histological investigation shows muscle fibres in various stages of degeneration.

8.2 Causative agents

Percids can exhibit severe disease symptoms caused by viruses, bacteria and fungi. Walker (1969a) detected three viruses from skin lesions in walleye from Oneida Lake, USA. These viruses were associated with lymphocystis, which was common, and with dermal sarcoma and epidermal hyperplasia which were locally common in the walleye from the lake (Bowser *et al.* 1988). The incidence of these viruses extends into central and western regions of Canada and may be widespread throughout walleye populations in North America (Yamamoto *et al.* 1976; Yamamoto *et al.* 1985). In central Canada a fourth virus (*Herpesvirus vitreum*) associated with a different form of epidermal hyperplasia has been isolated (Kelly *et al.* 1983). Lymphocystis is a common viral disease of walleye and may cause mortality (Margenau *et al.* 1988). Afflicted fish exhibit wart-like protuberances on the fins, jaws and opercula. It was first recognised in the ruffe, *Gymnocephalus cernuus*, (Weissenberg 1965) and is probably widespread in many of the percids (Walker & Weissenberg 1965; Amin 1979). Anders and Darai (1985) identified the genus *Cystivirus* as causing lymphocystis in percids and other fish families. Dermal sarcoma tumours (Bekesi & Kovacs-Gayer 1986) are similar in gross appearance to those of lymphocystis. Dermal sarcoma in walleye are probably caused by a virus from the family Retroviridae (Martineau *et al.* 1991, 1992; Bowser & Wooster 1994; Bowser *et al.* 1994). They are large, white growths often tinged with red. On more careful observation, they have a smooth appearance in contrast to the granular nature of the enlarged cells of lymphocystis. Bowser *et al.* (1990) investigated the effect of temperature on the transmission of dermal sarcoma in fingerling walleye. Transmission was most successful at 15°C compared to 10 and 20°C. Dermal sarcoma regresses seasonally in walleye (Bowser & Wooster 1991). Epidermal hyperplasia may be associated either with a retrovirus or a herpesvirus and has a clear slime-like appearance. However, the lesion with the associated herpesvirus has a more diffuse character and there is often a swelling of the underlying tissue (Yamamoto *et al.* 1985). Epidermal hyperplasia was identified by Walker (1969b) in yellow perch but no viruses were detected. Virus-containing skin lesions in walleye show some seasonality in incidence and are frequently common during the spawning period. Lymphocystis and dermal sarcoma appear to have the most effect on adult fish causing open wounds and loss of blood.

Walleye fry and fingerlings are susceptible to infectious pancreatic necrosis virus (IPNV) and infectious haematopoietic necrosis virus (IHNV), two viruses that cause severe diseases in salmonids (Schat & Carlisle 1980). Problems of cross infection may occur if walleye and infected salmonids are stocked together in the same pond.

The epizootic haematopoietic necrosis virus (EHNV) has been isolated from perch in Australia (Langdon *et al.* 1986; Eaton *et al.* 1991; Hedrick *et al.* 1992; Hengstberger *et al.* 1993; Gould *et al.* 1995; Whittington *et al.* 1996). The origin of the virus is unclear since the perch were originally introduced from England (Cadwallader & Backhouse 1983) and no isolation of the virus nor similar symptoms have been recorded in Europe.

Bacterial infections appear to be less important than viral infections in fishes. There are only a few recorded cases where bacteria have been identified as being the primary cause of a disease outbreak in percids. In many cases diseases have been aggravated by secondary bacterial infections (Bauer *et al.* 1973). Nordstrom *et al.* (1960) isolated gram negative bacilli which were similar to *Pseudomonas* from yellow perch found dead in large numbers in Dailey Lake, USA. The disease, which produced symptoms of multiple petechiae present in the skin and dorsal musculature and inflammation of the peritoneal cavity, appeared to be specific to the yellow perch. Granulomatous skin lesions and granulomatous peritonitis and peritonitis were observed in two yellow perch populations in Alberta, Canada. *Mycobacterium chelonae* was isolated from affected fish in both epizootics although the source and route of infection was not identified (Daoust *et al.* 1989). A well recorded high mortality of perch occurred in Windermere, England (Bucke *et al.* 1979). The perch population has been studied over many years (see Chapter 9) so it was possible to estimate that over 98% of the population, over one million fish, were killed by 'perch disease'. A wide variety of common water bacteria were isolated from both healthy and diseased fish. The bacteria included *Aeromonas*, *Pseudomonas* and some flexibacteria. The commonest bacterium found in infected fish was *A. hydrophila* subspecies *hydrophila*. These bacteria were considered to be secondary in nature and the causative organism was not isolated. However, symptoms resembled those of furunculosis and a form of *Aeromonas* not identified may have been the cause of the disease. This disease was specific to perch and no other species in the lake seemed to be affected, although the decline in the perch population had a dramatic affect on other species, particularly pike, the main predator of perch. Tentatively, a subspecies of *Aeromonas salmonicida* specific to perch may be the cause of a form of furunculosis which is normally found in salmonids. Furunculosis has been identified in perch by McCarthy (1975) and is probably more widespread than is realised.

The diseased Windermere perch also had secondary infections of fungi. These were mainly species from the family Saprolegniaceae. Six genera were isolated, *Achlya*, *Aphanomyces*, *Leptolegnia*, *Leptomitus*, *Pythiopsis* and *Saprolegnia* (Bucke *et al.* 1979). The occurrence of fungi was higher in the winter than in the summer. The seasonal differences may be caused by the preference of the fungi for lower temperatures compared to the bacteria found in the perch lesions. In both winter and summer up to four different species of fungi could be found on a single lesion. Spores of fungi are present in the water (the sewage fungus, *Leptomitus lacteus*, is more concentrated in the littoral than in the centre of the lake; Willoughby &

Roberts 1991) but infection of the fish does not occur until there is a break in the skin. It is of interest that a year before the outbreak of perch disease in Windermere, the parasitic branchiuran, *Argulus foliaceus*, was discovered for the first time in the lake. The acquisition and growth of fungi appeared to be fairly unselective. This was also the case when young-of-the-year perch in Windermere were found to be infected by fungi in an earlier study (Willoughby 1970). *Saprolegnia ferax*, common in Windermere water samples, did not occur on the perch in either study. Secondary infections of Saprolegniaceae fungi on damaged eggs can also cause serious losses in all species of percids. Disease-inducing fungi are composed of branched hyphae without septa and form dense, sometimes cotton-like growths in severe cases. The sporangia are located at the end of the hyphae, separated by septa, and in the sporangia large numbers of zoospores are formed. These spores are liberated into the water when ripe.

The number of zooparasite species associated with perch, yellow perch, walleye and zander is extensive; a comprehensive list is given in Table 8.1a (end of chapter). Protozoa and Myxosporida are parasites of the skin, gills, gut, eye, urinary system and blood. Many species of these groups are limited to a specific host of percid. However, *Icthyophirius multifiliis* is a common parasite found under the epidermis of perch, yellow perch and walleye. When suitable conditions occur it can become an epizootic and cause mass mortality. The mature trophonts break away from the host, sink to the bottom and attach to plants or other substrates. Each cyst divides into 1000–2000 daughter cells which are released into the water (Bauer *et al.* 1973). These mobile ciliates then search for a new host, penetrate the skin, grow and mature. If a new host is not found within two to three days the parasite dies.

Monogenea and Trematoda are common parasites of percids. Monogeneans are ectoparasites of gills, skin and body orifices and trematodes infect many of the body organs (Table 8.1a). The monogeneans have a direct life cycle with a single host. For example, *Urocleidus adspectus* lives on the gills of yellow perch. Its eggs may overwinter away from the host on the lake bottom (Tedla & Fernando 1969). Free-swimming larvae hatch from these eggs and infect a new host when contact is made. Trematodes may need one or two intermediate hosts for development and the definitive host can be a fish, bird or mammal. Apart from a few exceptions such as *Sanguinicola occidentalis*, it is usually the larval stage of the parasite which causes disease in fishes. The adult parasite of *S. occidentalis* lives in the blood vessels of yellow perch and walleye, and a snail, belonging to the family Limnaeidae, acts as the single intermediate host. *Diplostomum spathaceum* infects the eye of perch, yellow perch and zander. These fish are the second intermediate host to the parasite; the definitive host is a bird, often a gull. Eggs from the adult worm in the gut of the gull are released in the faeces and motile miracidia, which develop from these eggs, infect the first intermediate host, a limnaeid snail. Cercaria are released from the snail, penetrate the skin of the percid and reach the eyes through the blood system. Here the parasites develop into metacercaria and the life cycle is complete when a bird eats the fish.

Many cestodes are parasitic on percids. The adult tapeworms live in the gut of the definitive host but encyst in the body organs and muscle of fishes which act as intermediate hosts. Adult *Triaenophorus nodulosus*, for example, live in the gut of pike. Coracidia which develop from the eggs released in the faeces of the pike infect copepods such as *Cyclops bicuspidatus* Claus. Here they develop into infective procercoids and when eaten by a percid, penetrate the intestine of the fish. They enter the body cavity and eventually encyst as plerocercoids in the liver. If a pike does not eat the percid the parasites will eventually die and degenerate.

Adult nematode worms are found in the gut of percids and encysted larval stages often inhabit the liver. Adult *Raphidascaris acus* occur in the pyloric caecae and anterior regions of the intestine. The main definitive host appears to be pike although gravid female worms have also been found in the intestines of walleye (Poole & Dick 1986). Fertilised eggs are released with the host's faeces. A fish host may eat the eggs directly or a fish or an invertebrate may eat larvae hatching from the eggs. In the latter situation, the fish becomes infected when the fish or invertebrate is eaten. After the perch ingests the larvae they penetrate the intestine wall and enter the body cavity. They migrate to the liver where they encyst. The parasite is transmitted to the definitive host when the perch is eaten. Walleye can also carry larval worms in the liver and thus act as an intermediate host (Poole & Dick 1986).

Acanthocephala adult and larval worms live in the intestine. Mature worms produce shelled acanthors which are liberated into the water and are ingested by the intermediate host, usually an invertebrate. *Acanthocephalus lucii* is a common parasite of perch and zander; the intermediate host is normally the isopod, *Asellus aquaticus*. Infection of the fish host occurs when *A. aquaticus* is eaten. Pike have been reported to become infected with the parasite by eating perch which acts in this situation as an intermediate host (Moravec 1979).

Several leeches (Hirudinoidea) are external parasites living on the skin and gills of fish. *Piscicola geometra* commonly occurs on these sites particularly at the base of the fins and in the mouth of perch. It has been introduced into North America and has been found on walleye and yellow perch. When a fish swims within range of a leech attached to the substrate the leech moves onto the surface of the fish and finds a suitable location. Blood is drawn from the host by insertion of a proboscis and when engorged the leech drops from the fish. Sexual reproduction takes place away from the host on the river or lake bottom. The cocoons each containing one egg are produced in large numbers and are deposited on stones.

Larval stages of some molluscs are also ectoparasites of fish including percids. The fresh water clam, *Lampsilis radiata*, has a life cycle which is partly parasitic. The early development of the clam takes place within the marsupium of the female. Glochidia, released by the females in late spring and early summer, burrow into the gills of fish including yellow perch and encyst. In this stage they develop adult organs and eventually break away and settle to the bottom as young clams. The glochidia of *Anodonta piscinalis* infect perch and ruffe in the spring (Jokela *et al.* 1991).

Some copepod and branchiuran crustacea are parasitic on the body, gills and oral cavity of percids. The copepod, *Ergasilus sieboldi*, is a common gill parasite of perch and zander. The mature female is the parasitic phase of the life cycle. Embryonic development takes place in egg sacs which are paired and contain about 100 eggs each. There are three nauplius and four copepodid free living stages each proceeded by a moult. Differentiation of the sexes and copulation occur at the fourth copepodid stage. The males die and the females enter the gill cavity and finally attach themselves to the gill filaments where they live for about one year (Bauer *et al.* 1973). Some parasitic species of copepod are found on a number of freshwater fish groups, for example *Lernaea cyprinacea* is common in cyprinids as well as percids. Others, such as *Actheres percarum*, have specific hosts, perch and zander (Fryer 1982). Branchiurans, unlike copepods, moult several times after reaching sexual maturity and adults can be found in a wide range of sizes (Fryer 1982). Eggs of *Argulus foliaceus* are laid in clumps or rows on stones or other suitable substrate and vary in number but are usually in the range 250–300 per clutch. Those laid in the summer hatch in that year while those laid in the autumn may overwinter as eggs. Larvae hatch as late nauplii (Fryer 1982) and must find a host fish within about 48 h. All stages, other than the egg, are free swimming thus the parasite can move from host to host. The maxillules are elaborate hook-like structures in the larval stage, which are used to cling onto the host. After several moults the maxillules are transformed into suckers. On attaching itself to the host, *A. foliaceus* pierces the skin with its proboscis and sucks blood.

The last group of parasites to be considered belongs to the Acarina. Their life cycle is not completely clear. Eggs of water mites are usually laid on stones, vegetation or debris but in the case of *Hydrachna* they are deposited in the tissues of aquatic plants. Free swimming larvae emerge which eventually become attached to aquatic insects. A species of *Hydrachna* has been found on the gills of yellow perch (Tedla & Fernando 1969). The parasitic larvae develop into inactive nymphs called nymphophans. Nymphs break from the host and after an active free living period, pupate on rooted vegetation or bottom debris. This second pupal stage is called a teleiophan; from it emerges the sexually mature adult which is also free living.

8.3 Abiotic and biotic factors

As outlined in the introduction to this chapter, abiotic and biotic conditions of the environment play a major part in the population dynamics of parasites and the outbreak of disease. Abiotic factors not only influence ectoparasites and free living stages of endoparasites directly, but through their control of the physiological and/or hormonal state of the host (invertebrate or vertebrate), affect all parasites indirectly. Temperature is probably the single most important abiotic controlling factor (Bauer 1962; Chubb 1982). This is particularly the case with ectoparasites. In monogeneans for example, temperature is significant in all stages of life including the rate of ovipositor, the speed of development of the embryo, the length of life of the oncomiracidium, the speed of maturation of the juvenile and the longevity of the adult. An increase in temperature can increase activity and decrease longevity of these parasites. It can shorten embryonic development time and in some cases promote synchronised hatching. Larvae of *Ichthyophthirius multifiliis* hatch faster from the ciliospores at higher temperatures although they tend to be smaller (Bauer 1962). The optimum temperature for embryonic development is characteristic for a given species of parasite and these may be different for two or more parasites living off the same host. *Triaenophorus nodulosus* and *Diphyllobothrium latum* eggs can develop at 0°C but the optimum is 20°C and death of the embryos occurs at 35°C. Eggs of *Argulus foliaceus* will not develop below 10°C and optimum temperature is 25–26°C. This is the same optimum temperature for *Ichthyophthirius multifiliis*, *Ergasilus sieboldi*, *Lernaea cyprinacea*, *Ligula intestinalis* and others. This latter group will not develop below 14°C. Temperature limits restrict the distribution of parasites. The trematode *Clinostomum complanatum* is normally located within the annual 10°C isotherm and warmer regions. It has been found in a colder area where its host existed in warmed waters from a cooling plant and the artificially raised temperatures allowed the eggs of the parasite to develop (Grabda-Kazubska 1974). Temperature increases may not always benefit egg production. *Bunodera lucioperca* matures during the winter, is gravid in the spring and releases eggs in the summer. A cool period through the winter is required for its development (Andrews 1977). Egg release at an optimum temperature can produce epizootics in a number of species which cause fish kills at the hottest time of the year. This has been reported for *Ichthophthirius multifiliis* and *Diphyllobothrium latum* and probably occurs in many other disease-inducing parasites. Temperature influences such fish-parasite active invasive phases as protozoan ciliospores, monogenean free-swimming larvae, digenetic trematode miracidia and cercaria, cestode coracidia and various stages of crustaceans. The seasonal changes in prevalence and intensity of parasites (Margolis *et al.* 1982) are in many cases a reflection of temperature. Chubb (1977; 1979; 1980; 1982) has reviewed the subject of seasonality of helminth parasites extensively but the complex underlying controlling factors are still not elucidated. A study conducted on the helminths found in the intestine of yellow perch of Lake Opeongo, Canada, illustrates the complexity of seasonal abundance of parasites in the host (Cannon 1973). Two trematodes, *Bunodera sacculata* and *B. lucioperca*, and one cestode, *Proteocephalus pearsei*, showed their lowest incidence in the summer. One trematode, *Crepidostomum cooperi*, and one nematode, *Dactinitoides cotylophora*, were highest in incidence in the summer. The cestode *Bothriocephalus* sp., the nematode *Spinitectus gracilis* and the acanthocephalan *Leptorhynchoides thecatus* showed no seasonality. Some parasites may not show any seasonality in their overall abundance although the maturation state of the parasite may show marked seasonality. Larvae and adults of the nematode *Camallanus lacustris* are found in the intestine of perch throughout the year and their abundance is non-seasonal (Skorping 1980). However, the number of gravid females shows marked seasonality. The parasite is found mainly in the pyloric region of the intestine from May to October, but from December to March most of the worms are found in the posterior region.

Light, oxygen concentration, pH, salinity, and possibly water current and water chemistry help control the life cycle of parasites but have less influence than temperature. Only a few experiments have been carried out to test the importance of light. The eggs of *Triaenophorus nodulosus* were found to take longer to develop in the dark compared to those kept in diffuse light (Mikhailov 1951). The coracidia of this parasite are positively phototactic and thus, like the cyclopoids which ingest them, are active in the daytime. Although light may be important in the development of larvae from eggs, it can reduce time of the free living larval stage in some species. The ciliospores of *Ichthyophthirius multifiliis* die two to four times faster in diffuse light than in the dark (Bauer 1962). Light probably plays an important role in the seasonality of parasites. This influence can be direct, such as on ectoparasites and free living stages of endoparasites, or indirectly, through hormonal response of the host to light.

Many ectoparasites and free living stages of endoparasites show considerable variability in their tolerance to different concentrations of oxygen in the water. *Piscicola geometra* for example has a high oxygen demand and when it is free living, it is found on wave-washed shores of lakes or in fast-flowing streams and rivers (Elliott & Mann 1979). At the other extreme, *Argulus foliaceus* appears to be intolerant of high levels of oxygen and can carry out all stages of its life cycle at low oxygen concentrations. It is best suited to living in stagnant water. Oxygen dependence of parasites living in the fish will be determined by the physiological state of the fish. Those parasites living in oxygen rich areas of the body, for example the blood, heart and swim bladder, may carry out more aerobic respiration than those living in oxygen poor areas, such as the gut, where anaerobic respiration is probably obligatory. Endoparasites live in a fairly constant pH environment of 7.2–7.8. Ectoparasites and free living developmental stages will normally tolerate pH limits over a similar range to their fish hosts. *Icthyophthirius multifiliis* has a tolerance range from pH 5.5–10.1. Above or below these values the parasite cannot attach itself to the host and cysts are not formed (Bauer 1962). The coracidia of *Triaenophorus crassus* die at pH 5.0.

Many of the parasites found on percids are also tolerant to the same salinity ranges as their hosts (Chapter 6). *Triaenophorus nodulosus* is euryhaline and its eggs can develop in salinities up to 6 ppt and coracidia can live in salinities not exceeding 7 ppt. This parasite exists on perch and zander which live in brackish areas such as the Baltic Sea. They appear to be able to tolerate the transition between fresh and brackish water. Some ectoparasites, including leeches, molluscs and crustacean larvae, are sensitive to increases in salinity.

Other living organisms in the environment have a major influence on the abundance of fish parasites. The diet and rate of food intake of the fish host influence parasite transmission. If the incidence of the parasite in the infective stage can reach a peak at the same time as maximum feeding in the host then transmission of the parasite will be greatly enhanced. Chubb (1964) showed that the low incidence of *Triaenophorus nodulosus* in perch at Lake Bala, Wales, was caused by the absence of copepods, first intermediate hosts, in the diet of perch at the main period of infection. In a long term study at Heming Lake, Canada, on the occurrence of *T. nodulosus* in yellow perch and pike which were subjected to heavy fishing, Lawler (1969) discovered some interesting aspects concerning parasite loading and the state of the fish host populations. Young yellow perch had a higher prevalence of the parasite than the older and larger yellow perch because cyclopods were far more important in the diet of young fish. If the yellow perch were not eaten by the definitive host, in this case the pike, the encysted plerocercoids degenerated. This occurs after one and a half to three years in perch (Chubb 1964). At Heming Lake the reduction in the size of the yellow perch population by fishing allowed the remaining yellow perch to grow more quickly but the fishing also reduced the numbers of large old pike which formed a reservoir for *T. nodulosus* eggs. The smaller pike were not able to feed on the larger yellow perch and a reduction in the prevalence of the parasite took place. After fishing stopped *T. nodulosus* again became common in the yellow perch (Poole & Dick 1985).

Predators may eat a number of parasites. For example, some copepods feed on *Icthyophthirius multifiliis* and the coracidia of *Ligula intestinalis* (Bauer 1962) and perch feed on *Piscicola geometra* (Bauer 1962) and *Argulus* sp. (Craig 1974a). Bauer (1962) suggests that considerable numbers of eggs of helminth parasites may be eaten and digested by detritus feeders and this could have a significant effect on parasite abundance.

Some species of parasite may compete for similar sites of infection. The metacercaria of *Diplostomum gasterostei*, *Tylodelphys clavata* and *T. podicipina* are found in the eyes of perch, usually in the vitreous humour, although *D. gasterostei* is occasionally found in the lens. *Tylodelphys clavata* was accidentally introduced into Slapton Ley, England, where previously the only eye-fluke observed in perch had been *D. gasterostei*. The population of *T. clavata* in the perch host and its interactions with other eye-flukes were studied over the next seven years (Kennedy 1981). During the study *T. podicipina* was introduced into the system. The definitive host of the *Tylodelphys* species is the great crested grebe, *Podiceps cristatus* Linnaeus, which started a breeding population and is believed to have introduced the parasites. All three parasites infect the snail, *Lymnaea peregra* Muller, as the first intermediate host. The population of *T. clavata* increased slowly at first then rapidly and finally declined as the population level reached equilibrium. Kennedy (1981) considered a decline in the population of *D. gasterostei* was due to two independent events. Firstly, the numbers of the perch host carrying heavy infections declined and secondly, the recruitment of the parasites into young perch was reduced. Kennedy suggested that these processes related to changes in the population density of the other eye-flukes. The decline in the number of fish heavily infected with *D. gasterostei* was significantly related to the increase in *T. clavata* and the decrease in the recruitment into the young fish was related to the increase in *T. podicipina* in the eyes of the young perch. Kennedy (1981) suggested that the two species of *Tylodephys* interact negatively with *D. gasterostei* and in Slapton Ley partition the perch habitat between them. There are probably many other examples of competition between parasites and antagonistic relationships between them.

In many cases the density of the fish host population controls the abundance of the parasite. A large host population increases the chance of success at the invasive phase of the parasite. When the fish is living at the extreme of its range and densities are low, the number of species and density of parasites is minimal compared with the host in its normal range (Bauer 1962). Outbreaks of disease often occur when fish populations congregate together during spawning which enhances the chances of infection although fish may be more vulnerable to invasion by the parasite at this time due to their physiological state. The condition or fat content of the fish host may influence the susceptibility of the fish to infection. At the time of spawning the condition of fish is usually poor. The life cycle of parasites may be linked to the hormonal condition of the fish. The actions of host hormones on the parasite are probably indirect though there is no real indication of the mechanisms involved. Some parasites make some selection in the sex of the host. In Lake Opeongo, Canada, no species of parasite was more prevalent in the male than in the female yellow perch but the cestode *Proteocephalus pearsei* and the acanthocephalan *Leptorhynchoides thecatus* were more prevalent in female fish (Cannon 1973). Prevalence of parasites increased with size of the yellow perch although the increase in some parasites was influenced by sex. For example, *Bunodera lucioperca*, *Bothriocephalus* sp., *Spinitectus gracilis* and *L. thecatus* only increased in abundance with increase in size of female yellow perch.

Long term climatic and biotic changes in the environment can lead to alterations in the parasitic fauna. During the period 1928 to 1936, 26 parasites were recorded in yellow perch from Oneida Lake, USA. In 1966, 24 parasites were found (Noble 1970a). Some rare or accidental parasites identified in the earlier period were not recorded in 1966. The only common species in 1928–31 not found in the yellow perch in 1966 was the trematode *Clinostomum marginatum*. The trematodes, *Tetracotyle* sp., *Apophallus brevis* and *Urocleidus adspectus* appeared for the first time in the 1966 investigation. Cooper *et al.* (1977) compared the long term changes in the helminth parasites of yellow perch in western Lake Erie, North America, between the period 1927–9 and the 1970s. They discovered that some parasites had increased in prevalence (*Triaenophorus nodulosus* and *Eustrongylides tubifex*), some had decreased in prevalence (*Crepidostomum cooperi*, *Bothriocephalus* sp., *Proteocephalus* sp. and *Dacnitoides cotylophora*) and some remained about the same (*Philometra cylindracea* and *Camallanus oxycephalus*). Care should be taken in interpreting the results of long term studies since different methods are often used over the time span.

8.4 Disease and mortality

The effect of parasites and diseases on the mortality of the host is not easy to quantify and mortality is usually only recorded when it is dramatic. An area of future research must be the part played by parasites in the natural mortality of fish populations (Chapter 3). Sublethal effects of parasites need further investigation. Do cysts of flukes in the eye of percids cause vision impairment? Do gut parasites reduce absorption in the gut and thus reduce feeding efficiency? How do parasites affect oxygen uptake and ion balance? Do some of these effects make the host more susceptible to predation? When large kills of percids have been recorded it has usually been difficult to identify the causal agent and often little is known about the fish populations which have died. It has long been recognised by fish culturists that transfer of fish between ponds or from one hatchery to another could induce outbreaks of disease often leading to heavy mortalities. Water quality was and is considered of fundamental importance in the successful rearing of fish (see Chapter 11). It has also been observed that the resistance of fish to disease is a function of their age, and exposure of fish to an epizootic results in more resistance to subsequent attacks.

Immunology is the study of the host's ability to make itself exempt from a disease by defending itself against attack (Anderson 1974). The most effective defence against pathogens, other than against those entering the gut after ingestion, is the skin. The protective layers include the mucous, scales, epidermis and dermis. The constant sloughing of the mucous removes foreign particles attached to it. The mucous may also contain antibodies. A break in this protective barrier by wounding, including penetration by parasites such as leeches and branchiurans, allows entrance of bacteria, viruses and fungi. Inflammation may occur at the wound. Inflammation allows repairs to tissue damage and prevents further harm by the accumulation of polymorphonuclear leucocytes and lymphocytes. This reaction and other immune responses of fishes are influenced by environmental factors, in particular temperature. Those fishes living near an optimum temperature will have better responses to parasite invasion and disease than those living in colder or warmer areas.

Although there has been recent development of intensive percid culture (see Chapter 11), the future of reliable sources of fish for recreation must lie in the enhancement and rehabilitation of their natural habitat (see Chapter 10). One method of doing this is by releasing reared larvae and fry into the water body. Little is known about the rate of recruitment of the indigenous parasitic fauna by the introduced fish and whether mortality occurs as a result. Walleye fry introduced into Heming Lake, Canada, rapidly acquired parasites that were non-specific and common to indigenous populations of walleye and yellow perch (Poole and Dick 1985). Some parasites, in particular *Raphidascaris acus* and *Ergasilus luciopercarum*, were acquired at a higher prevalence and mean intensity than those found in the indigenous stocks of walleye. These two parasites are known to be pathogenic. This type of study is clearly important in any introductions of fishes and must become necessary procedure in the future if successful management of percid stocks is to be achieved.

Appendix

Table 8a A list of zooparasites of *Perca fluviatilis* (P), *P. flavescens* (YP), *Stizostedion vitreum* (W) and *S. lucioperca* (Z). (For authorities see the references given.)

Parasite	Host	Site	Author
Protozoa and Myxosporida			
Ambiphrya ameiuri	Z	Skin, gills	Rahkonen 1994
Apiosoma sp.	P		Kennedy 1974
Balantidium sp.	YP	Intestine	Hoffman 1967 Margolis & Arthur 1979
Dermocystidium percae	P	Skin	Bykhovskaya-Pavlovskaya *et al.* 1964
Eimeria laureleus	YP	Intestine	Margolis & Arthur 1979 Molnar *et al.* 1974
E. percae	P		Bykhovskaya-Pavlovskaya *et al.* 1964
E. rivierei	P	Liver	Markevich 1963
E. tedlai	YP	Intestine	Margolis & Arthur 1979
Glossatella campanulata	P		Bykhovskaya-Pavlovskaya *et al.* 1964
Glossatella sp.	P		Campbell 1974
Henneguya acerinae	Z	Gills	Markevich 1963
H. creplini	Z, P	Gills	Petrushevskii 1963 Haaparanta *et al.* 1994
H. doori	YP	Gills	Hoffman 1967 Margolis & Arthur 1979
H. percae	YP	Gills	Hoffman 1967 Margolis & Arthur 1979
H. psorospermica	P	Gills	Bykhovskaya-Pavlovskaya *et al.* 1964
H. wisconsinensis	YP		Hoffman 1967
H. wolinensis n. sp.	P	Skin	Romuk-Wodoracki 1990
H. zschokkei	P		Bykhovskaya-Pavlovskaya *et al.* 1964
Henneguya sp.	P, YP	Gills	Kennedy 1974 Margolis & Arthur 1979
Ichthyobado necator	Z	Skin, gills	Rahkonen 1994
Icthyophthirius multifiliis	P, YP W	Skin, fins, gills	Bauer 1962; Hoffman 1967 Margolis & Arthur 1979
Myxidium percae	YP	Sub-dermal	Hoffman 1967
Myxobilatus asymmetricus	W	Urinary bladder	Hoffman 1967
Myxobolus carassi	P	Body cavity, liver, intestine, gills	Markevich 1963
M. dispar	P		Bykhovskaya-Pavlovskaya *et al.* 1964
M. ellipsoides	P	Gills, body cavity, gall bladder	Bykhovskaya-Pavlovskaya *et al.* 1964 Markevich 1963
M. macrocapsularis	Z		Petrushevskii 1963
M. minutus	P		Bykhovskaya-Pavlovskaya *et al.* 1964
M. mulleri	P	Gills	Kennedy 1974
M. percae	YP	Fin base, eye	Hoffman 1967 Margolis & Arthur 1979
M. permagnus	P	Gills	Markevich 1963

Cont.

Table 8a Cont.

Parasite	Host	Site	Author
M. phyriformis	YP		Hoffman 1967
M. sandrae	P	Vertebrae	Lom *et al.* 1991
M. volgensis	Z	Gills, eye (cornea), skin, fins, muscle	Markevich 1963
M. wegeneri	P		Bykhovskaya-Pavlovskaya *et al.* 1964
Myxobolus sp.	YP, W		Hoffman 1967
Myxosoma anurus	P		Bykhovskaya-Pavlovskaya *et al.* 1964
M. branchialis	P	Gills	Markevich 1963
M. dujardini	P	Gills	Markevich 1963
M. neurophila	YP	Optic tectum	Hoffman 1967
M. scleroperca	YP	Sclera of eye	Hoffman 1967 Margolis & Arthur 1979
Myxosporidia gen. sp.	W, Z	Various	Margolis & Arthur 1979
Spironucleus sp.	YP	Intestine	Molnar *et al.* 1974 Margolis & Arthur 1979
Thelohanellus piriformis	YP	Spleen	Hoffman 1967
Trichodina domerguei	P, Z	Gills	Bykhovskaya-Pavlovskaya *et al.* 1964
T. megamicronucleata	Z	Gills	Bykhovskaya-Pavlovskaya *et al.* 1964
T. meridionalis	P		Bykhovskaya-Pavlovskaya *et al.* 1964
T. nigra	P, Z	Gills	Bykhovskaya-Pavlovskaya *et al.* 1964
T. urinaria	P	Urinary bladder, ureters	Bykhovskaya-Pavlovskaya *et al.* 1964 Markevich 1963
Trichodina sp.	YP, W, Z	Gills, urinary bladder, ureters	Hoffman 1967 Margolis & Arthur 1979 Rahkonen 1994
Trichodinella epizootica	P		Bykhovskaya-Pavlovskaya *et al.* 1964
Tricophyra intermedia	P		Bykhovskaya-Pavlovskaya *et al.* 1964
T. piscium	YP, Z	Skin, gills	Hoffman 1967 Rahkonen 1994
Trypanosoma acerinae	P		Jastrzebski 1984
*T. lucioperca*e	Z	Blood	Markevich 1963
T. percae	P, YP	Blood	Bykhovskaya-Pavlovskaya *et al.* 1964 Hoffman 1967 Markevich 1963 Margolis & Arthur 1979
Monogenea and Trematoda			
Allocreadium isoporum	P		Bykhovskaya-Pavlovskaya *et al.* 1964
A. lobatum	W	Stomach	Hoffman 1967 Fischthal 1952
Ancyrocephalus creplin	P	Gills	Markevich 1963
A. percae	P		Chubb 1977
A. paradoxus	P, Z	Gills	Bykhovskaya-Pavlovskaya *et al.* 1964 Markevich 1963
Ancyrocephalus sp.	YP		Hoffman 1967
Apophallus americanus	YP, W		Hoffman 1967

Cont.

Table 8a Cont.

Parasite	Host	Site	Author
A. brevis	YP	Skin, fins, gills, musculature	Noble 1970a Margolis & Arthur 1979
A. itascensis	YP		Hoffman 1967
A. müehlingi	P, Z		Bykhovskaya-Pavlovskaya *et al.* 1964 Chubb 1979
A. venustus	YP, W	Musculature	Hoffman 1967 Margolis & Arthur 1979
Apophallus sp.	YP		Hoffman 1967
Ascocotyle coleostoma	P		Bykhovskaya-Pavlovskaya *et al.* 1964
Aspidogaster limacoides	P		Bykhovskaya-Pavlovskaya *et al.* 1964
Asymphylodora sp.	YP		Hoffman 1967
Azygia acuminata	YP, W		Hoffman 1967
A. angusticauda	YP, W	Stomach, intestine	Hoffman 1967
A. bulboa	W		Hoffman 1967
A. longa	YP, W	Stomach, intestine	Hoffman 1967
A. lucii	P, Z	Stomach, intestine	Bykhovskaya-Pavlovskaya *et al.* 1964 Markevich 1963
A. sebago	YP		Hoffman 1967
Azygia sp.	YP, W		Hoffman 1967 Margolis & Arthur 1979
Bolboforus confusus	P		Bykhovskaya-Pavlovskaya *et al.* 1964
Bucephaloide pusilla	W		Hoffman 1967
Bucephalopsis pusillum	YP		Hoffman 1967
Bucephalus elegans	YP	Gut (pyloric caecae)	Hoffman 1967 Margolis & Arthur 1979
B. markewitschi	Z	Intestine	Markevich 1963
B. polymorphus	P, Z	Intestine	Bykhovskaya-Pavlovskaya *et al.* 1964 Markevich 1963
Bunocotyle cingulata	P, Z	Intestine	Bykhovskaya-Pavlovskaya *et al.* 1964 Markevich 1963
B. luciopercae	YP, W, P, Z	Intestine	Dechtiar 1972a Margolis & Arthur 1979 Rahkomen 1994
B. nodulosum	YP		Hoffman 1967
Bunodera luciopercae	P		Scholz 1986
Bunodera sp.	YP		Margolis & Arthur 1979
Bunoderina sacculata	YP, W	Intestine	Hoffman 1967 Margolis & Arthur 1979
Centrovarium lobotes	YP, W	Intestine	Hoffman 1967 Margolis & Arthur 1979
Cleidodiscus aculeatus	W		Hoffman 1967
Cleidodiscus sp.	YP, W	Gills	Hoffman 1967 Margolis & Arthur 1979
Clinostomum complanatum	P, YP, W	Gills, musculature	Bykhovskaya-Pavlovskaya *et al.* 1964 Margolis & Arthur 1979 Markevich 1963

Cont.

Table 8a Cont.

Parasite	Host	Site	Author
C. marginatum	YP, W	Gill cavity, musculature, fins, mesenteries	Hoffman 1967 Margolis & Arthur 1979
Clinostomum sp.	YP, W	Musculature, viscera	Margolis & Arthur 1979
Coitocoecum skrjabini	P, Z	Intestine	Markevich 1963
Cotylurus communis	W	Mesenteries, liver	Hoffman 1967 Margolis & Arthur 1979
C. pileatus	P		Bykhovskaya-Pavlovskaya *et al.* 1964
Crassiphiala bulboglossa	YP, W	Musculature, skin	Hoffman 1967 Margolis & Arthur 1979
Crepidostomum cooperi	YP, W	Intestine, pyloric caecae, gall bladder	Hoffman 1967 Margolis & Arthur 1979
C. farionis	P, YP	Intestine, pyloric caecae, gall bladder	Hoffman 1967 Kennedy 1974 Margolis & Arthur 1979
C. isotomum	YP	Intestine	Margolis & Arthur 1979
C. laureatum	YP		Hoffman 1967
Crepidostomum sp.	YP, W	Intestine, gall bladder	Hoffman 1967 Margolis & Arthur 1979
Crowcrocoecum skrjabini	P		Bykhovskaya-Pavlovskaya *et al.* 1964
Cryptogonimus chili	YP	Intestine, pyloric caecae, stomach	Hoffman 1967 Margolis & Arthur 1979
Cyrodactylus longiradix	P		Bykhovskaya-Pavlovskaya *et al.* 1964
Dactylogyrus tenuis	P	Gills	Bykhovskaya-Pavlovskaya *et al.* 1964
Digenea gen. sp.	W	Skin, fins, eye, viscera, musculature	Margolis & Arthur 1979
Diplostomulum clavatum	P	Eye (vitreous humour and lens)	Bykhovskaya-Pavlovskaya *et al.* 1964 Markevich 1963
D. huronense	YP	Eye (vitreous humour & lens)	Hoffman 1967
D. scheuringi	YP, W	Eye (vitreous humour)	Hoffman 1967 Margolis & Arthur 1979
Diplostomulum sp.	YP, W	Eye (vitreous humour) brain, pharynx	Hoffman 1967 Margolis & Arthur 1979
Diplostomum adamsi	YP	Eye (retina)	Margolis & Arthur 1979
D. baeri	P	Eye (retina)	Hoeglund & Thulin 1992
D. gasterostei	P	Eye	Kennedy 1981
D. spathaceum	P, YP, Z	Eye (vitreous humour, lens)	Bykhovskaya-Pavlovskaya *et al.* 1964 Margolis & Arthur 1979 Molnar *et al.* 1974 Chubb 1979

Cont.

Table 8a Cont.

Parasite	Host	Site	Author
D. spathaceum huronense	YP	Eye	Margolis & Arthur 1979
D. volvens	P		Deufel 1961
Diplostomum sp.	P, Z	Eye (vitreous humour, lens)	Kennedy 1974 Rahkonen 1994
Distomum nodulosum	YP		Hoffman 1967
Echinochasmus donaldsoni	YP	Gills	Hoffman 1967
Euclinostomum heterostomum	P		Bykhovskaya-Pavlovskaya *et al.* 1964
Euparyphium melis	YP	Nares, urinary bladder	Hoffman 1967
Gyrodactylidae gen. sp.	YP	Gills	Margolis & Arthur 1979
Gyrodactyloidea gen. sp.	YP, W	Skin, gills	Margolis & Arthur 1979
Gyrodactylus freemani	YP	Fins	Margolis & Arthur 1979
Gyrodactylus sp.	P, YP	Skin, fins, gills	Hoffman 1967 Kennedy 1974
Hemiurus appendiculatus	P		Bykhovskaya-Pavolvskaya *et al.* 1964
Hysteromorpha triloba	P		Bykhovskaya-Pavlovskaya *et al.* 1964
Ichthyocotylurus cucullus	P		Kennedy 1974
I. pileatus	P, Z		Chubb 1979
I. platycephalus	P		Chubb 1979
Ichthyocotylurus sp.	P		Kennedy 1974
Leuceruthrus sp.	YP	Gut	Hoffman 1967 Margolis & Arthur 1979
Maritrema medium	YP		Hoffman 1967
Metagonimus yokogawai	P, Z		Bykhovskaya-Pavlovskaya *et al.* 1964 Chubb 1979
Metorchis conjunctus	YP	Musculature	Margolis & Arthur 1979
Microphallidae gen. sp.	YP		Margolis & Arthur 1979
Microphallus medius	YP		Hoffman 1967
M. opacus	YP	Intestine	Hoffman 1967 Margolis & Arthur 1979
Neascus brevicaudatus	P	Eye (vitreous humour and iris)	Bykhovskaya-Pavlovskaya *et al.* 1964 Markevich 1963
N. ellipticus	YP	Muscle	Hoffman 1967
N. longicollis	YP	Skin	Hoffman 1967
N. oneidensis	YP		Hoffman 1967
N. pyriformis	YP	Skin, fins	Hoffman 1967 Margolis & Arthur 1979
Neascus sp.	YP, W	Mesenteries, gills, skin	Hoffman 1967 Margolis & Arthur 1979
Paracoenogonimus ovatus	P, Z		Bykhovskaya-Pavlovskaya *et al.* 1964 Chubb 1979
Petasiger nitidus	YP		Hoffman 1967
Phyllodistomum americanum	YP		Hoffman 1967

Cont.

Table 8a Cont.

Parasite	Host	Site	Author
P. angulatum	P, Z	Urinary bladder	Bykhovskaya-Pavlovskaya *et al.* 1964 Markevich 1963
P. folium	P		Bykhovskaya-Pavlovskaya *et al.* 1964
P. pseudofolium	P		Bykhovskaya-Pavlovskaya *et al.* 1964
P. superbum	YP, W	Ureters, urinary bladder	Hoffman 1967 Margolis & Arthur 1979
Posthodiplostomum brevicaudatum	P		Chubb 1979
P. cuticola	P		Bykhovskaya-Pavlovskaya *et al.* 1964
P. minimum	P, YP, W	Kidney, liver mesenteries, spleen	Hoffman 1967 Margolis & Arthur 1979
P. minimum centrarchi	YP	Liver	Margolis & Arthur 1979
Prosorhynchoides pusilla	W	Stomach, pyloric caecae, intestine	Margolis & Arthur 1979
Ptychogonimus fontanus	YP	Stomach	Hoffman 1967 Margolis & Arthur 1979
Rhipidocotyle illense	P		Kennedy 1974
R. papillosa	YP	Musculature	Margolis & Arthur 1979
Rossicotrema donicum	P		Bykhovskaya-Pavlovskaya *et al.* 1964
Sanguinicola occidentalis	YP, W	Blood	Hoffman 1967 Margolis & Arthur 1979
Sphaerostoma bramae	P	Intestine	Bykhovskaya-Pavlovskaya *et al.* 1964 Markevich 1963
Stephanophiala farionis	YP		Hoffman 1967
Striegeidae gen. sp.	YP	Mesenteries	Margolis & Arthur 1979
Tetracotyle diminuti	YP	Mesenteries	Margolis & Arthur 1979
T. echinata	P	Peritoneum	Bykhovskaya-Pavlovskaya *et al.* 1964
T. intermedia	YP	Heart, mesenteries	Margolis & Arthur 1979
T. perca fluviatilis	P	Peritoneum, walls of swim bladder and surface of various body cavity organs	Bykhovskaya-Pavlovskaya *et al.* 1964 Markevich 1963
T. variegatus	P, Z	Peritoneal epithelium, walls of swim bladder and other body cavity organs	Markevich 1963

Cont.

Table 8a Cont.

Parasite	Host	Site	Author
Tetracotyle sp.	P, YP, Z	Heart, pericardium, kidney, wall of swim and urinary bladders, mesenteries, musculature	Hoffman 1967 Kennedy 1974 Margolis & Arthur 1979 Petrushevskii 1963
Tylodelphys clavata	P, Z	Eye (vitreous humour)	Kennedy 1974 Chubb 1979
T. podicipina	P	Eye (vitreous humour)	Kennedy 1974
Urocleidus aculeatus	W	Gills	Margolis & Arthur 1979 Suriano & Beverley-Burton 1981
U. adspectus	YP	Gills	Hoffman 1967 Margolis & Arthur 1979
Uvulifer ambloplitis	YP, W	Skin, musculature, fins, gills	Hoffman 1967 Margolis & Arthur 1979
Cestoda			
Abothrium crassium	W		Cross 1938
Bothriocephalus claviceps	P	Gut	Scholz 1986
B. cuspidatus	YP, W	Pyloric caecae, intestine	Hoffman 1967 Margolis & Arthur 1979
Bothriocephalus sp.	P, YP	Pyloric caecae, intestine	Kennedy 1974 Margolis & Arthur 1979
Cestoda gen. sp.	YP	Musculature, mesenteries, viscera, intestine	Margolis & Arthur 1979
Corallobothrium sp.	YP	Intestine	Margolis & Arthur 1979 Molnar *et al.* 1974
Cyathocephalus truncatus	P, YP, Z	Pyloric caecae, intestine	Bykhovskaya-Pavlovskaya *et al.* 1964 Dechtiar & Loftus 1965 Deufel 1961 Margolis & Arthur 1979
Diphyllobothrium latum	P, YP, W	Musculature, body cavity	Bykhovskaya-Pavlovskaya *et al.* 1964 Hoffman 1967 Margolis & Arthur 1979
Diphyllobothrium sp.	P, YP, W	Viscera, musculature, body cavity, blood vessels of heart	Kennedy 1974 Margolis & Arthur 1979
Eubothrium crassum	P	Gut	Bykhovskaya-Pavlovskaya *et al.* 1964
Eubothrium sp.	P, Z		Kennedy 1974 Rahkonen 1994

Cont.

Table 8a Cont.

Parasite	Host	Site	Author
Gryporhynchus cheilancristrotus	P		Bykhovskaya-Pavlovskaya *et al.* 1964
Ligula intestinalis	P, YP, Z	Body cavity	Hoffman 1967 Margolis & Arthur 1979 Markevich 1963
Proteocephalus ambloplitis	YP, W	Viscera	Hoffman 1967 Margolis & Arthur 1979
P. cernuae	P	Intestine	Bykhovskaya-Pavlovskaya *et al.* 1964 Markevich 1963
P. dubius	P	Intestine	Bykhovskaya-Pavlovskaya *et al.* 1964 Markevich 1963
P. exiguus	P		Hanzelova *et al.* 1996
P. luciopercae	W	Intestine	Hoffman 1967 Margolis & Arthur 1979
P. pearsei	YP, W	Intestine	Hoffman 1967 Margolis & Arthur 1979
P. percae	P, Z	Intestine	Markevich 1963
P. pinguis	YP, W	Intestine	Hoffman 1967 Margolis & Arthur 1979
P. stizostethi	W	Intestine	Hoffman 1967 Margolis & Arthur 1979
P. torulosus	P	Intestine	Markevich 1963
Proteocephalus sp.	P, W, Z	Intestine, pyloric caecae	Hoffman 1967 Kennedy 1974 Margolis & Arthur 1979 Rahkonen 1994
Schistocephalus solidus	YP	Abdominal cavity	Margolis & Arthur 1979 Molnar *et al.* 1974
Triaenophorus crassus	W	Intestine	Margolis & Arthur 1979
T. nodulosus	P, YP, W, Z	Liver and gut	Bykhovskaya-Pavlovskaya *et al.* 1964 Hoffman 1967 Margolis & Arthur 1979 Markevich 1963
T. stizostedionis	W	Intestine	Hoffman 1967 Margolis & Arthur 1979
T. tricuspidatus	YP		Hoffman 1967
Triaenophorus sp.	YP, W	Musculature	Margolis & Arthur 1979
Nematoda			
Agamospirura sp.	Z		Chubb 1980
Anguillicola crassus	P, Z	Swim bladder	Thomas & Ollevier 1992
Anisakis sp.	Z		Eslami & Mokhayer 1977
Camallanus lacustris	P, Z	Intestine	Bykhovskaya-Pavlovskaya *et al.* 1964 Markevich 1963
C. oxycephallus	YP, W	Intestine	Hoffman 1967 Margolis & Arthur 1979
C. truncatus	P, Z	Intestine	Bykhovskaya-Pavlovskaya *et al.* 1964 Markevich 1963

Cont.

Table 8a Cont.

Parasite	Host	Site	Author
Camallanus sp.	YP		Hoffman 1967 Margolis & Arthur 1979
Capillaria catenata	YP, W	Intestine	Hoffman 1967
C. salvelina	P	Intestine	Kennedy 1974
Contracaecum aduncum	P		Bykhovskaya-Pavlovskaya *et al.* 1964
C. brachyurum	YP, W	Stomach, intestine	Hoffman 1967
C. rudolphii	YP		Chubb 1980
C. spiculigerum	YP	Liver, mesenteries	Hoffman 1967 Margolis & Arthur 1979
C. squalii	P	Body cavity	Bykhovskaya-Pavlovskaya *et al.* 1964
Contracaecum sp.	YP, W	Intestine, stomach, viscera, mesenteries, musculature	Hoffman 1967 Margolis & Arthur 1979
Cucullanellus cotylophora	YP, W	Intestine	Margolis & Arthur 1979
Cucullanus marinus	Z	Intestine	Markevich 1963
C. truttae	P	Intestine	Kennedy 1974
Dactinitoides cotylophora	YP, W		Hoffman 1967
Desmidocercella numidica	P		Chubb 1980
Desmidocercella sp.	P, Z		Bykhovskaya-Pavlovskaya *et al.* 1964 Chubb 1980
Diohelyne sp.	YP		Bangham 1955
Eustrongylides excisus	P, Z		Bykhovskaya-Pavlovskaya *et al.* 1964 Chubb 1980
E. mergorum	P		Bykhovskaya-Pavlovskaya *et al.* 1964
E. tubifex	YP		Chubb 1980
Eustrongylides sp.	P, YP, W, Z	Viscera, musculature, body cavity, ovary	Hoffman 1967 Kennedy 1974 Margolis & Arthur 1979 Petrushevskii 1963 Chubb 1980
Gnathostoma sp.	P		Bykhovskaya-Pavlovskaya *et al.* 1964
Ichthyobronema conoura	P	Intestine	Markevich 1963
I. gnedini	P		Bykhovskaya-Pavlovskaya *et al.* 1964
I. tenuissima	P	Stomach	Markevich 1963
Nematoda gen. sp.	YP, Z	Viscera, musculature, intestine, stomach	Margolis & Arthur 1979 Rahkonen 1994
Philometra abdominalis	Z		Chubb 1982
P. cylindracea	YP, W	Peritoneum, abdomen, swim bladder	Hoffman 1967 Margolis & Arthur 1979
Philometra sp.	YP	Body cavity, intestine	Margolis & Arthur 1979

Cont.

Table 8a Cont.

Parasite	Host	Site	Author
Porrocaecum reticulatum	P		Bykhovskaya-Pavlovskaya *et al.* 1964
Raphidascaris acus	P, W, Z	Intestine	Bykhovskaya-Pavlovskaya *et al.* 1964 Margolis & Arthur 1979 Markevich 1963 Chubb 1980
R. cristata	P		Kennedy 1974
Raphidascaris sp.	YP, W	Liver, gut	Margolis & Arthur 1979
Rhabdochona canadensis	W	Intestine	Margolis & Arthur 1979
R. cascadilla	YP	Intestine	Margolis & Arthur 1979
R. ovifilament	YP	Intestine	Hoffman 1967 Margolis & Arthur 1979
Rhabdochona sp.	YP	Intestine	Hoffman 1967 Margolis & Arthur 1979
Spinitectus carolini	YP, W		Hoffman 1967
S. gracilis	YP, W	Intestine	Hoffman 1967 Margolis & Arthur 1979
Spinitectus sp.	YP, W	Gut	Hoffman 1967 Margolis & Arthur 1979
Spiroxys contortus	YP	Intestinal serosa	Margolis & Arthur 1979 Molnar *et al.* 1974
Spiroxys sp.	YP	Viscera, mesenteries, digestive tract	Hoffman 1967 Margolis & Arthur 1979
Thynnascaris brachyura	W		Margolis & Arthur 1979
Acanthocephala			
Acanthocephalus anguillae	P	Intestine	Bykhovskaya-Pavlovskaya *et al.* 1964
A. clavula	P		Kennedy 1974
A. jacksoni	YP		Hoffman 1967 Margolis & Arthur 1979
A. lucii	P, Z	Intestine	Bykhovskaya-Pavlovskaya *et al.* 1964 Markevich 1963
Corynosoma semerme	P, Z	Body cavity	Bykhovskaya-Pavlovskaya *et al.* 1964 Petrushevskii 1963
C. strumosum	P, Z	Intestine	Bykhovskaya-Pavlovskaya *et al.* 1964 Petrushevskii 1963
Echinorhynchus clavula	P, Z	Intestine	Markevich 1963
E. coregoni	YP		Hoffman 1967
E. lateralis	YP		Hoffman 1967
E. salmonis	YP	Intestine	Hoffman 1967
E. salvelini	YP		Hoffman 1967
E. truttae	P	Intestine	Campbell 1974
Echinorhynus sp.	P		Kennedy 1974
Leptorhynchoides thecatus	YP, W	Intestine	Hoffman 1967 Margolis & Arthur 1979
Metechinorhynchus lateralis	YP	Intestine, mesenteries	Margolis & Arthur 1979

Cont.

Table 8a Cont.

Parasite	Host	Site	Author
M. salmonis	P, YP, W	Intestine	Bykhovskaya-Pavlovskaya *et al.* 1964 Dechtiar 1972a Margolis & Arthur 1979 Oliver 1960
Metechinorhynchus sp.	W		Margolis & Arthur 1979
Neoechinrhynchus crassus	W	Intestine	Margolis & Arthur 1979
N. cylindratus	YP, W	Intestine	Hoffman 1967 Margolis & Arthur 1979
N. pungitius	YP	Intestine, stomach	Margolis & Arthur 1979
N. rutili	P, YP	Intestine	Bangham 1955 Margolis & Arthur 1979 Bykhovskaya-Pavlovskaya *et al.* 1964
N. tenellus	W		Hoffman 1967 Margolis & Arthur 1979
Neoechinrhynchus sp.	YP, W	Intestine	Margolis & Arthur 1979
Pomphorhynchus bulbocolli	YP, W	Intestine, mesenteries	Hoffman 1967 Margolis & Arthur 1979
P. laevis	P	Intestine	Kennedy 1974
Pomphorhynchus sp.	W	Intestine	Margolis & Arthur 1979
Pseudoechinorhynchus clavula	P		Bykhovskaya-Pavlovskaya *et al.* 1964
Hirudinoidea			
Actinobdella sp.	YP		Hoffman 1967 Margolis & Arthur 1979
Cystobrunchus verrilli	W		Hoffman 1967
Hemiclepsis marginata	P		Bykhovskaya-Pavlovskaya *et al.* 1964
Illinobdella alba	YP	Skin	Hoffman 1967 Margolis & Arthur 1979
I. moorei	YP, W		Hoffman 1967
Illinobaella sp.	YP	Fins	Hoffman 1967 Margolis & Arthur 1979
Macrobdella decora	W	Skin	Margolis & Arthur 1979
Myzobdella moorei	YP, W	Fins	Margolis & Arthur 1979
Percymoorensis marmorata	W	Skin	Margolis & Arthur 1979
Piscicola geometra	P, W, Z	Skin	Bykhovskaya-Pavlovskaya *et al.* 1969 Markevich 1963 Rawson 1957 Rahkonen 1994
P. punctata	YP, W	Skin, gills	Margolis & Arthur 1979 Oliver 1960
Piscicola sp.	YP	Skin	Hoffman 1967 Margolis & Arthur 1979
Placobdella parasitica	YP	Skin	Hoffman 1967 Margolis & Arthur 1979
P. picta	YP		Margolis & Arthur 1979

Cont.

Table 8a Cont.

Parasite	Host	Site	Author
Mollusca			
Anodonta anatina	P		Campbell 1974
Anodonta sp.	YP	Gills	Margolis & Arthur 1979
Elliptio complanata	YP	Gills	Margolis & Arthur 1979
Lampsilis radiata	YP	Gills	Margolis & Arthur 1979
Unionidae gen. sp.	YP, W	Fins, gills, skin	Margolis & Arthur 1979
Crustacea			
Achtheres lacae	YP	Gills	Hoffman 1967
A. percarum	P, Z	Gills, mouth	Bykhovskaya-Pavlovskaya *et al.* 1964 Markevich 1963
Argulus appendiculosus	YP		Hoffman 1967
A. biramosus	YP, W		Hoffman 1967
A. canadensis	W		Dechtiar 1972b
A. catostomi	YP		Hoffman 1967
A. foliaceus	P, Z	Oral and gill cavities, body surface	Bykhovskaya-Pavlovskaya *et al.* 1964
A. stizostethii	YP, W	Skin, fins	Hoffman 1967 Margolis & Arthur 1979
A. versicolor	YP, W	fins	Dechtiar 1972a Hoffman 1967 Margolis & Arthur 1979
Caligus lacustris	P	Gills, skin	Bykhovskaya-Pavlovskaya *et al.* 1964
Copepoda gen. sp.	W	Gills, fins, mouth	Margolis & Arthur 1979
Ergasilus briani	P	Gills	Bykhovskaya-Pavlovskaya *et al.* 1964
E. caeruleus	YP, W	Gills	Hoffman 1967 Margolis & Arthur 1979
E. centrarchidarum	YP, W	Gills	Hoffman 1967 Margolis & Arthur 1979
E. confusus	YP, W		Hoffman 1967 Margolis & Arthur 1979
E. luciopercarum	YP, W	Gills	Bykhovskaya-Pavlovskaya *et al.* 1964
E. sieboldi	P, Z	Gills	Bykhovskaya-Pavlovskaya *et al.* 1964
Ergosilus sp.	YP, W	Gills	Margolis & Arthur 1979
Lernaea cyprinacea	P	Skin	Bykhovskaya-Pavlovskaya *et al.* 1964
L. esocina	P, Z	Skin and gills	Bykhovskaya-Pavlovskaya *et al.* 1964 Markevich 1963
Parergasilus longidigitus	P		Tuuha *et al.* 1992
Thersitina gasterostei	P	Skin	Bykhovskaya-Pavlovskaya *et al.* 1964
Acarina			
Hydrachna sp.	YP	Gills	Margolis & Arthur 1979

Chapter 9
Population Dynamics and Community Structure

9.1 Introduction

Many percid populations are economically important supporting varied sport and commercial fisheries (see Chapter 10). A significant amount of effort has been made to appraise their potential biological production and yield. This involves estimating their numbers, biomass and growth and how and why these fluctuate with time. Bioenergetics models (Chapter 5) have proved to be useful in understanding these processes. Percid populations, unlike marine species, are normally confined within limited areas, which makes sampling and analysis a little easier. Population dynamic analysis involves estimating numbers of organisms and the rates at which these numbers change. Essentially it is involved with birth and death rates and indirectly with somatic growth since growth has a direct influence on both rates. Birth rate is the number of offspring produced per female per unit of time. As soon as eggs are laid, they and the resulting offspring are subject to a mortality rate which will normally vary at different life stages of the fish. As a rule mortality will be most intense in the early stages, declining as age and size increase. The annual instantaneous mortality rate, Z, is normally calculated where the number at age t+1, N_{t+1}, depends upon the number at t, N_t. Thus:

$$N_{t+1} = N_t \exp(-Z)$$

Z can represent two components, natural mortality, M, and fishing mortality, F, where $Z = F + M$ (Chapter 3). In order for populations to remain in a steady state, a balance must exist between birth and mortality rates. These rates depend on the density of the population. Thus, as the population increases, the death rate will tend to increase and the birth rate will tend to decrease. This is an example of density dependence and may result from factors such as shortage of food caused, for example, by competition, from predation by other species or from cannibalism. The effect of environmental factors such as temperature and oxygen concentration on birth and death rates does not depend on the size of the population and thus, is density independent. Both density dependent and density independent factors influence the abundance of a population.

In stable, percid populations unperturbed by humans, although there are very few examples, there is a tendency for populations to maintain a constant biomass (Craig 1982). This is brought about by synchronisation between abundance and growth. It is often assumed that abundance of fish populations is determined during the larval stage when small changes in larval survival can bring about large changes in population abundance. Jensen (1993) modelled the impact of applying random survival of eggs and larvae on population abundance, using the walleye population of the western basin of Lake Erie, North America, as an example, assuming food-limited growth and size-dependent mortality and age of maturity. In populations with size-dependent mortality, variations in larval survival have very little effect on the resulting adult numbers. If age of maturity is flexible, variations in early survival have a profound effect, causing large fluctuations in abundance with the occasional large year class. If age at maturity is not flexible and if mortality is not size dependent, variation in larval survival will cause population abundance to fluctuate widely and randomly. Jensen (1993) concluded that compensation during the larval life stage was not necessary for regulation of fish population size. Walleye and probably other percids can adjust growth, age at maturity, fecundity and longevity to produce a relatively constant lifetime cohort egg production per recruit over their range and the environmental conditions to which they are subject (Beverton 1987). The energy expended for this constant egg production tends also to be constant even in populations with quite different life history strategies (Craig *et al.* 1995).

9.2 Population dynamics

Numbers and biomass

Percid population abundance varies between and within species, by geographical location and by the biotic and abiotic conditions under which the fishes live. A number of methods have been used to calculate percid population abundance and biomass. These have included virtual population analysis (VPA) (Hartmann 1994, 1995), mark-recapture (Craig 1974a; Guti 1992) and egg census (Newsome & Aalto 1987). Serns (1986) found that mark-recapture estimates were close to those derived by cohort analysis. Walters and Punt (1994) used a yellow perch fishery to illustrate the importance of undertaking well designed surveys even when age composition data for VPA are available.

In unexploited percid populations oscillations of population size probably are cyclic and regular in the long-term. In the short-term these fluctuations may not appear to be regular. Year class strengths of walleye in West Blue Lake, Canada, vary greatly and it is suggested that abundance of the year class depends on spawning success (Kelso & Ward 1977). In the long-term this unexploited population has low egg to fry survival and in the short-term, seasonal abundance is regulated by mortalities in the autumn.

Intensive fishing can upset the self-regulating mechanisms (see Chapter 10). A classic example is the blue pike population in Lake Erie, North America. These fish were abundant from about 1860 (time of commencement of records) to the early-1900s but then numbers began to fluctuate widely. Regier *et al.* (1969) suggest that these fluctuations were due to over-intensive fishing which eventually upset the self stabilising mechanisms of the blue pike population. The changes which occurred and reasons for the changes are very similar in some respects to those of the perch population in Windermere, England. A reduction in adults led to a population explosion and then the production of very poor year classes due to excessive cannibalism by the older fish. As these older fish died out another population explosion would occur. Eventually these oscillations would be expected to dampen out. However, the oscillations in the blue pike population indicated that the main controls were not only cannibalism but also the predators themselves, which were oscillating in numbers due to the fishery (Parsons 1967). A continuation of intensive fishing even when they were scarce resulted in the extinction of blue pike in Lake Erie.

Recruitment

The main factors controlling percid year class strength and the number of fish that are recruited to the population include temperature, wind velocity, water level of lakes or flow rate of rivers, light and water transparency, oxygen concentration, spawning stock size, prey availability, predation and cannibalism. An important factor controlling year class strengths of percids in many populations has been water temperature during the first summer of growth as growth is driven by water temperature. Thus, climate and climate change can have a significant effect on recruitment (Böhling *et al.* 1991; Lappalainen *et al.* 1995; Lehtonen & Lappalainen 1995; Pereira *et al.* 1995). As stated in Chapter 3, it is important for the young percids to reach a certain size by the end of the first summer as overwinter starvation is size dependent (Post 1988). Starvation can occur at the start of exogenous feeding and daylength may be critical for visually feeding larvae, as it is in older fish during the winter. Water clarity can also affect feeding and many percids show adaptations to varying light regimes (Chapter 2). Winds can have an effect, for example, by producing internal seiches which subject embryos and possibly larvae to thermal shock (Newsome & Aalto 1987). They can also disturb feeding habitats such as macrophyte beds in shallow water areas (Hartmann 1993) and make the water turbid. In some populations factors such as water level (Henderson 1985) or river flow (Belogolova 1987) may be the most significant in controlling year class strengths.

Although fecundity can be highly variable between individuals within a population, there is usually no relationship between the number of eggs laid and the strength of the subsequent year class. However, larger eggs produced by larger females may produce larvae which have a better survival than small eggs (Chapter

4). Egg and embryo size is related to yolk volume. Larvae hatched from small eggs are smaller, have more body deformities and higher mortality rates than larvae hatched from larger eggs (Moodie *et al.* 1989). Usually there is no relationship between the size of the parental stock and the size of the resulting year class although Craig and Kipling (1983) fitted a stock–recruitment curve to Windermere long-term perch data which showed the negative effect on recruitment at the highest levels of stock biomass. This was probably due to cannibalism. Bronte *et al.* (1993) also fitted a stock and recruitment curve to yellow perch data from western Lake Superior, North America. They found that age specific mortality increased at sexual maturity. Kjellman *et al.* (1996) concluded that when recruitment is dependent on a variety of abiotic and biotic factors, the population effect 'may frequently be overruled thus breaking down any single stock–recruitment curve'.

Mortality

Intrinsic and extrinsic factors and the interactions of both can cause mortality of fishes. Intrinsic factors include genetic deaths, physiological breakdown, caused for example by spawning or by senescence in adult fish, and diseases such as cancer. Extrinsic factors include lethal levels of temperature, oxygen, salinity, turbulence, pH and chemicals, parasites and infectious diseases, predation including cannibalism, starvation, caused by for example lack of light or insufficient prey of a suitable size and probably makes the fish susceptible to predation, and fishing. Slow growing species or populations of fish often have a lower natural mortality rate (M) than faster growing species or populations. Craig (1982) found that M was positively related to K, a growth function of the von Bertalanffy (1938) growth equation (Chapter 3).

Immigration and emigration

Apart from spawning migrations, which can be extensive for some species and populations, (Chapter 7) particularly in large lakes, rivers and coastal areas, immigration and emigration may be unimportant in percid population dynamics.

9.3 Role in the community

Trophic status

Fish communities are structured in relation to different forces acting on them including the abiotic environment, predation and competition. Interactions are complex and are often indirect. Abiotic factors may affect one species which undergo changes which then affect another species (Persson 1997). For example roach, *Rutilus rutilus*, are more sensitive to changes in pH and conductivity than perch. Acidification can alter roach populations which then has an impact on perch populations. Percids are important components of aquatic food webs and have been used in biomanipulation experiments to reduce nuisance algal growths (Chapter 5). Their trophic position in relation to other consumers can be estimated by using, for example, stable nitrogen isotope ratios (Vander-Zanden *et al.* 1997) or by carrying out enclosure experiments (Broenmark 1994; Diehl 1995). The position or trophic status of percids can change temporarily or permanently for example during periods of low prey abundance when changes occur from interspecific competition. Low prey density and high predation can lead to resource partitioning and a shift in resource use even in small water bodies (Keast & Fox 1990). Fishing pressure often leads to a reduction in species diversity (Cohen *et al.* 1993).

Predation and competition

Predation and competition are the major factors controlling the structure of fish communities. Predation is much easier to measure than competition. Predator–prey relationships and competition are a function of abiotic factors such as temperature and light which can affect the density and foraging activity of predators, competitors and prey. Many predators are flexible and can switch prey type (Chapter 5) in response to change; walleye, for example, fed on darters and minnows in Sparkling Lake, USA, when yellow perch abundance was low but returned to the yellow perch when they became abundant again (Lyons & Magnuson 1987). Cannibalism is a special type of predation which can have a major impact on population dynamics (see 9.4 Case histories). Cannibalism can occur early in life leading to a bimodal length distribution within a year class (Brabrand 1995). Normally predation of percids on their siblings occurs between year classes, in particular cannibalism on young-of-the-year by older fish.

Predation is usually size selective and can be limited by the gape size of the predator (Post 1988; Post & Evans 1989b; Chapter 5). Thus, in many cases predation changes with ontogenetic development. Alewives prey on yellow perch soon after hatch in late April and early May in Lake Ontario, North America (Mason *et al.* 1986; Brandt *et al.* 1987). As the yellow perch grows beyond a size to be preyed upon by the alewife, the alewife affects the state of the yellow perch population by competing for the available food resources. Size selective predation can limit recruitment of the prey allowing the survivors to increase growth due to reduction in intraspecific competition, and reach a size that is invulnerable to predation (Tonn *et al.* 1992; Persson *et al.* 1996). Indeed, prey sizes eaten by different piscivore species of similar length are similar and the variation in prey size is due to differences in the size of the predators (Mittelbach & Persson 1998). In small, species-poor lakes in northern Wisconsin, mudminnows, *Umbra limi* (Kirtland), compete unfavourably with young yellow perch. This reduces the food intake, growth and condition of the mudminnows which then makes them more susceptible to predation by larger, older yellow perch (Tonn *et al.* 1986). However, mudminnows show flexibility which allows them not to be totally eliminated from the lakes (Tonn & Paszkowski 1987). They tend to be less dense (12% of mudminnow only lakes) and larger when found with perch (Tonn & Paszkowski 1986).

The introduction of a predator can result in a significant change in the community structure (see Chapter 10). After the introduction of zander into Lake Gjersjoeen, Norway, the abundance of juvenile roach declined in the pelagic to be replaced by larger roach. The loss of the pelagic as a refuge for juvenile roach made them more vulnerable to perch predation in the littoral. In the littoral there was a shift from roach to perch dominance (Brabrand & Faafeng 1993).

The variability, structural complexity and number of habitats play an important role in predator–prey relationships and thus the structuring of fish communities (Christensen & Persson 1993). Pike can influence habitat distribution of the prey. Perch are less vulnerable in open water than in weedy littoral areas (Ekloev & Hamrin 1989) as they are more flexible in their behavioural response to predators than fish such as roach who take refuge in macrophytes (Ekloev & Persson 1996). Walleye are more successful than muskellunge (*Esox masquinongy* Mitchill) in anti-predatory behaviour, taking refuge in vegetated areas (Wahl 1995). Minnows, *Phoxinus phoxinus*, seek out shallow water when preyed upon by perch (Eklöv *et al.* 1994). However, taking refuge may have its disadvantages, for example, perch hiding in vegetation from pike or older perch were found to exhibit reduced individual growth. This may have been due to the physical complexity of the submerged macrophytes interfering with benthic feeding and an increase in the density of fish in the limited vegetated areas (Diehl & Ekloev 1995).

Large populations of perch are usually slow growers and a reduction in their numbers results in improved growth, an indication that there is intraspecific competition (Alm 1946; Le Cren 1958). Competition may arise if the same food supply is sought. In Lake St George, Canada, two sequential year classes of yellow perch differed in density by a factor of three. The low density cohort undertook the typical ontogenetic shift in behaviour from an early presence in the pelagic to a complete migration to the littoral by midsummer. The second, high density cohort split, probably due to intraspecific competition, into a fast growing group living in the littoral and a slow growing group living in the pelagic. The mortality of the high density cohort was twice that of the slow growing cohort (Post *et al.* 1997). One species may be more successful at competing for this food supply than another. For example Hanson and Leggett (1985) showed that yellow perch and pumpkinseed, *Lepomis gibbosus* (Linnaeus), kept separately in enclosures in the littoral zone of Lake Memphremagog, Canada–USA border, gained weight at low densities and lost weight at high densities. Yellow perch and pumpkinseed feed on similar food items in the littoral zone of lakes. When yellow perch and pumpkinseed were placed in the enclosures together, growth of yellow perch was suppressed but pumpkinseeds grew at the same rate as if they had been on their own. It would appear that the pumpkinseed was superior to yellow perch in competing for the food. Although different species may be feeding on the same food source, in some situations they may be separated into different habitats in time or space, or they may use the same habitat but feed on different prey (Crowder *et al.* 1981). Timing of ontogenetic development may prevent competition between species feeding from the same food resource (Roseman *et al.* 1996). In Lake Ros, Poland, juvenile roach occupy the upper layers while perch and smelt inhabit deeper water (Jachner 1991).

Observations of species dominance and community structure have been extensive both in the natural state and in artificial situations such as species removal or introductions or in artificial water bodies such as reservoirs. However, in most cases, the causal mechanisms underlying species dominance have not been explained quantitatively. A survey of Swedish lakes was made to study the interspecific dominance in fish communities (Svärdson 1976). The perch is the most widely distributed fish in Sweden but its populations do not always compete favourably with other species. Where perch and whitefish, *Coregonus* sp. Lacépède, compete, the perch is more successful than the whitefish in small lakes where the littoral area is large in relation to the open water. However, the whitefish is a better plankton feeder than the perch and is more successful in large, oligotrophic lakes. The whitefish is better adapted also in colder lakes since its physiological optimum temperature is lower than that of perch. The zander prefers warmer temperatures than other percids and has been successfully introduced into the warm eutrophic lakes of Sweden. In these lakes it competes against perch and pike and causes a decline in these two species. The zander may be more successful because it is a 'wandering' predator making use of both the pelagic and benthic habitats. The perch is less successful as a pelagic fish and the pike tends to stay in the littoral zone where it stalks its prey.

The most detailed studies on competition in freshwater fishes have been on perch and roach in Sweden. The roach spread throughout Sweden during the Pleistocene period but its distribution was less extensive in Norway as mountains formed a barrier to its distribution. In addition to preferring warmer water than the perch, the roach is not very tolerant of low oxygen concentrations or extremes of water hardness or low pH. It cannot reproduce in waters below pH 5.5. When lakes containing perch and roach are subjected to increased acidification the roach may be eliminated and the perch population may increase in size, both in numbers and growth of individuals. However, where roach lives sympatrically with perch it dominates; the effects are highly asymmetrical in favour of roach. The roach is very plastic in its feeding habits; in particular, it utilises zooplankton better than juvenile perch. In their first year perch feed on cladocerans but in the presence of roach they move to copepods and macroinvertebrates (Persson 1987b). Roach will feed on algae and detritus when other foods are limited and can become a pelagic algal eater. If juvenile perch are forced to move to a diet of benthic macroinvertebrates they may compete with older perch. A decrease in size of older (1+ perch) may be an indirect result of roach abundance. This has been demonstrated in highly productive (Persson 1986; Persson & Greenberg 1990a, 1990b) and unproductive (Persson 1997) lakes. Perch undergo ontogenetic diet shifts from plankton to benthic macroinvertebrates to fish. They may decrease in abundance and biomass especially with increasing productivity due to competition with successful planktivores such as roach and benthivore specialists such as ruffe (Bergman & Greenberg 1994). Yellow perch in North America also undergo similar shifts in diet and may undergo competition with other specialists in a particular habitat, for example white suckers, *Catostomus commersoni* (Hayes *et al.* 1992).

Yellow perch and walleye can thrive in the same lake as was illustrated for Oneida Lake (this chapter) and has been shown in West Blue Lake, Canada (Kelso & Ward 1977). When sauger are present there is usually a reduction in the proportion of yellow perch and walleye in the community (Clady 1978). Species diversity appears to be negatively related to the importance of yellow perch and positively related to the number of sauger. Species diversity does not appear to be related to the relative abundance of walleye. Sauger tend to inhabit more southerly lakes than the yellow perch and these waters contain more diverse fish faunas. However, indications are that where yellow perch exists in North American lakes and rivers it competes very well against other species.

In northern Canadian lakes walleye commonly coexist with pike. In a study of Ontario lakes the existence and composition of four common species was determined (Johnson *et al.* 1977). The most common combination was walleye and pike. Walleye combinations with lake trout, *Salvelinus namaycush*, and smallmouth bass, *Micropterus dolomieui* Lacépède, were extremely rare. Walleye and pike tend to feed on the same prey (Fig. 9.1) but their habitat is different since pike is a daytime feeder and walleye is a twilight and night feeder. Thus, they are able to live

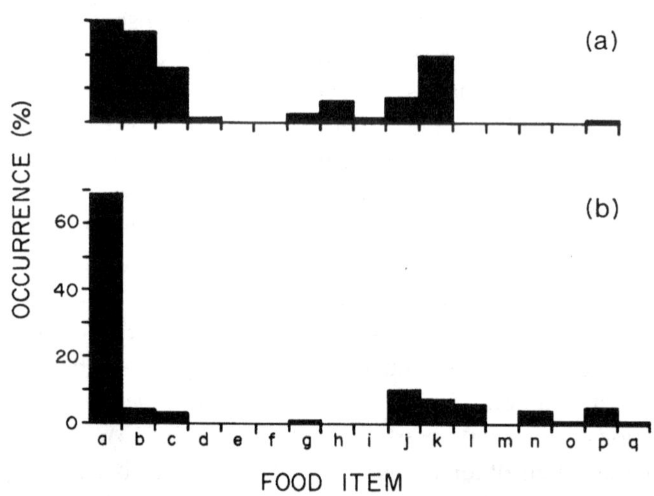

Fig. 9.1 The occurrence of prey items in the stomachs of: (a) pike and (b) walleye in Wolf Lake. *a.* empty, *b Perca flavescens*, *c Notropis hudsonius*, *d Coregonus clupeaformis*, *e Catostoma commersoni*, *f Pimephales promelas*, *g Lota lota*, *h Esox lucius*, *i Couesius plumbeus*, *j* fish remains, *k* dragonfly nymphs, *l* damselfly nymphs, *m* dipteran larvae, *n* dipteran pupae, *o* leeches, *p* gammarids and *q* macrophytes. (After Craig & Smiley 1986.)

sympatrically. In small lakes walleye are less common and the only predatory fish is usually the pike. This may be due to the increased proportion of the littoral compared to the pelagic.

Evidence suggests that fish populations are limited by food availability and species within a water body must compete for that available food. Percids seem to be quite successful and labile as is illustrated by their common occurrence and high biomass in many bodies of water. However, their requirements are fairly specific and perturbations to their niche can alter their balance in the ecosystem.

9.4 Case histories

Oneida Lake

Oneida Lake is situated in New York State, USA. It is a relatively eutrophic lake and has been used extensively by sports fishermen. The lake has been studied for many years and the following account will deal with the period from 1900 to 1975. It is of particular interest because of the relationship between yellow perch and walleye (Forney 1977a). From 1900 to 1975 the pike and eel, *Anguilla rostrata*, declined and the chain pickerel, *Esox niger* Lesueur, became nearly extinct. The walleye established itself as the main predator in the lake. In surveys carried out from 1915–1927 and from 1960–1970 the yellow perch was found to be the most abundant species in the littoral regions of the lake. The lake became dominated by these two percids and in the period 1958–1973 they made up 85% of the catch in experimental gill nets. The reduction of other predators in the lake resulted in increased numbers of walleye (Fig. 9.2) but not an increase in their size (Forney 1977a). Thus, although recruitment of walleye was improved there was increased intraspecific competition for food as growth was a function of prey density. This was illustrated by Forney

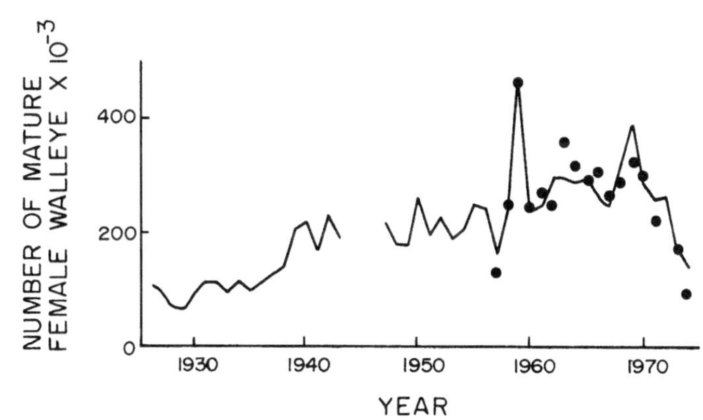

Fig. 9.2 Estimated number of mature, female walleye in Oneida Lake from 1928–1974. Values based on mark-recapture are also given (•). (After Forney 1977a.)

(1977a) who found a close relationship between yellow perch abundance and length increments of one to four year old walleye.

In 1971, first year yellow perch were abundant and the weights of three-year-old walleye increased rapidly in July and slowed towards September. The walleye fed on invertebrates from spring until mid June when yellow perch fry were about 18 mm and became susceptible to the walleye. The invertebrate diet gave no anabolic energy for growth but the yellow perch diet gave the walleye excess energy for growth above that required for catabolic processes. In 1973 the yellow perch were less abundant and there were insufficient of these even for the maintenance diet required by the walleye. This was particularly apparent in five-year-old walleye which decreased in weight in July. The walleye started to gain weight in August and this gain was rapid through September and October. During this time the consumption of white perch, *Morone americana*, increased. The white perch year class in 1973 had been more successful than that of the yellow perch but the walleye did not prey on them until the white perch changed from a pelagic to a demersal habit. The walleye consumed a high proportion of the prey production and the stable size of the adult stock was brought about by equilibrium with the prey. Thus, growth and recruitment in the walleye stock changed in response to prey abundance. High prey density increases walleye growth and recruitment and it also reduces cannibalism. The year class produced in this situation soon increases its biomass through rapid growth and eventually this leads to intraspecific competition and reduction in growth. The large year class, both in number and individual size, intensifies cannibalism and suppresses succeeding year classes, a similar situation to large year classes of perch in Windermere, England (see *Windermere* case study below). Cannibalism can be a major factor in controlling recruitment in walleye. Fast growth and the attainment of *c.* 175 mm by the end of the summer results in greater recruitment than slow growth which exposed fish to a prolonged period of cannibalism (Forney 1976, 1980).

The yellow perch have improved their growth over the period of study in Oneida Lake due to reduced competition with other prey species. The populations of walleye and yellow perch were less varied in abundance in the 1930s when other predators were present in the system. The walleye may have been able to feed on several age groups rather than relying merely on the abundance of first year yellow perch as the yellow perch were smaller at that time. The present single predator and prey relationship is probably less stable than a mixed one.

Older yellow perch in Oneida Lake are cannibalistic on yellow perch fry but the walleye affect the mortality of the yellow perch to a much greater extent. The density of yellow perch fry declines exponentially through the late summer and autumn. The number of yellow perch consumed by walleye was estimated from examination of walleye stomachs caught by otter trawls (Forney 1977b). The abundance of pelagic yellow perch fry was estimated by Miller samplers (Noble 1970b) in June and demersal yellow perch were estimated by trawls. Numbers of yellow perch consumed by walleye between mid June and October ranged from

18 000 ha^{-1} in 1973 to 243 000 ha^{-1} in 1971. The number eaten was found to be approximately proportional to the fish available. The numbers eaten by three-year and older yellow perch were 56 000 ha^{-1} in 1971, 24 000 ha^{-1} in 1975 and were negligible in other years.

A predator–prey model was formulated by Forney (1977c) to predict the effect of different levels of walleye biomass on year class strengths of yellow perch (in their first year). Assumptions on which the model was based were: (1) all mortality of young yellow perch was caused by walleye predation and (2) mortality of yellow perch was proportional to walleye biomass on days when ration size was less than the satiation ration. Number of yellow perch (P) on day t was described by:

$$P_t = P_0 - \sum_{t=1}^{t_i} 0.000186\ (9.5) B_t$$

where P_0 was the initial density of yellow perch ha^{-1} and B_t was the biomass of walleye in kg ha^{-1}. The daily mortality of yellow perch, 9.5, was described by the product of average attack and average duration of nocturnal feeding. The ration, R, consumed kg^{-1} by walleye was given by:

$$R_t = P_t \hat{W}_t 0.00176$$

where $\hat{W}_t$ was the mean weight of yellow perch on day t. The ration to satiate walleye was assumed to be 0.06 body weight. Biomass of walleye (B_t) was calculated from the product of average weight and numbers of individuals. The numbers declined at an exponential rate Z and individual daily weight increment was calculated as 0.14 of the ration in excess of 0.01 body weight. Thus:

$$B_t = N_0 \exp(\ Z_t)\ [\bar{w}_0 + \sum_{1}^{t} (R - 0.01w)(0.14)]$$

The model was used for predicting yellow perch fry biomass using an initial density of 240 000 and growth data collected by weekly sampling from June to October 1971 (Fig. 9.3). A biomass of 22 kg ha^{-1} was estimated for the walleye in the spring of 1971. Yellow perch biomass at a walleye biomass of 18–30 kg ha^{-1} peaked in late June and early July and declined in the late summer. The predicted values did not fit the observed values very accurately. Forney (1977b) suggested that the rate of walleye predation may have been lower or biomass of yellow perch was sufficient to satiate walleye and thus, the effective rate of predation was reduced. Observed rate of predation did not exceed 0.06 body weight but in practice it may have been lower than this due to physiological factors. A minimum biomass of walleye of 10–18 kg ha^{-1} is needed to control the year-class strengths of yellow perch otherwise the amount of available food must limit yellow perch recruitment. Recruitment is probably a function of parental stock size. Ricker (1954, 1975) derived the resulting curve thus:

$$R = \alpha\ S \exp(\beta P) \qquad \text{Equation 9.1}$$

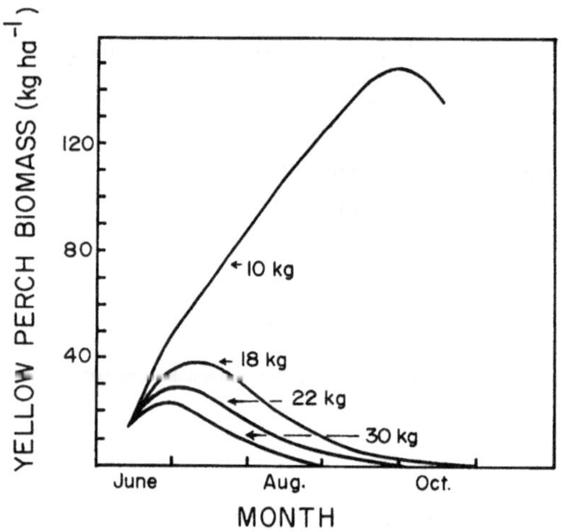

Fig. 9.3 Simulated values of young-of-the-year yellow perch biomass in relation to walleye biomass in June ranging from 10–30 kg ha^{-1} in Oneida Lake. (After Forney 1977c.)

where R is the number of recruits, S is the size of the parental stock, α is a dimensionless parameter and β a parameter with dimensions $1/P$. In the next case history the recruitment of walleye is described as a function of parental stock and spring spawning temperatures.

Western Lake Erie

Lake Erie, North America, has a long history of commercial fishing (see Chapter 10). The walleye has been a major species exploited in this fishery. From the late nineteenth century there have been many habitat changes brought about by human influence. Particularly important to the walleye has been the reduction in water quality and spawning areas (Koonce *et al.* 1996). The walleye has also been over-exploited and fishing probably exceeded sustainable yields in the mid 1950s (Shuter & Koonce 1977). A reduction in walleye density over the period of observation was apparent from catch per unit effort data and year class strengths were found to decline. In response to this reduction in numbers, younger fish increased in growth and some walleye became sexually mature earlier. Shuter and Koonce (1977) showed that growth was related to population density which implies that forage density was the controlling factor. In 1927 about 40% of three-year-old female walleye were mature and this increased to about 80% by 1965. Recruitment (R) was related to stock size (S) and spring temperature (t, the rate of water temperature increase in the spring during spawning) by:

$$\log R = \alpha + \beta \log s + \delta \Delta t$$

where α, β and δ are constants. The number of yearlings (R_i) was related to the adult spawning stock and spring temperature in the year before ($i-1$). There were indications that these stock and recruitment relationships were working in the period 1947 to 1967. From these basic principles, a model for the walleye population was developed (Shuter & Koonce 1977; Shuter *et al.* 1979). An assumption was made that the annual natural mortality was 20% and this acted on all ages from one to eight years. A minimum length of 152 mm at age one year was assumed in the modelling. Different regimes of fishing mortality were used. The resultant predictions are very close to the observed values for both recruitment (Fig. 9.4) and stock abundance (Fig. 9.5). Thus, information on relative abundance, year class strength and growth could be combined into a model which appeared to be a good predictor for western Lake Erie walleye from 1947–1967.

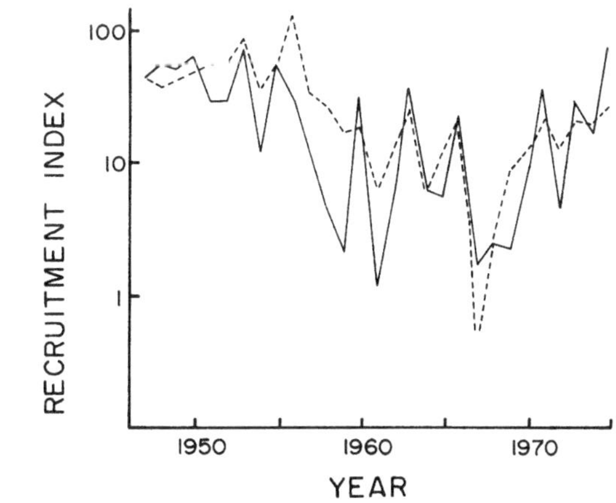

Fig. 9.4 Observed (solid line) and simulated (broken line) values of the recruitment index (number of one year olds per trap lift) of walleye from 1947–1975 in western Lake Erie. (After Shuter & Koonce 1977.)

The average age of maturity of female walleye is now five years (Henderson & Nepszy 1994). Age of maturity is related to the physiological state of the fish; sufficient energy must be acquired for spawning. Growth of walleye declined from 1965 to 1984 which may have been a result of food shortage and a reflection of prey type and abundance. Gizzard shad, *Dorosoma cepedianum*, emerald shiner, *Notropis atherinoides*, and spottail shiner, *N. hudsonius*, were the main prey of walleye when they were abundant but walleye consumed more white perch, *Morone americana* and yellow perch when the preferred prey were less abundant (Hartman & Margraf 1992). From the late 1960s to the late 1980s the relative abundance of prey species was fairly constant but fishes such as emerald shiner, spottail shiner, alewife and white bass, *Morone chrysops* declined, significantly after 1977. Increased predation by walleye is thought to be responsible for the change in species composition in the lake (Knight & Vondracek 1993). Madenjian *et al.* (1996) developed a recruitment model for walleye for the period 1981–1993 which accounted for 92% of the variation in recruitment of two year olds. Unlike Forney (1976, 1980) they found that recruitment was not related to first year growth and the effects of cannibalism. Their best fit model based on equation 9.1 took the form:

Fig. 9.5 Observed (solid line) and simulated (broken line) values of abundance (number of fish per trap lift) of walleye from 1947–1975 in western Lake Erie. (After Shuter & Koonce 1977.)

$$R = 1.53P \exp(-0.0736P + 0.00460SHAD + 5.23WMRT)$$

where *SHAD* is the abundance index for age 0 gizzard shad in the autumn prior to walleye spawning in the following spring and *WMRT* is the spring warming rate (°C day^{-1}). The authors suggest that gizzard shad are important in supplying appropriate lipids during the winter and spring to maturing females. Parental stock size and spring warming temperature also influenced walleye recruitment.

As in many other systems, western Lake Erie 0+ yellow perch undergo an ontogenetic diet shift from zooplankton to benthos, normally during June to August (Wu & Culver 1992), a time when zooplankton abundance declined. Yellow perch >50 mm were able to make the switch. Those <30 mm were unable to feed on benthic organisms. Slower growing fish are probably prone to starvation and more vulnerable to predators. Walleye can play a major role in influencing the year class strength of yellow perch through predation even in a lake as large as Lake Erie (Hartman & Margraf 1993). In 1988 it was estimated that small walleye (two years or younger) ate 28–90% of the yellow perch reaching 18 mm TL.

Lake Ijssel

The importance of temperature in growth and recruitment of percids is illustrated by the zander in Lake Ijssel. Lake Ijssel is a shallow eutrophic rather turbid lake in the Netherlands (Willemsen 1977b). It was created in 1932 by damming a marine bay and forming a freshwater lake. The lake contains mainly zander, perch, pike, ruffe, smelt, bream, *Abramis brama*, roach, and eel, *Anguilla anguilla* (Linnaeus). The gill net fishery for perch and zander is characterised by large variation in the yield. The variation is due to the unstable nature of the system brought about by the intensive fishing and the large differences in year-class strengths (Buijse 1992). Up to 1970 an eel trawl was used in the lake but this was discontinued when it was realised that it caused heavy mortality on other desirable species. Willemsen (1977b) found that the food of perch and zander was very similar and there was little predation or competition for food between or within species and growth did not appear to be limited by food availability. Smelt were the most important food for both zander and perch when zander were > 50 mm and perch > 100 mm. Cannibalism was only slight, 0.14% in zander and 2.00% in perch. In perch it increased with increase in density.

During the period 1966–1989 the year class strengths of zander and perch, measured as the abundance of 0+ fish caught by autumn survey trawls, varied by a factor of ×70 and ×400 respectively. For zander 38–86% of the variation in recruitment could be explained by water temperature, wind velocity and year of sampling. For perch 36–71% of the variation was taken up by stock biomass of piscivorous fish, water temperature and year of sampling. Stock and recruitment relationships could not be found for either species. Length in November of 0+ zander was also correlated, 72.5% of the variation, with water temperature (Buijse & Houthuijzen 1992). Depending on factors such as water temperature and food availability the initial unimodal length frequency of 0+ zander developed into a positively skewed bimodal or negatively skewed distribution towards the end of the summer. Non-piscivorous zander decreased in condition over time while those becoming fish eaters increased. High temperatures favoured the onset of piscivory which positively affected growth and survival.

Windermere

Windermere, the largest lake in England, was formed by glaciation and is situated in the Lake District, north-west England. It is divided into north and south basins by a series of islands and shallows. Windermere contains a relatively simple fish fauna consisting mainly of perch, pike, charr, *Salvelinus alpinus*, brown trout, and eels. The fish in this lake have undergone investigation since the start of World War II (Le Cren 1987). Particular emphasis has been placed on factors that control the abundance of perch and pike. Historically, the lake was fished commercially but no commercial fisheries are in operation today, apart from a small plumb-line fishery for charr. In contrast to Oneida Lake, the sport fishery is not extensive.

In a wartime fishery (Worthington 1950), from 1941 to 1947, over 90 tonnes of perch were removed from the lake reducing the biomass of the perch from 136 to 55 t. Large-scale removal of perch continued in the north basin of Windermere until 1948 and in the south basin until 1964. Unbaited fish traps were used to catch fish for the fishery and these traps have continued to be used for sampling the population ever since. The pike depends on the perch as its main source of food, a similar, single predator–prey relationship to that found in Oneida Lake. There were fears that the large scale removal of perch would alter the feeding habits of the pike, specifically that the pike would feed heavily on more desirable species such as charr. In 1944 an experimental winter gill net fishery was started to remove pike >550 mm and this fishery has continued on an annual basis until the present. About one third of the pike population >550 mm is removed by the fishery each year.

Trends in the perch population were closely followed by that of pike; similar trends occurred in both basins (Fig. 9.6). There was an initial fall in number as a result of the fisheries and numbers of perch remained low until 1957 in the south basin and until 1961 in the north basin. Pike recovered more quickly although they were still being fished. From 1949 to 1971 there were more pike than before the wartime fishery, but these were younger fish. After 1971 the pike population declined; Kipling (1984) describes five stages of population changes during the study (Table 9.1). Up to the end of the 1950s, year classes of perch and pike were dependent on summer temperatures in their first year of life. The slow growth of perch made them available as food for both young-of the-year and older pike. In the early 1960s the populations were dominated by the very strong 1955 and 1959 year classes of both species. Year classes of perch produced during this period were heavily preyed upon by pike and also by older perch. The year classes were already

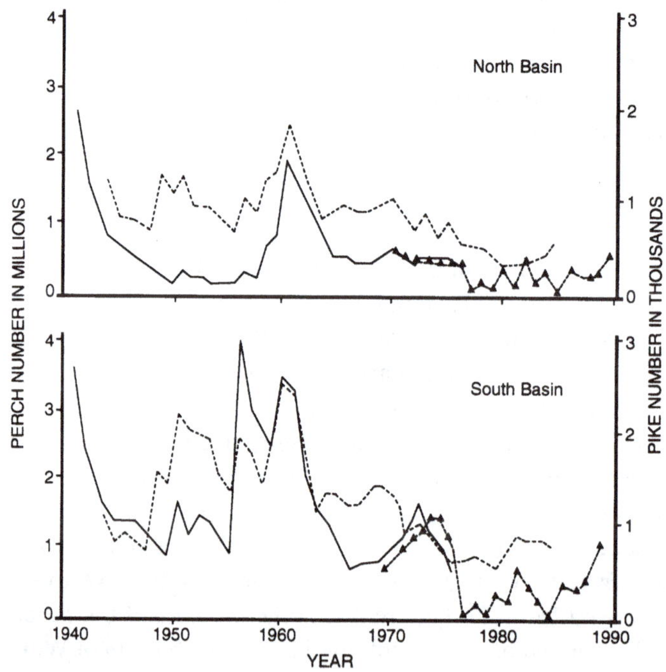

Fig. 9.6 The estimated numbers of perch (solid line) and pike (broken line), aged two years or older, in the north and south basins of Windermere. The triangles connected by broken lines (1970–1989) show the unadjusted catch per unit of effort data from perch traps, scaled to the same axis, to show the recovery of the population following the 1976 disease outbreak. (After Mills & Hurley 1990, with kind permission from Blackwell Science.)

Table 9.1 A summary of summer temperatures, growth and year class strengths of perch and pike during five periods in Windermere, England. (After Kipling 1984.)

Period	Summer temperature	Growth of perch	Growth of pike	Year class strength of perch	Year class strength of pike
Pre-exploitation	Variable	Very slow	Slow	Not known	Not known
Late 1940s and 1950s	Variable	Increasing	Variable	Variable	Variable
Early 1960s	Low	Slow	Slow	Weak	Medium
Late 1960s and early 1970s	High	Fast	Fast	Strong	Weak
1976–1982	Variable	Fast	Fast	Weak	Weak

weak as a result of cold summers in the years of hatch. The very poor growth of old pike at this time indicates that they were short of food. In the late 1960s the summers were warmer and perch year classes were stronger and they grew faster. Perch in this period were seven or eight times heavier than they had been at the start of the perch fishery. They rapidly outgrew the size at which young pike could feed on them. Pike became more cannibalistic which reduced their numbers, although growth improved. In 1976 a disease killed over 98% of the perch population and the pike had lost almost completely their previous main source of food. The pike continued to decrease in numbers but grew fast by obtaining their food from cannibalism and probably charr. Large pike were no longer found in the summer in the littoral zone but they were caught from time to time by charr fishermen in deep water. Since 1976 the perch have shown signs of recovery.

Although numbers of perch declined as a result of the fishery, the perch biomass did not decline proportionally because the growth rate increased. In the south basin numbers and biomass increased in the mid 1950s, due to the strong 1955 year class, but did not start to increase in the north basin until the early 1960s when the 1959 year class made a major contribution to the population. This year class dominated the population in the north basin for most of the 1960s and it made up most of the adult biomass (Fig. 9.7). After reaching a peak in 1962 the total perch biomass in the north basin fell slightly until 1964 but then remained almost constant (49 ± 3 t) until the disease of 1976. Year classes from the early 1960s made little contribution to the total biomass (note ordinate in Figure 9.7 is on a logarithmic scale). As the 1959 year class declined, later year classes made an increasing contribution to the biomass. It may be that the warm summers of the mid and late 1960s and early 1970s allowed stability in the population through density dependent factors. However, it is unlikely

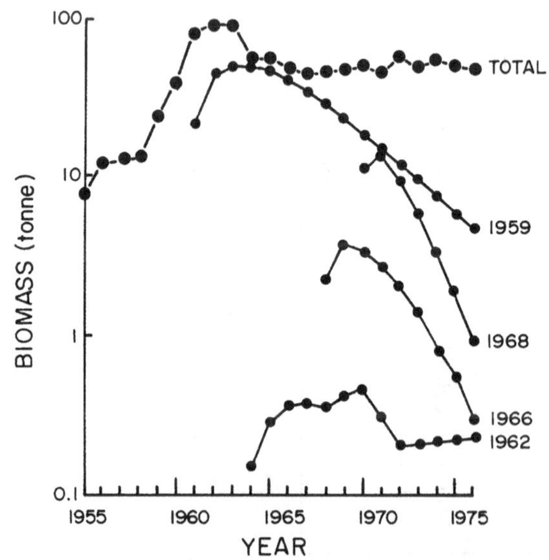

Fig. 9.7 Total biomass of perch (aged two years or older) in the north basin of Windermere from 1955–1976. Biomass data are also given for the 1959, 1962, 1966 and 1968 year-classes. (After Craig 1982.)

that this population, with its rapid growth and high mortality, was as stable as pre-1941 (before the fishery).

Some of the causes for the changes described above can be elucidated from the principles of population dynamics. An increase in mortality rate, either natural (*M*) or fishing (*F*), with resulting decrease in population density, should result in increased birth rate. Perch responded to decrease in numbers by increasing their rate of egg production. In the period 1941 to 1960 a female perch of 200 mm length spawned on average 13 100 ± 700 eggs compared to the period 1979 to 1981 when a female of the same length laid 19 300 ± 1800 eggs, a 47% increase. Before the disease outbreak in 1976 the youngest age of sexual maturity in the female perch had been three years but after 1976 some mature two-year-old fish were found. However, egg numbers laid had no correlation with the number being recruited into the population at age two years (Fig. 9.8). One of the lowest adult stocks produced one of the largest year classes. Therefore, the factors controlling year class strengths must be affecting the survival of perch in the first two years of life. The chief factors have been found to be summer water temperature in the year of hatch, the biomass of adult perch and the year class strength of pike of the same year. Not all these factors have been significant throughout the whole study.

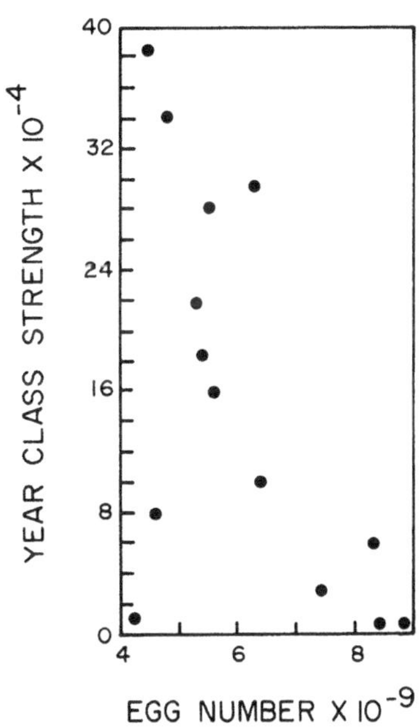

Fig. 9.8 Perch year class strength (number at age two years) in relation to total number of eggs laid from 1961–1974 in the north basin of Windermere. (After Craig & Kipling 1983.)

During the period from 1961 to 1975, summer water temperatures could account for 73% of the variation in year class strength of perch. Over the whole period from 1942 to 1975 the correlation between year class strength and summer water temperature was still significant but only accounted for 28% of the variability. Between 1959 and 1974 there was a significant negative correlation between year class strength and the biomass of adult perch in the year of hatch. Before 1959 the intensity of predation by pike on young perch may have made cannibalism by adult perch insignificant. From 1959 to 1974 the negative correlation between perch year class strength and pike year class strength in the year of hatch continued to be highly significant. From these relationships a simple model was produced to predict recruitment from 1959 to 1974:

$$\ln R = C_0 + C_1 T + C_2 S + C_3 \ln Y \qquad \text{Equation 9.2}$$

where R was the year class strength at age two years, T was temperature measured in degree-days over 14°C, S was biomass of the adult stock in tonnes and Y was the year class strength of pike. C_0, C_1, C_2 and C_3 are all constants; C_2 and C_3 are negative.

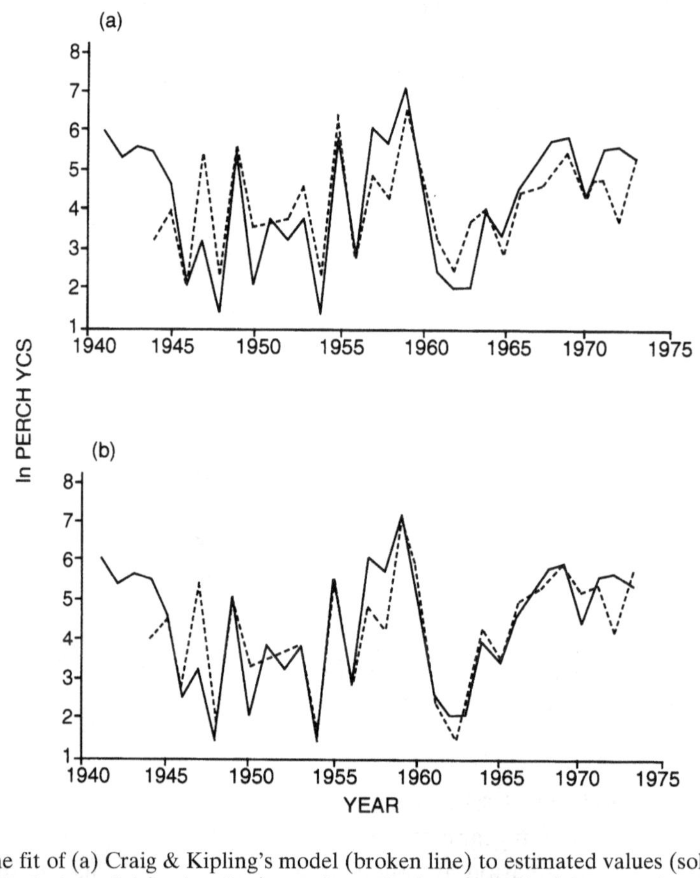

Fig. 9.9 The fit of (a) Craig & Kipling's model (broken line) to estimated values (solid line) and (b) Mills & Hurley's model (broken line) to estimated values (solid line) for year class strengths (at age two years) of perch in the north basin of Windermere. (After Mills & Hurley 1990, with kind permission from Blackwell Science.)

The model showed how an increase in temperature, a density independent factor, increased year class strength but an increase in predation both by cannibalism and predation by pike, density dependent factors, decreased year class strength. The model does not allow for much variability caused by competition between fish of the same year class but perhaps this is not important in Windermere except at extreme density levels. Mills and Hurley (1990) modified Equation 9.2 to incorporate a stock and recruitment expression (Equation 9.1) which allowed for a density dependent component, $C_{21} \ln - S\text{-}C_{22}S$:

$$\ln R = C_0 + C_1T + C_{21} \ln S - C_{22} + C_3 \ln Y$$

However, the predictive capabilities only showed a slight improvement (Fig. 9.9).

Since the 1976 disease outbreak that killed most (98%) of the adult perch in Windermere, there has been considerable variation in recruitment although the general trend has been one of increase (Winfield *et al.* 1998d). The increase probably results in part from reduced cannibalism by adults. A significant positive relationship was found between zooplankton biomass and perch recruitment (27% of the variability was explained), covering the years 1966–1990, although the authors found no relationship with water temperature. However, the relationship between zooplankton abundance, growth and overwintering of 0+ perch is not clearly understood in the Windermere study. The problem is that recruitment is not measured until after two winters. The critical period is probably the first summer of life. Although temperature has proved to play a major role in determining year class strengths its direct influence has not been elucidated. It certainly affects growth and survival but not directly. It is probably involved in the control of timing of food production in the food chain. The correct timing of this production, especially the optimum food particle size at the right time, is critical to the survival of the young fish. This is an area where much research is required.

Chapter 10
Fisheries and Economic Importance

10.1 Introduction

Biological production is the change in biomass with time and includes reproductive products which may be released. Production is normally higher during the juvenile stages than in adults (Fig. 10.1). For example, in only one year class (1959) did perch adult production exceed that of juveniles in Windermere, England, and on average adult production was 61% of that of juveniles. Yield is the part of production used by humans. Over about the last 50 years management of both inland and marine fisheries has been based on the maximum sustainable yield (MSY) of populations of single species applying the production model of Schaefer (1954; 1957) or the analytical model of Ricker (1958; 1975) and Beverton and Holt (1957). Some scientists consider that it is unfortunate that the concept of MSY was conceived. Fisheries scientists have been unable to control the technique, distribution and amount of fishing effort which has led to the collapse of several important fish stocks (Ludwig *et al.* 1993).

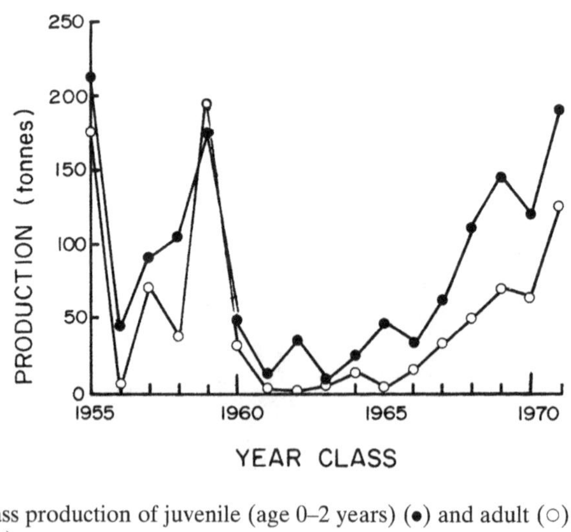

Fig. 10.1 Year class production of juvenile (age 0–2 years) (•) and adult (○) Windermere perch. (After Craig 1980a.)

Consider a simple model relating fish stocks, commercial fishing effort and annual yield (Fig. 10.2). At a low level of fishing (*A*), the yield will be small and the stocks will remain high (A^1) as the population makes adjustments to this added mortality. Point *C* is the level of MSY. At this point annual yield is balanced with stock abundance and the ability of the population to sustain itself. If fishing increases past this point, the yield (*E*) and the stocks (E^1) will be reduced, the latter possibly to a point of collapse. Fisheries have a tendency to exceed MSY. This may be explained in the following economic terms. At a low level of fishing the cost of production, including profit, is below the return and a super profit or economic rent is generated. This encourages the existing fishermen to expand operations and other fishermen to enter the fishery. If a single fisherman operated the fishery, expansion would continue until a point was reached (*B*) where the costs of production were still marginally below the revenue generated. At this level of fishing the stocks would still be capable of responding to the mortality imposed on them. However, individual fishermen, both commercial and recreational, compete against each other

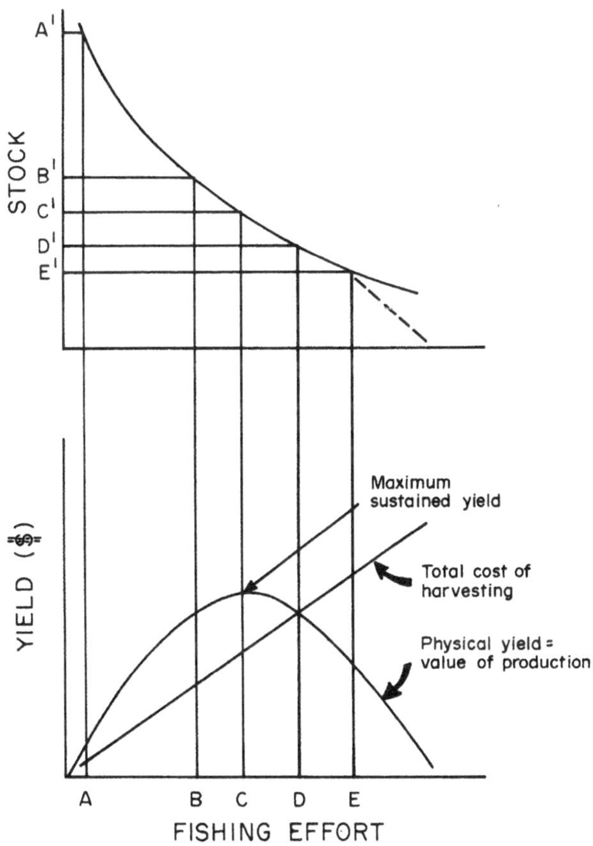

Fig. 10.2 The relationship between fish stocks and yields to fishing effort. See text for description.

but individually only consider their own costs and returns. No consideration is given to the overall yield, fishing effort and cost of production, and fishing increases until costs equal revenue (*D*). If governments subsidise the fishery, effort will increase to point *E* and the possible collapse of the fishery. Management policy aims to control fishing effort around point *B*. At this level the excess profits are extracted as payment for the resource. This type of policy applies equally well to recreational and commercial fisheries.

In 1976 the concept of an optimum yield was formalised in Canada and the United States (Healey 1984). It took into account biological, economic and social factors. For example, as well as considering the conservation of the fished species, the maximum net income of the fisherman and the maximum employment of people in the industry were also used in determining the yield from the fisheries.

Percids normally exist as part of a community (Chapter 9) and single species methods of managing them can be inappropriate. There has been progress in developing single species virtual population analysis (VPA) and cohort analysis, into multi-species VPA. Declining catches often cause fishermen to switch to species further down the food chain leading to fundamental changes in the ecosystem. An understanding of ecosystem functioning and the role of fisheries has been furthered by the development of suitable models such as ECOPATH, based on the principle of mass balance and ECOSIM which simulates the effects of different harvest rates (Pitcher in press).

Yields of percids have shown considerable spatial and temporal variation. Several models have been developed for predicting sustainable yields and for deciding management policy (e.g. Christie & Regier 1988; Buijse *et al.* 1992; Tyutyunov *et al.* 1993).

10.2 Fisheries

Remains of percids have been discovered in archaeological excavations of human habitations. Perch and zander were fished along the south coast of the Baltic Sea 6000 to 3000 years BP (Tsepkin 1984a). The mean size of the zander caught in the fishery at that time was greater than the present-day average. Perch and zander remains have also been found in excavations of human sites in the upper regions of the Dnieper River, Russia. Fossils there date between 2700 and 400 years BP (Tsepkin 1984b). Native North Americans made use of fish stocks including percids in inland waters, but it was probably not until the invasion of European settlers that demands on these fish stocks became significant.

Worldwide commercial landings of *P. fluviatilis*, *P. flavescens*, *S. lucioperca*, *S. vitreum*, *S. canadense* and *G. cernuus* are given in Table 10.1 for the period 1994–97. Finland has the highest perch catches, the third most important species by weight (9.9% of the total fish landed in 1997) after herring, *Clupea harengus*, and sprat, *Sprattus sprattus* (FAO 1999). The countries with the highest zander yields are

Kazakhstan, the Russian Federation, Turkey and Finland, representing 12.6, 0.1, 0.3 and 1.5% respectively of the total fish landed in 1997. Yellow perch and walleye fished in Canada have their biggest market in the central USA. The largest commercial catches of walleye are landed in Canada.

In economic terms recreational fisheries are extremely important. For example, in Canada in 1995 they involved over 4.2 million adult anglers who fished for 55.5 million days and caught over 254 million fish, of which 113 million were kept. In central Canada, including Ontario, Manitoba, Saskatchewan, Alberta and the Northwest Territories, the anglers caught about 61% of the total number (Table 10.2). Ontario has the largest and most valuable recreational fishery resource in Canada. The resource is used by recreational fishermen, commercial fishermen and by natives for domestic use. The total commercial catch of freshwater fish (about 100 species) from this region of Canada in 1995 represented over 94% of Canada's freshwater fish harvest and landed value. However, the landed value was only 2.5% of the monies spent directly on recreational fishing.

Three examples, the yellow perch fishery of Lake Erie, the walleye fishery of Manitoba and the zander fishery of Lake Balaton, illustrate the local economic importance of percids.

Lake Erie

Native North Americans living near the shores of Lake Erie used fish as a main source of food but their impact on the fish population was small. The early European settlers caught fish to supplement their diet but fishing was confined to inshore areas. The first commercial hook and line fishery began in eastern Lake Erie in 1795 (Regier *et al.* 1969). Then after the war of 1812, a commercial fishery using seines was established in western Lake Erie. During the nineteenth century, sauger, walleye and smallmouth bass, *Micropterus dolomieui*, were the main species caught. The most desired species were lake trout, *Salvelinus namaycush*, cisco, *Coregonus artedii*, and lake whitefish, *C. clupeaformis* (Mitchill). The lake sturgeon, *Acipenser fulvescens* Rafinesque, was at first considered a nuisance since it became caught and tore nets set for desirable species. As many as possible were removed. It was not until the 1860s that the sturgeon became commercially important and by that time its numbers had declined considerably. The intensity of fishing increased during the second half of the 1800s due to improved transportation (canals, lake, rail and road), improved preservation techniques and increases in the human population of the area. Fishing was made easier by the introduction of steamboats in the 1880s which also meant that the central region of the lake could be fished. Not only were the fish stocks subjected to heavy fishing pressure but they were also exposed to a number of other perturbations caused by human activities. These included eutrophication, introduction of non-indigenous species (such as the sea lamprey, *Petromyzon marinus* Linnaeus, through the Welland Canal in about 1921, and the rainbow smelt, *Osmerus mordax* (Mitchill) in 1931), tributary and shoreline development, including

Table 10.1 World catches of percids in metric tonnes of Eurasian perch (*Perca fluviatilis*), yellow perch (*P. flavescens*), zander (*Stizostedion lucioperca*), walleye (*S. vitreum*), sauger (*S. canadense*) and ruffe (*Gymnocephalus cernuus*) by country from 1994 to 1997.* = data not available; e = estimate; — = negligible; 0 = >0<1. (Data from FAO 1999. The data for Canada have been modified according to information provided by the Canadian Department of Fisheries and Oceans Statistical Services.)

	1994	1995	1996	1997		1994	1995	1996	1997
P. fluviatilis					*S lucioperca*				
Belarus	15e	10e	10e	12e	Albania	25e	5	10	4
Belgium	25	25	25	25	Azerbaijan	21e	12e	10e	9e
Bulgaria				63	Belarus	4e	3e	3e	4e
Czech Republic	38	31	33	34	Belgium	15	15	15	15
Denmark	104	90	63	83	Bulgaria	30e	22	26	45
Estonia	1 223	1 007	1 030	1 202	Croatia	12	12	12	14
Finland	16 713	16 876	17 647	17 860	Czech Republic	164	86	130	157
Germany	887	948	764	782	Denmark	9	14	12	51
Kazakhstan	728	736	670e	610e	Estonia	627	657	726	460
Latvia	54	64	56	56	Finland	2 272	2 346	2 476	2 630
Lithuania	115	114	83	114	Georgia	2e	3e	*	*
Netherlands	136	219	376	336	Germany	578	570	546	560
Norway	*	*	*	9	Hungary	277	121	259	226
Poland	864	952	1 169	2 006	Iran	95	10	6	4
Romania		39	38	8	Kazakhstan	5 436	6 089	5 570e	5 070e
Russian Federation	1 976	2 234	2 628	3 527	Kyrgyztan	17	7	6e	5e
Slovakia	30	32	18	12	Latvia	50	57	73	40
Slovenia	2	2	1	1	Lithuania	55	44	56	65
Sweden	356	343	267	386	Moldolva Republic	2e	3e	4e	5e
Switzerland	396	373	475	364	Netherlands	185	79	100	89
Ukraine	288	319	126	80	Norway	*	*	*	5
Species total	23 950e	24 414e	25 479e	27 570e	Poland	496	440	512	619
					Romania	859	353	94	30
P. flavescens					Russian Federation	4 078	4 221	6 215	6 104
					Slovakia	89	93	76	70
Canada	3246	2595	2648	2829	Slovenia	5	5	5	5
USA	1387	1246	711	622	Sweden	425	351	312	366
Species total	4633	3841	3359	3451	Switzerland	10	13	10	6

Tajikistan	87e	92e	90e	75e
Turkmenistan	78	49e	28	28
Turkey	5 181	5 877	8 042	1 500
Ukraine	508	512	811	955
UK	3	1	1	0
Uzbekistan	161	282	130e	117
Species total	21 856e	22 443e	26 366e	19 333e
S. vitreum				
Canada	7543	7966	7816	7773
USA	22	17	18	4
Species total	7565	7983	7834	7777
S. canadense				
Canada	*	*	*	*
USA	1	4	1	
Species total	1	4	1	
G. cernuus				
Denmark	5	15	13	5
Species total	5	15	13	5

Table 10.2 (a) Recreational fisheries: total number of fish caught, number of days fished by anglers (mean per angler shown in brackets), numbers of yellow perch and walleye caught (percentage of the total shown in brackets) and direct expenditure (Can. $). (b) Commercial fisheries: total, yellow perch and walleye yields (t) and landed values (Can. $) (percentage of the total shown in brackets), for the Provinces of central Canada in 1995. (Data kindly provided by the Canadian Department of Fisheries and Oceans Statistical Services.)

(a) Recreational fisheries

Province	Total number of fish caught	Number of days fished by anglers	Number of yellow perch caught	Number of walleye caught	Direct expenditure (Can. $)
Northwest Territories	712 143	168 922 (11)		97 693 (14)	15 633 274
Alberta	15 186 996	3 714 744 (16)	3 413 054 (23)	2 465 126 (16)	125 266 999
Saskatchewan	10 070 553	2 234 248 (13)	2 631 664 (26)	3 177 821 (32)	95 393 147
Manitoba	8 938 960	2 196 866 (14)	1 030 770 (12)	3 806 439 (43)	77 947 364
Ontario	120 506 506	23 436 102 (15)	22 750 464 (19)	28 769 167 (24)	1 027 311 660
Totals	155 415 158	31 750 882 (14)	29 825 952 (19)	38 316 246 (25)	1 341 552 444

(b) Commercial fisheries

Province	Total harvest (t)	Landed value (Can. $)	Yellow perch harvest (t)	Yellow perch landed value (Can. $)	Walleye harvest (t)	Walleye landed value (Can. $)
Northwest Territories	1 434	1 637 000	0	0	51	221 000
Alberta	1 578	1 687 000	3	8 000	89	316 000
Saskatchewan	3 032	3 845 000	0	0	492	1 997 000
Manitoba	11 757	20 986 000	414	1 744 000	2 419	9 384 000
Ontario	17 633	42 779 000	1 932	9 315 000	4 904	21 120 000
Total	35 434	70 934 000	2 349 (7)	11 067 000 (16)	7 955 (22)	33 038 000 (47)

drainage of wetlands, an increase in siltation and turbidity, and the release of toxic materials, including biocides from industry, vessels and vehicles (Regier & Hartman 1973).

The early development of the Lake Erie fisheries took place in the USA. In the 1880s when catches of lake whitefish and cisco were at their highest, the landings in Canada were only 10% of the total catch. However, in the early 1900s the Canadian fisheries gradually increased and in the last 50 years Canada has dominated the commercial fisheries. The main reasons for this include lower labour costs, government support to stabilise prices, a more innovative industry, greater stocks of marketable fish in Canadian waters, and a greater use in the USA of the lake for a sport fishery (Regier *et al.* 1969). The USA buys fish in Canadian ports and the fish are transported overland. Yellow perch are marketed in the form of fillets, breaded and cooked products. The fisheries of Lake Erie have always come under the jurisdiction of the four border states in the USA, Michigan, Ohio, Pennsylvania and New York, and the province of Ontario in Canada. These authorities have guarded jealously the regulatory powers of the fisheries. The great pressures put on the fish stocks in the Great Lakes in general led to the formation of the Great Lakes Fishery Commission in 1955. Its mandate was to implement a programme of sea lamprey control (the fish has never been a major problem in Lake Erie), formulate and co-ordinate research programmes and advise governments on measures to improve fish stocks. However, actual fishery regulations remained with the individual states and provinces.

From 1927–1936 landings of yellow perch from Lake Erie averaged 5700 t and most fish came from the USA. The decline in cisco led to increased pressure on the yellow perch by American trap netters. In the 1950s, the introduction of nylon gill nets resulted in a two or threefold increase in the efficiency of the nets over cotton or linen nets and a rapid decline was evident in the walleye and blue pike fisheries (Plate 3). The latter species became extinct (Chapter 1) and the walleye fishery collapsed. At this time increased pressure was placed on the yellow perch populations and Canada became the chief producer. The highest catch of yellow perch was recorded in 1969 at 13 546 t and this declined to about 5000 t in the mid 1970s. Further pressure was placed on the yellow perch at this time by the partial ban on white bass, *Morone chrysops*, and a complete ban on walleye. The catches of yellow perch have been strongly influenced by year class strengths. Strong year classes were recorded in 1952, 1954, 1956, 1959, 1962 and 1965 (Nepszy 1977). Relatively strong year classes were recorded in 1970 and 1975. The 1959 year class was particularly strong and was represented in large numbers during the late 1960s. The year classes of yellow perch in Lake Erie appear to be influenced by similar factors to those of walleye; in particular the spring temperature in the year of hatch (Chapter 9). The great fluctuations in year class strengths and the increased pressures put on the yellow perch populations was considered to indicate that they were under stress and were likely to deteriorate. In 1975 the Yellow Perch Committee was established by the Great Lakes Fishery Commission to monitor closely the yellow perch populations. From 1980–1984 the Canadian commercial catch of yellow perch in Lake Erie varied from 2564 to 5731 t with a mean of 4071 t. Over this period the yellow perch represented 55.5% of the value of all fish landed in Canada from Lake Erie. In the 1990s the yellow perch landings declined further, averaging 1960 t (range 1246–3345 t). The largest landings continued to be of the less valuable smelt. Although catches of white bass and walleye were much lower than historic levels they made a significant contribution to the value of the Lake Erie landings due to the demand put upon them by the consumer. In the early 1980s walleye catches increased significantly and the trend has continued in the late 1980s and the 1990s. Stocks of Lake Erie walleye recovered after a ban was placed on landing them, due to mercury contamination, in the early 1970s, and after an international management quota system was applied to reduce the fishing effort by 20% by 1990 (Hatch *et al.* 1990). The improvement in lake water quality and walleye abundance has resulted in an increase in recreational fishing (Lichtkoppler 1992). Lake Erie is an example of a lake which has undergone extensive perturbations with the resultant changes in the nature of the fisheries both in species composition and yields. A review of historical changes has revealed the complex nature of interactions between over-exploitation, introduction of exotics, habitat modifications and pollution (Koonce *et al.* 1996). The management of the fisheries depends on an understanding of the temporal and spatial dynamics of the stocks (Chapter 9) and the ecosystem of which they are a part.

Manitoba commercial fishery

Manitoba has been a major producer of walleye in Canada. Large scale commercial fishing started in the 1880s after the construction of the Canadian Pacific Railway and the arrival of Icelandic immigrants on the west shore of Lake Winnipeg (Gislason *et al.* 1982). Up to 1900 fishing was carried out in the summer, mainly with gill nets, although some pound nets and trap nets were also used. With the invention of the jigger (Fig. 10.3) gill nets could be fished under the ice and winter fisheries started in many of the Manitoba lakes. Other important developments included the introduction of petrol driven boats in the 1920s. In some of the large lakes, boats were used with net lifters for the first time in the 1940s and this brought about a reduction in crew numbers. In the 1950s nylon nets replaced cotton ones.

The viability of a commercial fishery is dependent on transport from source to consumer. Initially lake whitefish, goldeye, *Hiodon alosoides* (Rafinesque), and sturgeon were the most important species. The whitefish kept well when salted and stored in barrels. With the development of railways and roads and improvement in preservation techniques, including refrigeration, walleye became more important to the commercial fisheries. Today, walleye caught in northern lakes are flown to places which have good ground transport to Winnipeg where the fish are processed.

In the later part of the nineteenth century restrictions were introduced to protect future stocks. These included closed seasons and area, gear, mesh size, and licence

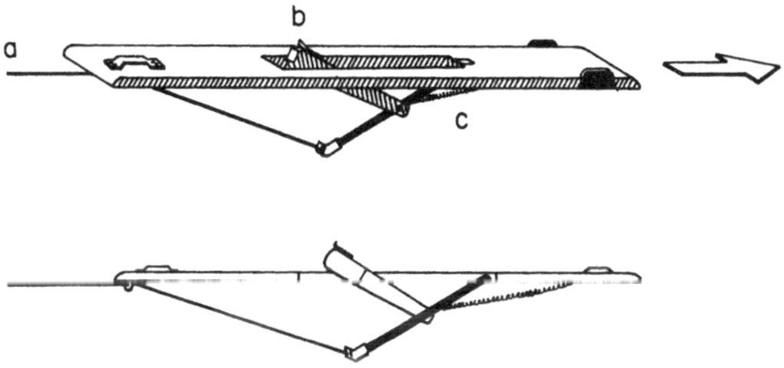

Fig. 10.3 A jigger for setting gill nets under the ice. When rope a is pulled, the metal claw *b* pushes against the ice and moves the jigger in the direction of the arrow. When *a* is released, the spring *c* takes the claw back ready for the next pull. (After Hamley 1980.)

restrictions. Initially, the fisheries were operated by private individuals of three types: owner operators, 'renters' who hired equipment from fish companies, and fishing labourers who were paid wages by a fish company or an owner operator or a renter. This system resulted in a number of complications including duplication of services by existing fish companies, fishermen's lack of bargaining power, lack of a basic price for fish, and poor quality of fish produced. These problems and the need to co-ordinate transportation led to the formation of the Freshwater Fish Marketing Corporation (FFMC) in 1969. FFMC has exclusive jurisdiction over inter-provincial and export trade in freshwater fish for a region extending from northwestern Ontario to Alberta and north to the Northwest Territories.

In 1997 the walleye landed weight was 2096 t, representing 21% of the Manitoba catch, and a landed value of Canadian $7 567 000. In 1983 the landed weight of walleye was 4237 t, representing 27% of the Province's total commercial catch, with a value of Canadian $6 316 000.

Lake Balaton

Lake Balaton in Hungary has been fished by man since 7000 BP (Sági 1974). Records show that the fishery rights were given to religious orders by the monarchy 1000 years ago and since then the lake has been fished extensively by seines and gill nets (Biró 1997). Further records show that the fishery yields declined during the 150 years of the Turkish occupation in the 1500–1600s. In the 1800s flood control measures were implemented and many zander spawning areas were destroyed. The intensity of fishing increased and, by 1880, 20 000 gill nets were in use in Lake Balaton. The regulation of the fisheries was effectively started in 1884 with the formation of the Balaton Fishery Co-operative, which was replaced in 1900 by the Balaton Fishery Corporation. These bodies rented the lake until it was taken over by the Hungarian Government in 1948. The Corporation was responsible for the collection of fishing data, the modernisation of gear, the protection of spawning sites and the introduction of a number of new species. Thirteen have been introduced including the catfish, *Ictalurus nebulosus*, largemouth bass, *Micropterus salmoides* (Lacépède), and the pumpkinseed, *Lepomis gibbosus*, from North America, carp, *Cyprinus carpio*, vendace, *Coregonus albula*, eel, *Anguilla anguilla* and others. Large numbers of eels were stocked during the 1960s. With increased human development, the nutrient loading of the lake increased and by 1930 the lake was becoming eutrophic. In the 1950s, the use of artificial fertilisers increased the eutrophication and biocides also entered the food chains. A large kill of 500 t of fish, including 200 t of zander, in 1965 was linked to biocides. There have been several large fish kills since, usually due to deoxygenation or toxins resulting from algal blooms (Fig. 10.4).

Zander are fished commercially by 1000 m long seines with a cod-end mesh size of 35–40 mm (Biró 1985). The annual yield of zander increased from about 80 t in the early 1900s to about 160 t by the late 1930s (Fig. 10.4). After the state took responsibility for the fishery, there was increased fishing pressure and the yield soon fell to about 120 t. In the early 1970s, the landings of zander represented 6–12% of the total, while 78% of the catch was bream, *Abramis brama*. Between 1970 and 1980 the zander yield fell from 119 to 45 t, but has recovered slowly since (Biró 1997).

Shifts in species composition and abundance have occurred as a result of reduction in wetlands and flooded areas, shoreline regulations, use of agricultural ferti-

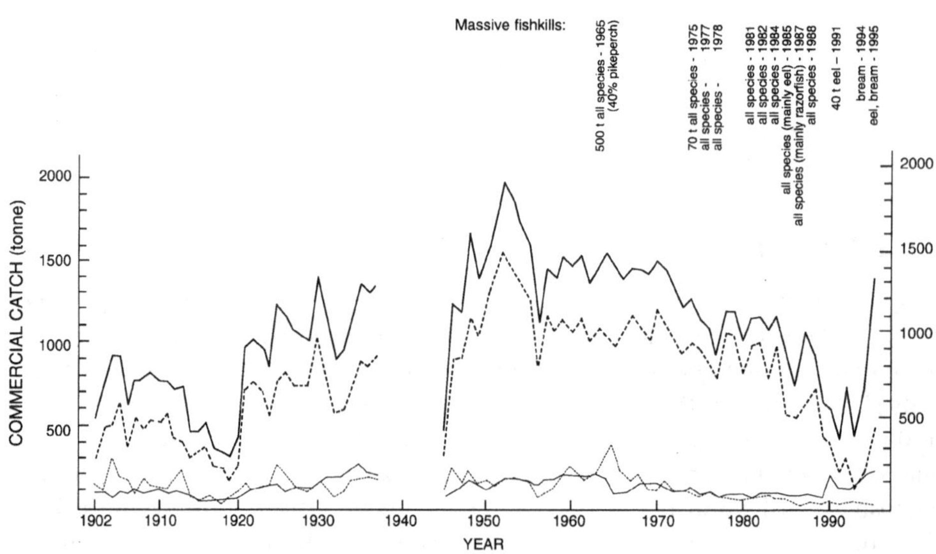

Fig. 10.4 Commercial catches from Lake Balaton. Massive fish kills between 1965–1995 are indicated. Thick solid line = total fish, large dashed line = *Abramis* sp., light dashed line = *Pelecus cultratus* and thin solid line = *S. lucioperca*. (After Biró 1997, with kind permission from Munksgaard International Publishers.)

lisers, urbanisation of the shoreline, sewage input, tourism, commercial and recreational fisheries, and stocking of indigenous and exotic species (Biró 1997). Management of the lake has included habitat reconstruction, rehabilitation of native fish species by stocking, improvement of water quality and removal of sediments by dredging. Biró (1997) indicates, however, the problem of water quality resulting in sudden massive algal blooms and fish kills. The large variations in primary production and the resulting oscillations in fish abundance make predictions about the lake and its fisheries difficult.

Fishing gears

A major factor determining the size, both in numbers and length (and/or condition), of the catch is the nature of the fishing gear. Most commercial freshwater fishing gear is passive and relies on the behaviour of the fish (gill nets, traps and hook and line) as compared to active gear which is moved through the water to catch the fish (trawls and seines). The main commercial fishing gears of economic importance used to catch percids include gill nets, trammel nets, seines, trawls, portable traps, fyke nets and hook and line for perch, gill nets, pound nets and trap nets for walleye and sauger and gill nets, fyke nets, trawls, seines, artificial lures and hook and line for zander (Fig. 10.5).

Mesh size will determine minimum size of the fish caught in both types of gear. In a survey carried out with Lake Winnipeg fishermen (Gislason *et al.* 1982), a number of characteristics about gill nets used to catch walleye were recorded. Green and blue nets caught more fish than white ones and brightly coloured nets such as yellow caught even fewer fish. The effect of colour was reduced, as the water became more turbid. Nylon nets caught more fish than cotton ones; a change from cotton to nylon nets was thought to have been a factor in the decline in zander in northern and central Finland (Colby & Lehtonen 1994). Also, the finer the ply of the net material the greater the catch although the finer nets had a shorter life. The position of the net is important. Nets set inshore caught more walleye in Lake Winnipeg than those set offshore, as did nets set perpendicular to the shore as compared to those set parallel. Craig and Fletcher (1982) found the position of nets in relation to the shore influenced the size of perch catches. More perch were caught in monofilament gill nets set perpendicular to the shore than those set parallel. The amplitude of gill net selectivity increases with mesh size but smaller mesh sizes are less efficient (Fig. 10.6). Selectivity curves for each mesh size are bimodal. The mode to the left represents walleye that are wedged in the mesh while the mode to the right represents fish that are tangled by teeth, maxillaries, preopercula and opercula. The difference in fatness or condition of the fish and thus size of girth determines the minimum size caught by a mesh size. This size will vary with time of year. Using a range of mesh sizes ensures that representatives of all fish within a certain size range are caught in the gear. Monofilament gill nets of stretched mesh sizes 10, 12.5, 16.5, 22 and 28 mm catch perch in the length range 90–210 mm with 72% efficiency

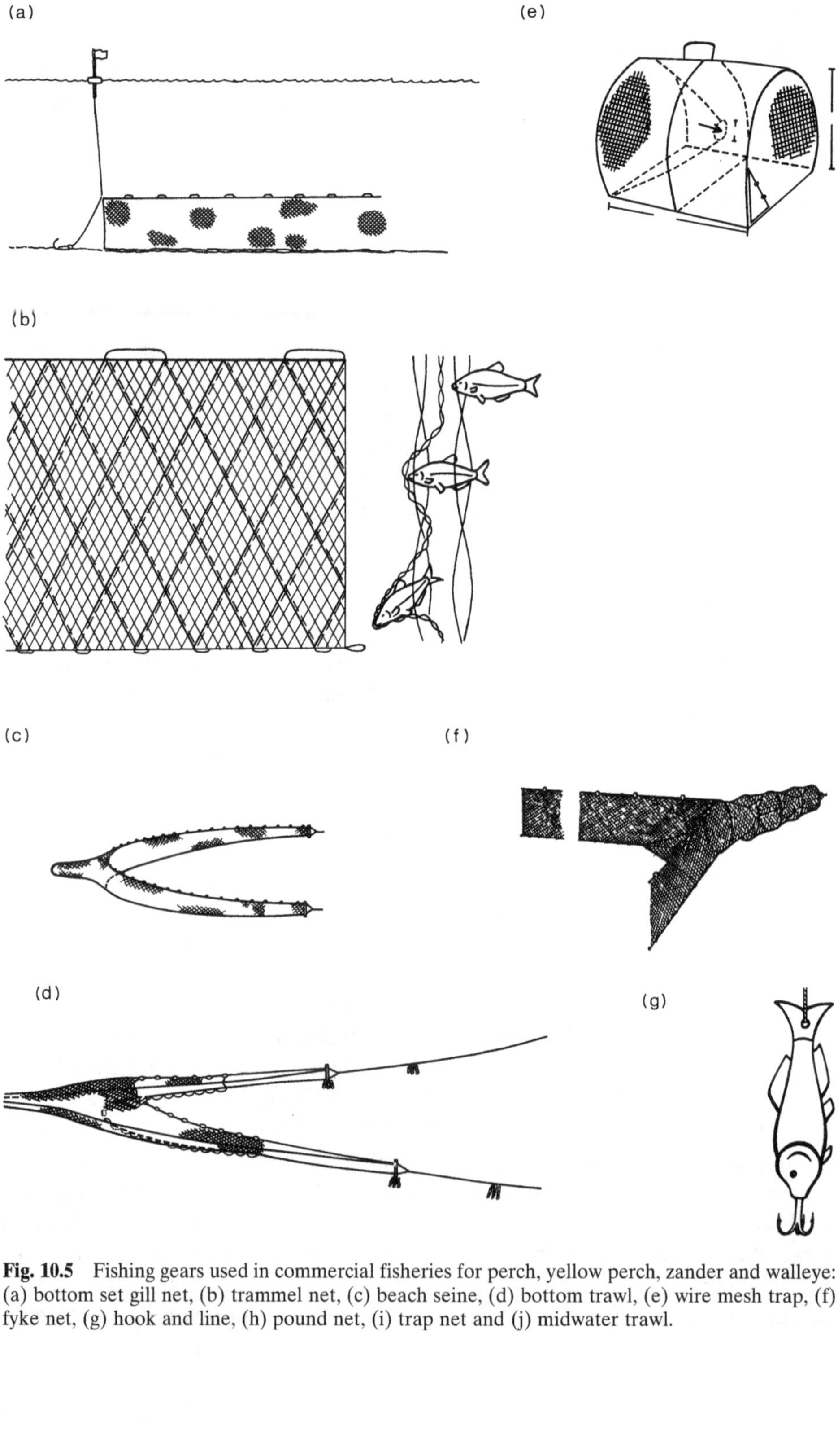

Fig. 10.5 Fishing gears used in commercial fisheries for perch, yellow perch, zander and walleye: (a) bottom set gill net, (b) trammel net, (c) beach seine, (d) bottom trawl, (e) wire mesh trap, (f) fyke net, (g) hook and line, (h) pound net, (i) trap net and (j) midwater trawl.

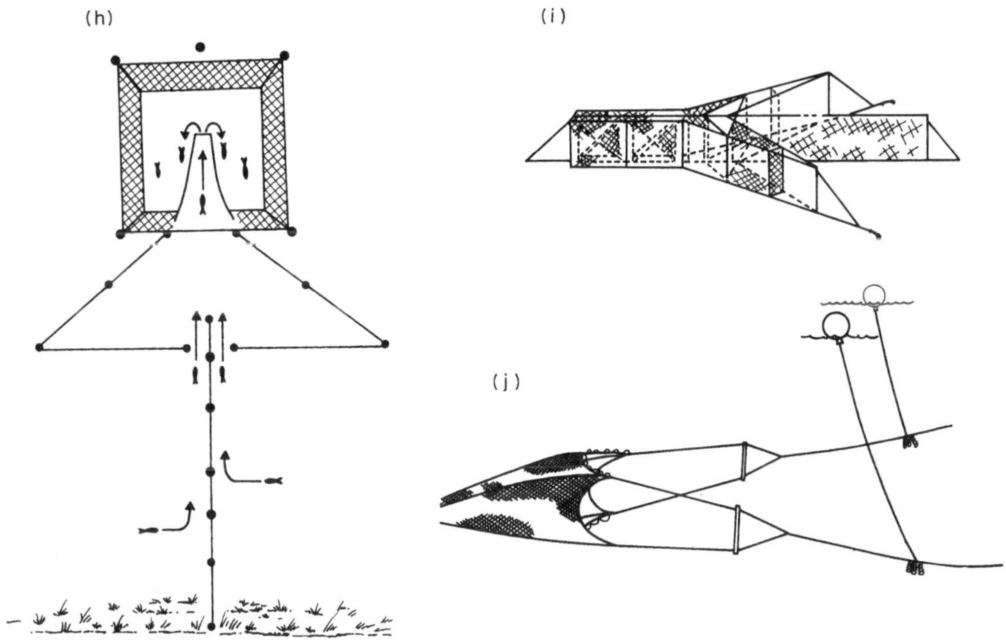

(Linlokken 1984). However, commercial nets are usually designed for high selectivity so that large marketable fish are caught and smaller fish are not netted. Net length and period of time set also affect the catch of fish, the scale of the effects depending on the species (Minns & Hurley 1988).

Factors influencing the number and size of percids caught in traps depend on the location, the length of time the trap is set, the size of the trap entrance, the size of the mesh, the season, the sex and the species (Craig 1980b). The variability in catches between traps is usually very large, as has been demonstrated for Windermere perch traps (Bagenal 1972) and trapnets used for yellow perch in Lake Erie (Hamley & Howley 1985). In the latter case, to obtain a geometric mean with confidence limits within 20–25% would require about 100 lifts of the trapnets.

Anglers tend to remove smaller walleye than commercial fishermen (Serns & Kempinger 1981; Craig & Smiley 1986) and many of these walleye are sexually immature. In Lac des Mille, Canada, during the period 1958–75, anglers removed walleye in the age range 4 to 6 years as compared to commercial fishermen who took 5 to 11 year old fish (Elsey & Thomson 1977).

Percid fisheries are important. Not only do they produce an economic return but also they provide hours of leisure time activity. However, many bodies of water containing percid fishes have and are undergoing significant changes including increased fishing pressure and increasing productivity. To sustain these fisheries for the future will require careful management and an appraisal of past management practices.

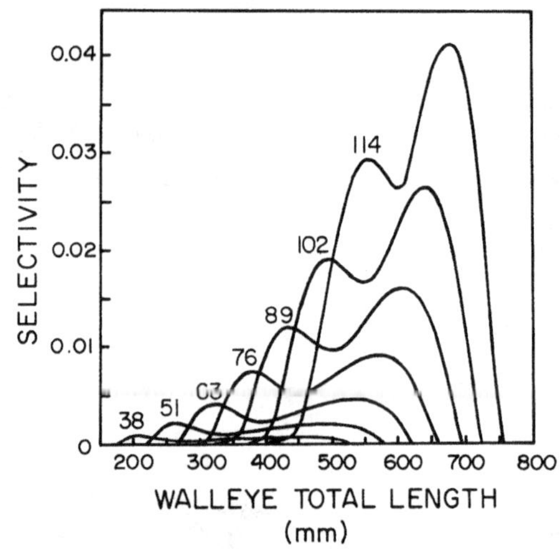

Fig. 10.6 Selectivity curves of 38–114 mm gill nets in relation to walleye fish length. Each curve is the sum of the wedging and tangling curves for that mesh. Selectivity is estimated from $S_{ij} = [R_{ijt}/X_{it}]/T_{jt}$ where S_{ij} is the selectivity of mesh *i* for size class *j*, R_{ijt} is the number of marked fish caught in day *t* by mesh *i*, X_{it} is the effort by mesh *i* in day *t* and T_{jt} is the number of marked fish of size class *j* alive at the start of day *t*. (After Hamley & Regier 1973.)

10.3 Management

The perch and yellow perch adapt very easily to new habitats within their range. In many areas attempts are made to suppress these species, especially when they are in supposed competition with salmonids. Areas that are affected include eastern Canada and British Columbia, western and north-eastern states of the USA, Scotland and Ireland. In contrast, in other more inland areas such as central Canada and the mid-western and central USA, and central Europe, perch species are held in high regard and are important in the fisheries. Some attempts, in most cases unsuccessful, have been made to manage perch and yellow perch fisheries by quota, minimum size and minimum mesh size. These policies are also applied to commercial fisheries for walleye. In the USA, walleye are extensively fished for sport and control measures include closed seasons, bag limits and minimum sizes.

The management of fish stocks has a long history. It is also very political and the degree to which the recommendations of biologists are taken into consideration depends on the local jurisdiction. Management has been improved by co-operation between riparian owners, (individuals, provinces, states and countries) that share the same stock (Geredeaux 1990; Francis *et al.* 1996). The managers of the resource have to consider the users, who in many cases are the voters who put the law makers in office. The manager of a fishery has six basic options to control the fishery. These are gear control, season control, limited entry, pricing the right to fish and non-transferable and transferable quotas. Control of fishing gear, for example, stipulating the mesh size to be used, is intended to permit sufficient escapement of young fish so that they may recruit later to the adult population and sustain the stocks. Gear controls are also used to limit fishing effort, for example, the number of nets set per day. Closed seasons, closed areas and shortened seasons aim to protect fish during critical periods in their life cycle such as spawning and also to control fishing effort. However, these measures only reduce the time in which the fish may be harvested and usually increase the fishing effort during the 'open' season. This is uneconomic as fishermen invest in more equipment to intensify fishing effort during a shorter period of time. The net effect is to increase harvesting, storage, processing and marketing costs. Limiting entry to a fishery may be a viable management option. Reduction in the number of boats usually leads to sophistication of the gear and vessels in the residual fleet exploiting the fishery. This system of management may be used in conjunction with a rent. In this way the beneficiaries of the resource pay for the benefits and these revenues can be used in the costs of management. Experience has shown that the use of quotas may be the best method of controlling the level of harvest and also in controlling investments commensurate with the value of the individual quota production (Johnson *et al.* 1992a). The major weakness of a non-transferable quota is its inflexibility in permitting the fishery to adjust to changing markets, costs and technology. The industry cannot consolidate quotas and rationalise the size of the fleet. These problems are overcome by an individual transferable quota system.

The removal of predators often results in an increase in smaller forage fish. This has already been illustrated in the Lake Erie fishery where the removal of blue pike, sauger and walleye led to an increase in yellow perch, smelt and white bass. Alternatively, an increase in predatory fish causes a decrease in overall yields as has been demonstrated in a number of Polish lakes (Bonar 1977) and in Lake Vortsjarv, former USSR (Pihu & Maeemets 1982). In Lake Vortsjarv the total yield in the early 1970s was about 300 t y^{-1} of which 80–90% was ruffe or small perch (Pihu & Maeemets 1982). The yield in 1979 declined to 184 t but the proportion of zander increased and with eels they represented 50–60% of the catch. The value of the total catch had risen by a factor of two to three. Careful management, including the introduction of elvers, thereby increasing the predatory pressure on forage fish, brought about the changes. The use of a fine mesh trawl, which caused high mortality of young stages of zander, was banned. Also, a closed season during spawning, a ban on fishing at spawning sites throughout the year, and size limits were introduced. Stocking with walleye and setting a minimum size limit of 15 inches (381 mm) tripled the predation rate of walleye on young perch in Lake Mendota which was sufficient to reduce recruitment of yellow perch (Johnson *et al.* 1992b). This had important implications for control of algae (Chapter 5). These are examples of effective management.

A limited walleye gill net fishery was established in the 1950s in the New York state waters of Lake Erie (Wolfert 1981). During the period 1950–78 four very large year classes were produced, 1959, 1962, 1964 and 1971. When a strong year class was produced, fishing intensity increased and small mesh, 70 or 76 mm stretched mesh, nets were used; 46% of the landings contained one year old walleye and many fish were removed before they could spawn. If the small gill nets had not been used but fishing had been restricted to large mesh nets, 121 or 127 mm, the fishery yields would have increased by 35% and many walleye would have reached the age of sexual maturity. In this example the fishery probably could have been better managed from a biological perspective.

An experiment was performed on the sport fishery of Escanaba Lake, USA (Churchill 1957). No restrictions were placed on hook and line fishing for any species and there was no evidence of depletion of fish stocks after ten years. In the period about 97 000 fish were angled of which about 80% were yellow perch or walleye. The fishing pressure was low during the spawning period, due to the remote location of the lake, so the effect of no closure at this time could not be evaluated. It was calculated that fishing regulations would have brought about a 50% reduction in the catch and minimum size limits would have more effect than bag or season limits.

For effective application of size limits the manager requires information on recruitment, growth, mortality and abundance of the stock (Brousseau & Armstrong 1987; Johnson *et al.* 1992a). Restrictions on size limits are intended to maximise the yield, increase the catch of larger fish and protect potential spawners. However, in many cases an increase in size limits has resulted in a decrease in the yield of large fish, and an increase in the number of young and smaller sized fish (Serns 1978). Taylor (1981) used a simple simulated model to determine the effect of size limits on the sport fishery for walleye in Lake McConaughy, USA. His model used the basic equation:

$$N_{t+1} = N_t - N_t(u + v) + f(g) + r \qquad \text{Equation 10.1}$$

where N_t and N_{t+1} are the numerical sizes of the population at the start and end of the time period (usually one year), u and v are the expectations of death from fishing and natural causes respectively, $f(g)$ is a growth function and r is the number of recruits during the time period t to $t+1$. Equation 10.1 was applied to each year class since values of u, v, g and r change with age. The long-term results of size limits on the walleye population illustrated that an increase in size limits resulted in an increase in the population but a decrease in the angler harvest (Table 10.3).

The object of stocking is to rear the fish to overcome an unfavourable period (high mortality) that they would naturally experience and time their introduction into a water body when conditions are at an optimum. Thus, a strong year class could be produced when a naturally produced year class might be weak. The chances of success must be weighed against the chance of selecting the optimum time to introduce the fish (later conditions may make the plantings a failure anyway), and the cost of rearing. Stocking is a common management practice and can be of three types: (1) introductory, in other words, the species stocked is new to the

Table 10.3 Estimated response of the Lake McConaughy, USA, walleye population to various length limits. (After Taylor 1981.)

Length limit (mm)	Population No. ha^{-1}	Population kg ha^{-1}	Angler harvest No. ha^{-1}	Angler harvest kg ha^{-1}	kg $(fish)^{-1}$	No. ha^{-1} caught and released
None	31.7	9.2	6.7	3.8	0.57	1.3
300	31.7	9.2	6.7	3.8	0.57	1.3
325	31.8	9.3	6.6	3.8	0.58	1.4
350	33.4	9.9	6.2	3.8	0.63	2.3
375	37.0	11.7	4.8	3.6	0.79	4.7
400	39.4	12.9	3.6	3.3	0.92	6.7
425	40.1	13.2	3.3	3.1	0.96	7.1
450	45.2	15.2	2.2	2.6	1.21	9.2

habitat; (2) maintenance stocking where the fish present do not reproduce naturally or if they do, their success is very limited; (3) supplementary stocking to augment natural reproduction of existing fish. In a study of walleye stocking in 125 bodies of water in the last 100 years, Laarman (1978) found that stocking success was very low. In type one there was 48% success, 32% in type two and only 5% in type three. Type one situations included stocking of newly formed reservoirs and in areas extending outside the species' range, providing conditions were still within the physiological range of the species. Type two included lakes which underwent occasional winterkills or where there was little spawning habitat. These stockings were often the basis for 'put and take' sport fisheries. Percids are normally fecund enough to produce large numbers of eggs at low population levels. Thus, if conditions are optimum for a good year class the planting of further larvae or fingerlings (type three) will make very little contribution to the total year class strength. However, it can affect the size of the fish. An examination of walleye stocking in Minnesota showed that stocking of type 2 in lakes decreased mean weight of the walleye (Li *et al.* 1996). In type 3, fry stocking decreased the mean weight of the fish although fingerling stocking had no effect. Li *et al.* (1996) found that type 3 stocking did increase the abundance of fish of the year but decreased the abundance of year classes one year older and one year younger.

Oneida Lake, USA, has a natural population of walleye. Stocking with larvae influenced the initial abundance of the year class but this influence did not persist to age one year and thus stocking of walleye larvae did not necessarily mean an increase in year class strengths in subsequent years (Forney 1976). The success of stocking larger walleye was measured in Michigan lakes by using fin clips as marks (Laarman 1981). Walleye were reared to a length of 110–170 mm, the average length of walleye in the lakes at the end of the first year and stocked at a rate of 25 ha^{-1}. The total (natural plus fishing) annual mortality from one to two years was 74% and from two to eight years was 56% for both native and stocked fish. The exploitation rate by anglers ranged from 3–35% with a mean of 17% but the anglers harvested only 3.5% of the stocked fish. The cost–benefit of stocking walleye was calculated in Dauphin Lake, Manitoba, which supports a winter commercial fishery. Walleye fry which were genetically distinct from the indigenous stock were introduced in the spring (Mathias *et al.* 1992). The walleye are fully recruited into the fishery at age three years. The survival of walleye from stocking as fry to capture in the fishery three plus years later was 0.04% and the introduced fish contributed *c.* 2.9% to the (1985) year class. The cost–benefit to the commercial fishery was 2.6:1. If the walleye had been fished for sport the ratio would have been improved to 1.2:1. Such a high cost as the one given in the example would not be practical for commercial fishing but may be for a sport fishery where much of the income is secondary in nature. Thus, not only is income derived from selling licences but is also derived from the sale of fishing equipment and hotel and travel costs.

At present, little is known about the benefits of stocking with larvae as compared to fingerlings. In an analysis of studies in which walleye stocking success was evaluated, 32% of fry stocking, 32% of small fingerling and 50% of advanced fingerling stocking were considered successful (Ellison & Franzin 1992). Indications are that unless conditions are suitable the exercise will be useless but this has not deterred many authorities from spending vast quantities of money in rearing fish for stocking. The success of walleye stocking to produce a self-sustaining population appears to depend on environmental and biological conditions of the water bodies rather than the numbers or sizes of walleye introduced. The most important of these are probably morphometry, water quality (Bennett & McArthur 1990), water temperature at spawning time, amount of potential spawning area and the biomass of potential predators feeding on the walleye. These examples provided for walleye are equally applicable to perch, yellow perch and zander, although conditions for success may vary. Methods of rearing percids have improved significantly in recent years (see Chapter 11).

10.4 Pollutants

Many bodies of water have elevated levels of substances harmful to fish and often to humans if fish are eaten. These increases are usually due to human activities. The toxicants can be heavy metals such as mercury, cadmium, lead, zinc, arsenic and copper, organic contaminants such as chlorinated insecticides and effluent from paper and pulp mills and coal mines. Many other substances produced in industrial processes can be liberated into freshwaters inhabited by percids such as chlorine in cooling water from power plants and accidental spills of radioactive materials. On a positive note, controls or ban on many of these substances has brought about a reduction in the tissue concentration in perch, yellow perch, walleye and zander (Frank *et al.* 1978; Armstrong & Scott 1979; Pfeifer *et al.* 1979) and often an improvement to the stocks (Weinbauer *et al.* 1980). The concentrations of organic contaminants in Lake Erie walleye declined between 1977 and 1982 but then remained fairly constant to 1990 (De Vault *et al.* 1996). The lack of a further decline may have been due to multiple contaminant pools or changes in the food web bringing about increased bioaccumulation.

Contamination may be brought about indirectly. After the creation of a reservoir on the Churchill River, Canada, mercury concentrations in walleye muscle rose from 0.2–0.3 $\mu g\,g^{-1}$ prior to flooding in 1978 to 0.5–1.0 $\mu g\,g^{-1}$ in the period 1978 to 1982 (Bodaly *et al.* 1984). Most of the mercury in the walleye was in the organic (methylated) form. There is no industrial development in this area and mercury was derived from the flooded terrestrial area by bacterial methylation of mercury occurring naturally in the soil. Bodaly *et al.* (1984) related the concentration of mercury to the area of land flooded. Johnston *et al.* (1991) developed a series of linear models relating mercury burden in walleye and other species to the physical characteristics of lakes and reservoirs in northern Manitoba. Ratios of flooded terrestrial areas to water volume of the reservoir and of inflowing waters were able to explain 76–84% of the variation in the mean mercury burden of the fish.

Bioenergetics models have been used to follow the pathway of heavy metal uptake (Harris & Snodgrass 1993; Korhonen *et al.* 1995; Post *et al.* 1996). Accumulation of heavy metals in the tissues is often age related (Berninger & Pennanen 1995). Heavy metals affect oxygen uptake, osmoregulation, carbohydrate metabolism and haematology. Fish exposed to heavy metals suffer from various pathological alterations in gill tissues. In the long term reproduction can be affected (Ellenberger *et al.* 1994; Friedman *et al.* 1996). Heavy metals in mine tailings can impact on walleye eggs and larvae (Leis & Fox 1994, 1996). Life-long exposure to contaminants such as polycyclic aromatic hydrocarbons, polychlorinated biphenyls (PCBs) and mercury can lead to exhaustion of the cortisol-producing endocrine system (Hontela *et al.* 1992, 1995).

Some metals exist in several forms depending on pH and the toxicity of different forms may vary substantially. Often fish are more susceptible to the metals in acidic waters. Mercury whole body burden is negatively correlated with pH (Cope *et al.* 1990; Suns & Hitchin 1990; Haines *et al.* 1992) and concentrations of dissolved organic carbon (Driscoll *et al.* 1995). In addition, synergism between acid and metal may occur. Mercury concentrations in fish can be reduced by liming acidic water bodies (Andersson & Haakanson 1991). Perch living in the effluent from a sulphide ore smelting plant accumulated metal residues in liver and muscle (Larsson *et al.* 1984). They showed signs of anaemia, hypocalcaemia, increase in muscle water content and reduced liver size, all of which were of transient nature, and disturbed chloride balance and hyperglycaemia, which were more persistent. Of interest was their reduced resistance to withstand stress compared to controls. Stress may increase the toxic nature of heavy metals.

Cadmium concentrates in the gills, kidney and liver of percids (Edgren & Notter 1980) as compared to mercury which is mainly deposited in the muscle (von Hegi & Geiger 1979). In the contaminated River Emaan, Sweden, perch exposed to cadmium had 40–100% more lymphocytes than controls, an indication of a stimulated immune system response. Contaminated fish showed signs of slight anaemia, changes in blood ion concentrations and disturbed carbohydrate metabolism. The physiological disturbance was not so great in perch from the river compared to laboratory fish. This indicates that the perch were able to develop resistance to cadmium in the natural situation or the cadmium was less available in the river. In other words it was bound or complexed (Sjoebeck *et al.* 1984) or the laboratory fish were more stressed than the river fish. At sublethal levels cadmium can affect growth rate in natural populations of perch (Kearns & Atchison 1979). Cadmium can be reduced by liming (Andersson & Holm 1995). Copper can also reduce growth rate in perch and at concentrations above 22 $\mu g\, l^{-1}$ of copper, food conversion efficiency is reduced (Collvin 1985). This is attributed to increased standard metabolism required for the detoxification of the copper. Unfertilised perch eggs also increase their respiratory rate when exposed to copper but the copper in this case dissipates an oxygen permeability barrier located at the chorion (Akberali & Earnshaw 1984). Perch are fairly insensitive to zinc (Ball 1967) and show little change in growth when exposed naturally to sublethal levels (Kearns & Atchison 1979).

Pulp and paper mill effluent adversely affects perch and walleye. Abnormalities have been observed in fish exposed to the effluent, such as deformities to the opercula and fin erosion in perch (Lindesjoeoe & Thulin 1990; Lindesjoeoe *et al.* 1994) and to blood cortisol levels and morphology of the pituitary–interrenal axis in yellow perch (Hontela *et al.* 1997). Untreated effluents can cause significant metabolic and physiological responses (Foerlin *et al.* 1995; Huuskonen & Lindstroem-Seppae 1995). They can also reduce recruitment to the population by affecting the reproductive systems of the parental stock and/or cause heavy mortality at the embryo or early larval stage (Karås *et al.* 1991). Effluents can affect fish 20–40 km from the discharge point (Balk *et al.* 1993). Perch held in kraft pulp mill waste water exhibited an inability to compensate in a rotary flow test when exposed to levels of 2 or above V/V concentrations (Lehtinen & Oikari 1980). Symptoms noted were: in the liver, hepatocytes with pycnotic nuclei, particularly near the central veins, increased cytoplasmic granularity and occasional swollen secondary lamellae, and parasitic cysts were noted between the secondary lamellae. These abnormalities were not observed in the controls.

Experiments with conifer groundwood fibre on walleye fingerlings showed that fish were shorter in length and lighter compared to controls when exposed to 50 $\mu g\, l^{-1}$ or higher concentrations of fibre (Smith *et al.* 1966). As for many other pollutants the standard metabolic rate went up, at 150 $\mu g\, l^{-1}$ of effluent it increased by 18% and the active metabolism went down 15%, compared to controls. The number of mucous cells also went down and at 150 $\mu g\, l^{-1}$ were 38% less than controls. Percids may be more vulnerable at certain life stages to pollutants. The eggs of perch are less affected by the insecticide mexacarbate than juvenile or adult fish (Mauck *et al.* 1977). The eggs of yellow perch are the most resistant to cyanide ions whilst the juveniles are the most sensitive especially at low temperatures and low oxygen concentrations (Smith *et al.* 1978). This may be the result of increased demands on the respiratory system to deal with the toxicant. Cyanide ions interfere with oxygen utilisation and the lowered oxygen available in the water may be an additional stress.

Perch exposed to chlorinated insecticides accumulate them in the body tissues. The main storage areas in the body for PCBs which are soluble in lipids are the viscera. The skin, scales and skeletal muscle are only minor sites (Guiney & Peterson 1980). Smith *et al.* (1992) found that contaminants of this type did not increase the susceptibility of spawning walleye to skin diseases such as lymphocystis and dermal sarcoma (Chapter 8). Many organic toxicants are concentrated in lipid. When lipid stores are saturated, toxicants may be released and cause stress. Usually the concentration of the chemical increases with size of fish (Mathews & Dolan 1983) and perch found in heated industrial discharges often have higher concentrations than those found in cooler areas (Edgren *et al.* 1981). This may indicate the effect of bioaccumulation in the food chain and faster food uptake in warmer water. Perch and yellow perch are able to break down several insecticides in the liver. These include carbaryl (Chin *et al.* 1979) and malathion (Gantverg & Rozengart 1984).

Radioactive caesium from the Chernobyl accident (1986) has been monitored in lakes containing perch (Saerkkae *et al.* 1995, 1996; Koulikov 1996; Zalewski *et al.* 1996). Perch, as piscivores, normally had higher levels of contamination than other fish species further down the food chain. Concentrations in sediment and fish tissue tended to be higher in oligotrophic lakes compared to eutrophic lakes which may be due to greater dilution of the caesium in the higher biomass of the latter systems. Perch in humic lakes also tended to have lower levels of the contaminant which may be due to absorption of the caesium by humus particles.

10.5 Conclusions

As this chapter has illustrated, humans have influenced and will continue to influence the abundance and quality of percid fishes. Many fish communities and stocks are depressed, collapsed, or extinct as a result of severe stresses brought about directly or indirectly by humans (Loftus 1976). Discrete stocks of percids in the same body of water have evolved over thousands of years and some of these stocks are now being lost. Loftus (1976) believes the main reasons for this are: (1) excessive exploitation; (2) incidental catches, by-catches by gears directed at adjacent stocks; (3) genetic drift imposed by selective fishing gear, often removing early maturing and/or fast growing individuals from the stock; (4) hatchery plantings may survive to mature and mix with the native stock thus diluting the gene pool of the well adapted stock (often plantings are not successful because they are not well adapted to the new environment); (5) environmental change may force the breakdown of segregation between formerly discrete stocks.

Rehabilitation of percid stocks can be achieved but further research is required in many areas including hatching, rearing, selective breeding, exploitation control and the precise determination of water productivity and quality (see Chapter 11). Only careful management, conservation and control of the fishes' environment may halt the decline in percid stocks. Guidelines should be developed for advising commercial and recreational fishermen on practical steps to be taken to conserve stocks.

The introduction of exotic fish either intentional or by accident needs to be carefully controlled. The outcome of introductions such as the ruffe into the Laurentian Great Lakes (Cangelosi 1997; see Chapter 13) cannot be predicted. Genetic markers are a useful tool in determining the impact of introductions (Skaala *et al.* 1990).

Temperature is a major controlling factor of percid populations and special care should be taken with alterations that will affect the temperature regime of their habitat. Attempts have been made to model the distribution, thermal habitat, and production of percids as a result of climate warming (Magnuson *et al.* 1990; Minns & Moore 1992). Further research is required.

Chapter 11
Aquaculture

Patrick Kestemont[1] & Charles Mélard[2]
[1]Unité de Recherches en Biologie des Organismes, Facultés Universitaires N.D. de la Paix, 61, rue de Bruxelles, B-5000 Namur, Belgium.
[2]Centre de Formation et de Recherches en Aquaculture, 10, Chemin de la Justice, B-4500 Tihange, Belgium.

11.1 Introduction

Among the Percidae family, four main species have been identified as candidates for aquaculture: two North American, the yellow perch and the walleye, and two Eurasian, the perch and the zander. There has also been interest in the culture of the sauger and the saugeye (hybrid between *S. canadense* male and *S. vitreum* female) though relatively less information is available on these fishes.

Historically, the yellow perch has been a highly valued food fish in the Great Lakes region of North America. The market demand for yellow perch has consistently been high, tied to the long-standing regional Friday night fish fries (Lesser & Vilstrup 1979). However, commercial harvest of the yellow perch from the Great Lakes and other parts of Canada have failed to keep pace with market demand (Chapter 10). Significant consumer demand and the associated high retail value have generated interest in the development of aquaculture techniques for yellow perch (Garling 1991). Apart from its economic value as a food fish, the species has many biological characteristics that make it an excellent candidate for commercial aquaculture. These include its acceptance of formulated feeds, its unaggressive and non-cannibalistic behaviour and its tolerance to crowding and other conditions associated with intensive culture (Calbert 1975; Heidinger & Kayes 1986). The walleye is also considered to be very valuable in northern USA and Canada, both for sport fishing and as a food fish. For many years there have been major efforts by provincial, state and federal US and Canadian agencies to supplement natural walleye populations (Chapter 10) and expand their range with fingerling stocking programmes (Conover 1986). More recently, the high demand of walleye as a food fish has generated development of efficient technology for food fish production. Like yellow perch, walleye exhibit many of the biological features required for aquaculture (Summerfelt 1996).

Compared to the research undertaken to overcome the technical constraints which limit the development of North American percid aquaculture, perch and zander have received little attention. For centuries perch and zander have been propagated by extensive culture where they have represented *c.* 2–5% of production in polyculture cyprinid ponds. Tamazouzt *et al.* (1993) documented the production and market of perch in Europe. Their overview indicated that the interest for perch varied according to the country or region.

Capture of perch in Scandinavia and alpine lakes has declined in recent years and consequently, the consumer demand has progressively exceeded the supply from the fisheries. High retail prices and the search for new candidates for freshwater aquaculture in Europe have increased interest in the production of this species and considerable improvements in intensive culture technology have recently been achieved. Compared to perch, the interest in zander production has had a much longer history. Schäperclauss (1961) suggested three major reasons for the cultivation of zander as a secondary fish in common carp, *Cyprinus carpio* pond culture:

(1) to eliminate undesirable prey fish
(2) to produce additional fish besides carp for the table and
(3) to produce fingerlings for restocking natural waters especially for recreational fisheries.

The zander is an attractive game fish, a key species in the biomanipulation of certain types of eutrophic lake (Chapter 5) and a fish valued by the consumer (Hilge & Steffens 1996). Unfortunately, although the extensive to semi-intensive propagation of zander has led to considerable improvements in the production of juveniles in ponds, intensive culture still requires additional research.

This chapter describes the major features of percid aquaculture, with a special emphasis on recent improvements. These are based on research efforts which have helped solve technological constraints, so that culture of all life stages of percids has progressively become more effective, leading to greater profitability.

11.2 Reproductive biology and spawning

Reproductive cycle and out-of-season maturation

As stated by Calbert (1977) in an introduction to a conference dealing with perch fingerling production for aquaculture, the development of any sound system of aquaculture depends on many factors. One of the most important of these is a dependable source of young organisms that can be grown and marketed at the proper time. Primary steps in the production of percid fingerlings are the development of reliable propagation techniques, either natural brood stock maturation and spawning during the normal breeding season or manipulated rearing which induces out of season spawning. The development of out of season spawning can improve the efficiency of fry culture systems, e.g. by allowing double or triple cropping, and provide more reliable and increased availability of fingerlings (Malison *et al.* 1998a).

The reproductive cycle of percid fish is characteristic of annual spawners with group-synchronous development of the ovary, although some differences exist between *Perca* and *Stizostedion* species (Chapter 4). In perch and yellow perch, ovarian development and oocyte growth are rather similar and can be divided into two successive phases: a rapid increase in oocyte diameter from late July to November–December, due to an accumulation of yolk vesicles, and a slower oocyte growth from winter to early spring, corresponding to the progress of exogenous vitellogenesis under steroid (oestradiol-17β, E_2) stimulation. The increase of temperature in March usually induces the completion of vitellogenesis resulting in a sharp increase of the gonadosomatic index (GSI = ovary mass/body mass) before spawning (Fig. 4.5) (Turner 1919; Breder & Rosen 1966; Hokanson 1977b; Kayes 1977; Treasurer 1981; Treasurer & Holliday 1981; Heidinger & Kayes 1986; Dabrowski *et al.* 1996; Sulistyo *et al.* 1998). According to latitude, the spawning period extends from March to late June at a temperature ranging from 7–20°C (usually 10–13°C).

In walleye and zander ovarian development and spawning season differ markedly. Walleye have a relatively short post-spawning quiescent period. By mid summer yolk vesicles are observed in most of the oocytes and the GSI rises rapidly during the autumn, coinciding with a corresponding increase in mean oocyte diameter and the highest serum E_2 levels. Vitellogenesis of walleye is nearly complete by mid January but spawning usually occurs from early spring in the southern extremes of the walleye's range (*c.* 35°N) to July in the northern extremes (*c.* 55°N), when water temperatures warm to 4–7°C (Hokanson 1977b; Becker 1983; Malison *et al.* 1994a; Malison & Held 1996a, b). Zander require much higher water temperatures for final maturation and spawning. Spawning occurs generally from April to mid summer, according to latitude, at a temperature of 10–14°C in the south of France but at only 8–10°C in northern Europe (62–64°N) (Salminen *et al.* 1993; Ruuhijärvi & Hyvärinen 1996; Schlumberger & Proteau 1996; Steffens *et al.* 1996). Gonadal recrudescence starts much later in zander than in walleye as does vitellogenesis.

Control of the reproductive cycle and out-of-season spawning are currently under investigation in percid fishes, with variable success according to species. In temperate fish species temperature and photoperiod are the main environmental cues involved in the regulation of the reproductive process (Chapter 4). Most studies have investigated the involvement of these two proximate factors in the regulation of percid reproduction cycles. Environmental control of gametogenesis and spawning of yellow perch for aquaculture purposes was studied two decades ago in Wisconsin, USA, by Kayes and Calbert (1979) while analyses of perch, walleye and zander have been conducted more recently.

In perch and yellow perch, photo-thermal conditions during the coolest seasons (autumn and winter) are critical for gonadal maturation and spawning. Manipulation of temperature in spring seems to have a limited effect on altering spawning. Kayes & Calbert (1979) demonstrated that perch captured from a lake in February at a temperature <5°C and warmed in the laboratory to 12°C, spawned within the same four day period as the fish in the lake, regardless of whether the fish in the laboratory were under a light:dark (LD) 13.5:10.5 or 10.5:13.5 photoperiod. Fish also spawned under photoperiods of LD 18:6 or 6:18 when warmed from 5–12°C. It was concluded that the onset of spawning in yellow perch depends more on the intrinsic maturation state of the gonads than on specific spring-time photoperiod–temperature cues, these factors having a modulatory influence on spawning. In contrast to the post-vitellogenic oocytes which seem to be resistant to photo-thermal manipulations, early stages of ovarian development are largely dependent on temperature and photoperiod. In female yellow perch, a period of low temperature (<10°C) in winter, during the principal ovarian growth phase, was found to be crucial for reproductive success (Hokanson 1977b; Heidinger & Kayes 1986; Dabrowski *et al.* 1994a) but the length of this cool period is not definitely known. According to Hokanson (1977b) temperatures no high than 6°C are required for 100% spawning participation and the cold duration needs to be at least 185 days. Recent results demonstrated that the compression of the photo-thermal cycle accelerated spawning in female yellow perch but caused disturbances in patterns and levels of steroids, diminished ovarian and hepatic growth, and reduced the quality of eggs (Dabrowski *et al.* 1996; Ciereszko *et al.* 1997). Similar responses were also observed in perch by Dlugosz (1983). Exposure of perch females to thermal effluents resulted in an asynchronous growth of oocytes and acceleration of spawning by 2–3 months. Under indoor rearing conditions, Sulistyo (1998) demonstrated that a shift of the photo-thermal cycle can be successfully applied to perch females and spawning can be obtained three months after the normal breeding season. If the cooling period is extended to about five months, plasma steroid levels and ova size of perch are similar to those females reared under an ambient photo-thermal regime. If the photo-thermal regime is compressed to three months the steroid levels and quality of ova are affected. Out-of-season spawning is obtained but the eggs are not fertilised by the males. All these results suggest shifting the entire reproductive cycle, instead of condensing it or shortening some chosen phases, seems to be a more promising tool to induce out-of-season spawning of perch and yellow perch. Both perch and yellow perch males subjected to delayed photoperiod and or temperature did not differ with regard to most reproductive characteristics from those males exposed to ambient photo-thermal cycles (Dabrowski *et al.* 1996; Ciereszko *et al.* in press; Sulistyo 1998). Thus, the male reproductive system responds to altered photo-thermal regimes in a more predictable and satisfying way than that of the female.

Out-of-season spawning of walleye (up to 10 weeks prior to the normal reproductive season) has been achieved successfully by Malison & Held (1996a) and

Malison *et al.* (1998a) by applying simple environmental and hormonal treatments. Wild walleye, captured in autumn and held in earthen ponds, were transferred to indoor tanks in late January, February and March (approximately ten, six and three weeks prior to natural spawning). Fish were submitted to an early increase in temperature (from 2 to 10°C) and photoperiod (from LD 8:16 to LD 12:12) and injected with different hormones, human chorionic gonadotropin (hCG), des-Gly^{10} (D Ala^{6}) LHRH ethylamide (LHRH-a) or 17α,20β-dihydroxy-4-pregnen-3-one (17,20-P). In January hCG was the most effective treatment at inducing out-of-season ovulation. In February and March all hormonal treatments induced ovulation while no control fish (injected with saline solution) ovulated at any time. In females treated with hCG and LHRH-a, the quality of eggs was comparable to those from walleye spawned during the normal reproductive season while eggs collected from 17,20-P-treated fish were small and had a very low survival. Due to out-of-season spawning of walleye, fingerling production can be initiated as early as January, allowing hatcheries to produce larger first year walleye fingerlings for stocking than would otherwise be possible (Malison *et al.* 1998a). So far, no control of zander reproductive cycle has been reported in the literature.

Induction of ovulation and spawning

Spontaneous ovulation and spawning of percid fishes in captivity, with or without spawning substrates, vary according to species and culture conditions. Capture of wild brood stock a few weeks or days before ovulation is currently practised and is extensively used by most hatcheries, although the use of captive broodstock can be successful. Synchronisation of ovulation and spawning can be induced by hormonal treatment in all percid species used in aquaculture.

Perch and yellow perch spawn naturally in tanks when both sexes are present without a spawning substrate (e.g. branches) and fertilisation rate is usually high, averaging 65–70% but reaching 90–100%. However, according to Kayes (1977) yellow perch sometimes spawn without any fertilisation of the eggs and a female can produce eggs without a male present. A detailed description of the natural release of eggs is described in Chapter 4.

The quality of perch eggs is not constant throughout the spawning season but usually declines during the second half of the season. Decrease of egg and larval quality are characterised by a reduction in fertilisation rate, a change in the aspect and firmness of the egg strands and premature hatching associated with mortality of newly hatched larvae (Kestemont *et al.* 1996). Several hypotheses have been made to explain the progressive decline in the fertilisation, embryonic and early larval development success. These include changes in the biochemical composition of the ova and semen quality from early to late spawning. Recent studies by Fiogbé (1996) and Abi-Ayad (1998) indicate that the morphology (size and mass) and biochemical composition (mainly essential amino acids and fatty acids) of perch eggs and larvae remain constant throughout the spawning season. Moreover, preliminary analysis of semen quality (concentration and motility of spermatozoa) and results of ova fertilisation with spermatozoa collected and cryopreserved at different periods of the reproductive season, suggest that this phenomenon is not linked to a decrease in the quality of semen (Kestemont & Cooremans unpublished data). Kestemont *et al.* (1999a) suggest the possible role of the lysosomal system in yolk reserve mobilisation during embryogenesis and endogenous feeding of larvae. They demonstrated the involvement of cathepsin L in the precocious degradation of yolk reserves and its potential consequences on the quality of perch eggs and larvae. In practice, it is thought to induce, hormonally, the ovulation and spawning of perch females early in the breeding season in order to avoid the possible ageing of non-ovulated mature oocytes.

In both *Perca* species, several hormonal treatments (LHRH-a, carp pituitary extract, hCG) have been used successfully to synchronise spawning during the normal reproductive season, followed by natural or artificial fertilisation (Kayes 1977; Dabrowski *et al.* 1994b, 1996; Kucharczyck *et al.* 1996a; Kouril *et al.* 1997). Time between injection and ovulation depends on type of hormonal treatment and temperature ranging from 14–16 h after the final injection in perch held at 10–11°C and injected with carp pituitary extract (two injections at 24 h intervals) to 4–7 days in perch and yellow perch reared at 13–15°C and injected with a single dose of hCG (0.23 IU g^{-1}), carp pituitary extract (1 $\mu g\, g^{-1}$) or LHRH-a, with or without a priming dose of pimozide (Table 11.1). Artificial fertilisation is generally carried out by the 'dry' method, in which the ribbons of eggs are covered with milt before water is added, but the wet method has also been performed successfully (Kayes 1977) with comparable fertilisation rates (80–97%). Males produce copious amounts of milt, but extender for walleye sperm can be used successfully, by facilitating the dispersal of semen into the whole egg mass. Walleye extender (Moore 1987, 1996) is prepared from 1920 ml distilled water, 0.234 g $CaCl_2.2H_2O$, 0.267 g $MgCl_2.6H_2O$, 0.472 g Na_2HPO_4, 3.744 g KCl, 13.155 g NaCl, 20 g glucose, 0.2 g citric acid monohydrate, 40 ml NaOH solution (1.27 g NaOH 100 ml^{-1} water) and 40 ml bicine solution (5.3 g bicine 100 ml^{-1} water). Methods for cryopreservation of yellow perch semen were investigated by Ciereszko *et al.* (1993) who obtained a fertilisation rate of 23.2% when semen was cryopreserved in an extender containing 8% dimethyl sulphoxide supplemented with 10% egg yolk, significantly higher than that obtained with glucose-glycerol (0.3%), but less than fresh semen (53.4%). Egg ribbons are not adhesive, thus treatment to prevent clumping is not required.

In zander several propagation methods have been developed: natural spawning under controlled conditions, induced spawning in tanks and artificial fertilisation of eggs (Ruuhijärvi & Hyvärinen 1996; Schlumberger & Proteau 1996; Steffens *et al.* 1996). Natural but controlled reproduction is induced using spawning nests (1600–3600 cm^2 area), made of various materials (e.g. willow or alder roots, synthetic wool fixed on wire mesh, rice roots attached to plastic wire and juniper branches) and set up in small earthen ponds (central Europe), concrete tanks or floating cages in lakes (Finland) (Chauderon 1970; Huet 1970; Horvath *et al.* 1984; Hofmann *et al.* 1987;

Table 11.1 Effects of various hormonal treatments (CPE = carp pituitary extract, hCG = human chorionic gonadotropin, LHRH-*a* = luteinising hormone-releasing hormone superactive analogue) on ovulation and fertilisation rates of percid species.

Species	Treatment and Dose	Temperature (°C)	Latency time[1] (degree days)	Ovulation (%)	Fertilisation (%)	References
Perca fluviatilis	CPE (4 μg g^{-1})[2]	10–11	6–8	100	62.5	Kucharczyk *et al.* 1996a
	hCG (5.7 IU g^{-1})	10–11	6–35	100	84.5	Kucharczyk *et al.* 1996a
	hCG (0.5 IU g^{-1}) + CPE (4 μg g^{-1})[2]	10–11	6–8	100	60.3	Kucharczyk *et al.* 1996a
	LHRH-*a* (0.125 μg g^{-1})	15.4	74.4	56.8	94.1	Kouril *et al.* 1997
Perca flavescens	CPE (1 μg g^{-1})	14	56–70	100	58.6–85	Kayes 1977
	hCG (0.23 μg g^{-1})	14	56–70	100		Kayes 1977
	LHRH-*a* (0.1 μg g^{-1}) + pimozide (10 μg g^{-1})	13-15	50–70	100	84.4	Dabrowski *et al.* 1994a
	LHRH-*a* (0.3 μg g^{-1}) + pimozide (10 μg g^{-1})	13–15	50–70	100	78.3	Dabrowski *et al.* 1994a
Stizostedion lucioperca	CPE (3–5 μg g^{-1})[2]	14	28	80	80–95	Antalfi 1979
	hCG (0.1–0.2 IU g^{-1})[3]	19–20	10			Steffens *et al.* 1996
	LHRH-*a* (0.1 μg g^{-1}) + pimozide (10 μg g^{-1})	14–20	25–30		> 80	Schlumberger & Proteau 1996
Stizostedion vitreum	CPE (7.7–22 μg g^{-1})	7.2–10	30			Lessman 1978
	hCG (0.11–1.1 IU g^{-1})	12.2	29–48	93–96	31.4–48	Heidinger *et al.* 1996
	hCG (0.5 IU g^{-1})	9–10	36–40	40	> 90	Barry *et al.* 1995
	LHRH-*a* (0.1 μg g^{-1})	9–10	24–50	60	> 90	Barry *et al.* 1995
	hCG (0.15 + 0.5 IU g^{-1})[4,5]	10	50	60–100[5]	41–72	Malison *et al.* 1998a
	LHRH-*a* (0.035 + 0.1 μg g^{-1})[4,5]	10	50	20–100[5]	63–31	Malison *et al.* 1998a
	hCG (0.15 IU g^{-1}) +17,20-P (2 μg g^{-1})[4,5]	10	50–80	40–100	0–1.4	Malison *et al.* 1998a

[1] After the final (resolving) injection. [2] Dose divided into two injections (10% + 90%) at 24 h interval. [3] Dose divided into 3 to 5 injections at 12 h intervals. [4] Two injections at 48 h interval. [5] Out-of-season spawning (January–March).

Salminen & Ruuhijärvi 1991). Cage spawning practices differ between countries, e.g. small 6 m^3 cylindrical cages in Finland stocked with three mature females and four males and 30–50 m^3 cages in Germany, with three to five females and four to 10 males. The average time zander females are held in the cages before spawning varies from six to 13 days depending on weather conditions (Salminen *et al.* 1993). The nests are checked daily and when spawning is detected, the nest with eggs is transferred into an empty cage. In small ponds and concrete tanks, the advised male–female ratio is 1:1–1.2 with one nest per male 10 m^{-2}.

Hormonally induced spawning in tanks usually requires a precise determination of the female maturation stage by biopsy. However, this is sometimes responsible for gonad regression before ovulation. Optimally, hormonal treatment (carp pituitary extract, hCG, LHRH-a) must be performed 4–5 days before ovulation (Table 11.1), when the oil droplet has formed in the ovule and the germinal vesicle is adjacent to the ovule membrane (Schlumberger & Proteau 1991). Hormonal treatment efficiency is improved by environmental manipulations such as a rise in water temperature from 10–14°C (Antalfi 1979; Steffens *et al.* 1996). Artificial fertilisation requires the same hormonal treatment but water temperature is usually increased to 19–20°C to synchronise female maturation and reduce latency time between hormonal injection and stripping. About 2 ml of milt are used to fertilise 100 g of eggs (semi-dry technique) and a fertilisation solution (0.3% NaCl) is used to prevent rapid sticking of eggs. Adhesiveness is then neutralised by washing the eggs in an alkaline protease solution (0.5%) or with a mixture of 100 g talcum and 20 g NaCl 10 l^{-1} water (Steffens *et al.* 1996).

Propagation methods for walleye include the capture of wild brood fish just prior to or after ovulation or the use of captive broodstock. Females captured one or two days prior to ovulation will often spawn in captivity without further treatment. Less mature females can be induced to spawn using various hormones similar to those described for the other species. Injection of hormones including hCG, LHRH-a, carp pituitary extract and several maturational steroids have been used successfully to induce final oocyte maturation in walleye (Goetz 1983; Pankhurst *et al.* 1986; Heidinger *et al.* 1990, 1996; Barry *et al.* 1995; Malison & Held 1996a). As previously mentioned, the combination of environmental manipulations and hormone injections can be used to greatly advance spawning in walleye (Malison & Held 1996a; Malison *et al.* 1998a) (Table 11.1).

Stripping and fertilising walleye eggs are performed by both the dry and wet methods. In the latter method the eggs or sperm are stripped into water and the other gametes are quickly added to accomplish fertilisation. Eggs must be fertilised within about two minutes of contacting water as spermatozoa are viable for less than one minute after contact (Malison & Held 1996a). As previously reported for zander, eggs of walleye become highly adhesive shortly after exposure to water. Prevention of egg adhesiveness includes the treatment of eggs with bentonite (166 $g\,l^{-1}$), protease (100–1000 $mg\,l^{-1}$) or tannic acid (400 $mg\,l^{-1}$) solutions (Colesante & Youmans 1983; Krise *et al.* 1986; Krise 1988). As males often become ripe several days before females they may be held in hatcheries for two to three weeks. This period of captivity may lead to a decrease in semen quality. Short-term storage (up to 14 days at 0.6–2.2°C) for walleye semen has been developed to enhance propagation efficiency and is currently utilised in hatchery operations (Moore 1987; Satterfield & Flickinger 1995; Koupal 1996). According to Koupal *et al.* (1995), semen extender ratios of 1:2, 1:4 and 1:6 were found to compare favourably with fresh semen. Preliminary cryopreservation of walleye semen has been investigated by Brown (1996) and the fertility rate, using semen that had been stored for one year was 31.3%.

11.3 Incubation and hatching

Egg strands of perch and yellow perch eggs are commonly incubated in horizontal trays or stretched on supporting wires in tanks. Recently an experimental incubation unit for perch eggs was described by Kucharczyk *et al.* (1996b) in order to provide identical incubation conditions for numerous separate batches of eggs. About one day before hatching, the egg strands are transferred to the larval rearing tanks. Estimation of egg numbers is made by weighing the whole strand after water hardening and counting several weighed sub-samples of eggs. Egg strands from large perch (1.5 kg body mass) may reach up to 250 mm in length (personal observation) with more than 120 000 eggs. In yellow perch, West (1977) reported strands of 213 mm in length.

Walleye eggs are usually counted by volumetric displacement, after water hardening, with sub-samples of eggs from different batches measured by hand or with calibrated egg troughs that provide an estimate of the number of eggs per litre (Malison & Held 1996a). Walleye eggs range in volume from 85 000–160 000 eggs l^{-1}. They are incubated in jars supplied with an upwelling flow of water. At hatching, fry are usually allowed to swim out of the jars and into a central collection tank. Malison & Held (1996a) stressed that the accumulation of egg shells, oil droplets and dead eggs can quickly lower water quality and block fine mesh filters that are used to contain the hatched fry. Solutions to this problem include screens, self-cleaning devices (e.g. bubble curtains adjacent to screens) and good husbandry. Zander eggs are incubated directly on the spawning nests up to the eyed stage and then transferred to incubation tanks. Hungarian farmers incubate zander eggs using 'mist-chambers' where the nests are hung vertically in a small room and sprayed with water (Horvath *et al.* 1984).

Incubation time is very similar for all percid species, ranging from 80–110 degree-days at a recommended temperature of 15°C (Fig. 11.1; see also Fig. 4.10). The relationship between temperature and median time of development of yellow perch embryos was derived by Hokanson & Kleiner (1974). According to these authors, the synchronisation of mouth and opercular movements in a regular breathing fashion indicates that hatching will occur very shortly. However, hatching of eggs

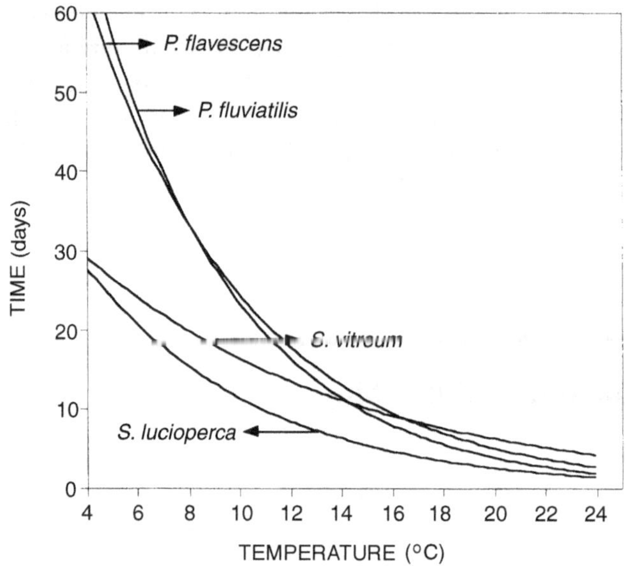

Fig. 11.1 Relationship between water temperature and incubation time (from fertilisation to hatching) in some percid fishes.

from the same female is temperature dependent and can often extend from several hours to five days, this period being shortened by increasing temperature a few degrees. In perch, five days separate the early and late hatching from the same egg strand at 15°C while it takes only three days at 19°C (Kestemont *et al.* 1999b).

11.4 Rearing of early life stages

Despite a carnivorous feeding habit at the juvenile and adult stages (Chapter 5), the size of newly hatched percid fishes is closer to that of cyprinid species than to that of predacious species such as members of the pike family. Size of hatchlings varies with species, population origin and size of breeders, but whatever the species, it is < 10 mm (total length) for a body mass < 1 mg (range usually between 0.4–0.8 mg) e.g. the total length of newly hatched walleye may range from 4.8–9.5 mm (Priegel 1970). Several factors, more or less related to the small size and fragility of percid larvae, have limited the development of efficient intensive larviculture methods for these species for several decades. Various culture systems have been investigated in percid fishes, from the most extensive production of fry and juveniles in undrainable ponds to the semi-intensive culture of larvae in green-water tanks. However, major constraints to the economical intensive culture of fry have been progressively resolved for walleye but remain under investigation for the other percid fishes. Some major features are given below of extensive to semi-intensive fry rearing systems in fertilised drainable ponds or concrete tanks and intensive production, which usually include recirculating systems.

Production of percid fry and juveniles in fertilised ponds or concrete tanks

The most common method of mass propagating yellow perch, walleye and zander fry in monoculture is to use a fertilised pond. Zander is also extensively cultivated in polyculture with carp and other cyprinids in several European countries, mainly in Germany and central Europe. The pond production of perch fry has received very little attention.

Development of walleye fry production in ponds, in particular fertilisation and stocking strategies, was stimulated by the unacceptable variation in survival and growth of fry to stockable-sized fingerlings in ponds (Culver 1996). Apart from the problems of low dissolved oxygen and high ammonia and pH occurring regularly in fertilised systems, the fish ponds developed blue-green phytoplankton blooms with few or no suitable algae for zooplankton, particularly if low N:P ratio (< 7:1) fertilisers were used (Helal & Culver 1991). However, it has been shown that appropriate zooplankton abundance largely determines the success of larval culture of percid fishes in ponds as these fish are planktivorous during their larval stages (Smith & Moyle 1943; Dobie 1956; Mills *et al.* 1989; Awaïss *et al.* 1992; Fox *et al.* 1992; Qin & Culver 1992; Wang 1994; Qin *et al.* 1995; Chapter 5). A few days after hatching, yellow perch larvae can consume prey or food particles as large as 0.19 mm (Schael *et al.* 1991) while mouth opening size of perch, walleye and zander is larger (about 0.35–0.45 mm) (Tamazouzt 1995; Kestemont *et al.* 1996; Chapter 5). Several taxonomic groups of zooplankton such as rotifers, copepod nauplii, cyclopoid copepods or small soft-bodied cladocerans are potential prey items as starter live food. As they grow, fry progressively switch to larger cladocerans and aquatic insect (mainly chironomids) larvae and pupae.

In percid fish culture several fertilisation regimes are used, based either on organic or inorganic fertilisers alone (Fox 1989; Qin *et al.* 1994), or in combination (Fox *et al.* 1992; Qin & Culver 1992). Qin *et al.* (1995) investigated the efficiency of inorganic and organic fertilisers in walleye larval culture in ponds and showed that, compared to organic fertiliser, inorganic fertiliser increased the levels of primary productivity, algal abundance, dissolved oxygen and total phosphorus, and raised the pH. The maintenance of adequate abundance of copepods and cladocerans for growth and survival of walleye larvae was achieved by applying liquid inorganic fertiliser with high N:P ratio (20:1) once per week to raise the inorganic concentration to $600\,\mu g\,N\,l^{-1}$ and no more than $30\,\mu g\,PO_4\text{-}P\,l^{-1}$. The same levels of inorganic fertiliser were used by Fiogbé *et al.* (1994) to enhance pond productivity, survival and growth rate of perch fry. Fox *et al.* (1992) also examined the effects of supplementary inorganic fertilisation on juvenile walleye growth, production and food availability in organically fertilised rearing ponds. These authors reported that walleye in organically fertilised ponds grew faster and were almost twice the mass of those in the control ponds at harvest. Improved walleye growth in the ponds, given an inorganic supplement, was apparently due to higher production of *Daphnia*, the major planktonic organism eaten by the fish.

Apart from inorganic and organic fertilisation, other procedures of plankton enhancement have been evaluated such as a short flooding of the pond to bring in plankton or the use of insecticides to establish communities of appropriate composition (Verreth 1984; Verreth & Kleyn 1987; Hilge & Steffens 1996).

Fish are commonly stocked in ponds 10–15 days after fertilisation. Pond size ranges from 0.2–0.8 ha for yellow perch and walleye but are usually larger in zander polyculture. Stocking densities vary from one to three million fertilised eggs ha^{-1} (one to two days prior to hatching) in yellow perch, from 60 000 to 600 000 larvae ha^{-1} in walleye and up to 100 000 larvae ha^{-1} in zander.

In drainable ponds, yields of two to three month old fingerlings are highly variable, ranging from 50 to 150 kg ha^{-1} of 35–65 mm perch and yellow perch, 68–133 kg ha^{-1} of 30-50 mm walleye and averaging 50 kg ha^{-1} of 0.5 g zander. Harvest characteristics are not correlated with stocking density. Typically at carrying capacity (which can be enhanced by appropriate fertilisation), fish size at harvest is inversely related to initial fish stocking density and number of fish harvested per unit area. According to Qin *et al.* (1995) the best growth of walleye fry was obtained in inorganically fertilised ponds stocked with 250 000 fish ha^{-1}, but the highest yield (up to 92.2 kg of 38-day-old fry ha^{-1}) was obtained at 500 000 fish ha^{-1}. Similar results were obtained with larval saugeye, a hybrid of female walleye and male sauger (Qin *et al.* 1994; see *Performances of hybrid percids* below). Survival in fertilised ponds is highly variable but as high as 40–60% in yellow perch, near 50% after four to five weeks in zander and up to 75–85% in walleye (Qin *et al.* 1995; Culver 1996; Flowers 1996; Steffens *et al.* 1996). Generally, survival rate is affected by high stocking density, probably because of a rapid exhaustion of adequate zooplankton during early life stages, and, as a corollary, the emergence of cannibalism. Indeed, on the basis of work carried out in Minnesota, USA, on the pond culture of walleye by Smith and Moyle (1943) and Dobie (1956), Cheshire & Steele (1972) reported that cannibalism caused by food shortages in mid summer was the main obstacle to rearing young walleye to fingerling size in natural rearing ponds. Walleye may turn cannibalistic at two stages during their early life. The first stage occurs within a week after hatching if zooplankton is scarce in ponds, and the second stage develops about mid summer, when the young walleye are ready to feed on fish. With respect to stocking density in ponds, Madon (1993) found no cannibalism in over 1000 larval and juvenile walleye stomachs within a stocking density range from 100 000–500 000 fish ha^{-1}. However Culver *et al.* (1993) recommended caution when stocking density increased beyond 500 000 fish ha^{-1} and fish size was not uniform. Occurrence of cannibalism was also suggested in perch by Kestemont *et al.* (1996). The bimodality in the size of a perch population, sampled one month after hatching, and the very small size heterogeneity at the time of harvesting, 75 days after larval stocking, indicated that extensive cannibalism occurred during the second half of the pond-rearing period.

Larval stocking density of percids in ponds may be increased if a partial harvesting of small fingerlings (20–30 mm) is undertaken during fingerling production. This combination is sometimes called the 'tandem pond–tank technique' (Moore *et al.* 1994). Because yellow perch and walleye are strongly attracted to light before they reach 50 and 32 mm respectively, 30–45 day old juveniles can be attracted to a floating light, combined with a lift net or surrounded with a soft, fine mesh seine (Manci *et al.* 1983) (Fig. 11.2). After removal from the ponds these juveniles are trained to accept formulated feeds, in hatchery tanks, by frequently offering them trout or salmon starter dry diets or semi-moist feeds. In walleye, training to feed in this way is more successful for juveniles harvested at 30 mm to fish harvested at

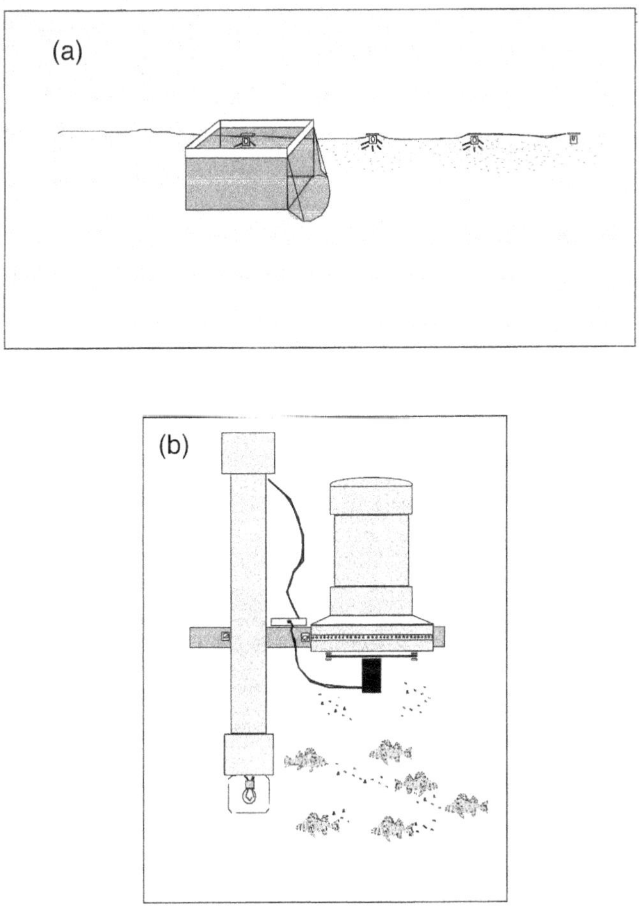

Fig. 11.2 Procedures for (a) harvesting yellow perch juveniles from ponds using light attraction and (b) training yellow perch juveniles to take artificial foods in ponds by the combined use of a vibrating feeder and light attraction. (Figures provided by Malison & Held, with thanks.)

20 mm in length (Table 11.2) (Malison & Held 1996c). In perch Kestemont & Mélard (1996) also showed that the success of conditioning of pond-reared juveniles to formulated feeds in intensive rearing tanks depended on the period at which the transfer was performed. When transferred to tanks at 35 mm, pond-reared fish were rapidly weaned to a dry diet by mixing the pellets with frozen chironomid larvae. However, dominant fish (usually the largest ones) were highly aggressive and most of the mortalities were related to attacks between conspecifics and cannibalism. When fish were transferred from ponds to tanks at a larger size (*c.* 100 mm), conditioning to dry diet required a longer transition period but no aggressive or cannibalistic behaviour occurred between fish (Table 11.2). Nagel (1978, 1996) and Malison & Held (1992) reported that in walleye and yellow perch respectively, an appropriate distribution of pelleted food combined with internal tank lighting improved the conditioning rate of juveniles to an artificial diet when transferred from nursery ponds to grow-out tanks. The sinking of pellets mimicked the movement of live prey and consequently induced predatory behaviour (Table 11.2). Compared to external lighting, internal lighting significantly increased the conditioning of pond-reared yellow perch to artificial food and reduced cannibalism. These authors observed that shadows and movement in the vicinity of the externally illuminated tanks alarmed the walleye and yellow perch fingerlings and caused erratic swimming, followed by an interruption of feeding for several hours, while the use of internal tank lighting minimised such disturbances. Malison *et al.* (1994b) developed a method of training yellow perch fry to accept formulated diets in ponds without moving them to hatchery tanks. By placing lights, either floating or submerged, and vibrating automatic feeders in the ponds, three week old yellow perch (initial size of 15–17 mm) were trained to accept formulated feed, allowing a production of up to 500 000 yellow perch fingerlings ha^{-1}, with more than 90% of fish trained to accept dry food (Fig. 11.2). Light has also been used to attract zoo-

Table 11.2 Percentage of perch, yellow perch and walleye juveniles conditioned to formulated feeds, or lost to starvation or cannibalism after harvesting from ponds at different sizes and reared under intensive culture conditions.

Size (mm)	% conditioned	% starvation	% cannibalism	References
Perca fluviatilis				
35.0	29.9	n.d.	14.3	Kestemont *et al.* 1996
97.0	78.8	21.2	0	
100.4	79.2	20.8	0	Fontaine *et al.* 1995
Perca flavescens				
16.9	54.5	26	20	Malison & Held 1992
32.5	67	14	18	
42.6	57	23	23	
Stizostedion vitreum				
20.0	40	55.2	5.8	Malison & Held 1996c
30.0	67.7	14.7	17.4	

plankton in lake cage culture of zander. In Germany production of 30–40 mm fry was obtained after 4–6 weeks, with survival rates of about 40%, by rearing zander in cages (8 m^3, 15 000–20 000 larvae per cage) submerged at double the Secchi disk depth in lakes of appropriate productivity. However, ectoparasites and bacterial diseases can cause serious problems during this sensitive stage of zander development (Jäger *et al.* 1984).

As previously mentioned, little attention has been given to the production of perch in fertilised ponds. Recent investigations have focused on the semi-intensive production of perch in fertilised concrete tanks (10 m^2). The method combines the advantages of newly hatched perch larvae first feeding on suitable zooplankton (mainly rotifers) during three to five days after hatching with the complementary supply of live and dry food. It can be considered as an efficient method of fingerling production (Kestemont *et al.* 1996; Mélard *et al.* 1996a). Tanks are organically fertilised with chicken or quail manure seven days before the stocking of eyed eggs (ready to hatch) in order to stimulate the development of phyto- and zooplankton. From hatching onwards the tanks are partly (80%) shaded to prevent any further development of filamentous algae that may trap fish larvae. During the first three to five days, the water is not renewed to prevent the flushing of rotifers. Tanks are then progressively changed to flow-through systems with water renewal rates from 1.0 to 1.8 $m^3 h^{-1}$, depending on stocking density (Mélard *et al.* 1996a). Two to three days after hatching, larval food is supplemented with *Artemia* nauplii. At one week a high quality dry diet, originally formulated for marine fish larvae, is provided together with *Artemia* nauplii. The latter is proportionally decreased until the end of the first production phase, six to seven weeks after hatching. Dry food is given only in the daytime using automatic feeders in the non-shaded area of the tank. Larvae are progressively trained to accept formulated diets during this period and all fry are conditioned to ingest dry feed at harvest.

The success of various stocking densities, ranging from 400–4000 eggs m^{-2}, were investigated at different water temperatures during a 44 day rearing period. Despite large year to year variation, mainly due to factors such as climatic conditions and their impact on phyto- and zooplankton development, origin of brood stock and egg quality, water quality and presence of parasites, the growth of perch was clearly enhanced at high temperatures: mean body mass ranging from 370–860 mg at 23°C and from 190–280 mg at 17°C (Fig. 11.3). However, the higher temperature negatively affected survival rate (means: 18.7% at 23°C and 38.5% at 17°C (Fig. 11.3)). The negative effects of higher temperatures on perch survival include an increase in fish size heterogeneity and, consequently, a higher proportion of cannibals within a sibling population, and an increase of pathogens (parasites and bacteria; Chapter 8). Regardless of temperature, high stocking density produced significantly more homogenous growth within the cultured stocks which decreased the overall mortality rate and the emergence of cannibals. Greater densities of cannibals are usually observed at the highest densities, e.g. from 2.6 to 14.7 cannibals m^{-2} at 500 and 4000 fish m^{-2}, respectively, but the semi-logarithmic nature of the

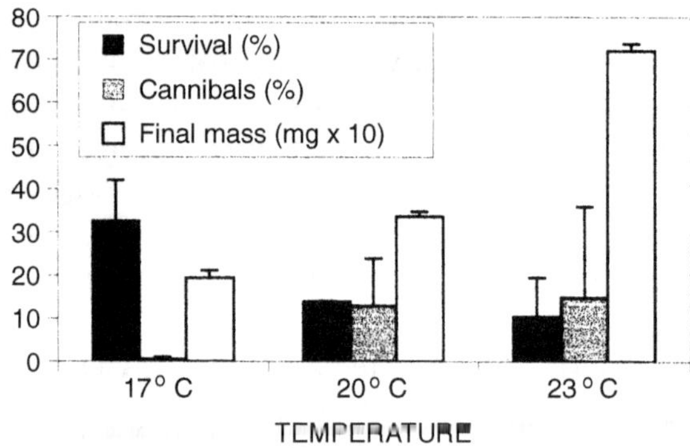

Fig. 11.3 Effects of temperature on survival (%), cannibalism (‰ population) and growth in perch larvae and juveniles semi-intensively reared in fertilised, concrete tanks, with additional supply of *Artemia* nauplii and dry food. (Data from Mélard & Kestemont 1994 and Mélard *et al.* 1996a.)

relationship between the two variables suggests that increasing stocking density proportionally reduces the impact of cannibalism. A model of perch survival rate (*SR*) after 44 days of rearing in fertilised concrete tanks was given by Mélard *et al.* (1996a):

$$SR(\%) = -58.20 + 26.65 \log (ISD) - 1.81 (10(SGR - 15))$$

where *ISD* = initial stocking density (fish m^{-2}) and *SGR* = specific growth rate (% day^{-1}). Predictive models of emergence of cannibalism and survival rate, depending on stocking density and growth patterns, clearly suggest that stocking density be increased as much as possible (within the tolerance range of perch larvae to deterioration in water quality) to reach the optimum growth–survival combinations.

Intensive larval rearing in recirculating systems

Since intensive culture of fry has many advantages (e.g. fairly stable culture conditions, rearing of fry produced by out-of-season spawning, more predictable production of juveniles and easier monitoring and control of cannibalism) compared to culture in ponds and cages, research has been directed towards the determination of optimal rearing conditions and nutritional requirements of percid fishes in these systems. There are several critical factors affecting success of percid intensive culture including small mouth gape (particularly in yellow perch larvae), dependence on live food organisms, non-feeding behaviour, non-inflation of the gas bladder (NGB), clinging behaviour (mainly in walleye) and cannibalism. Some of these constraints have now been overcome. Despite the availability of high quality feeds for small larvae (pellet size <200 μm), mainly formulated for marine species, the acceptance, growth and survival of percid fishes fed formulated diets as starting food are still highly variable and rather unsatisfactory, except walleye fed with high quality larval dry food.

In perch a relationship between mouth gape size (according to the formula of Shirota 1970) and the total length of the larvae (Treasurer 1990; Tamazouzt 1995; Chapter 5), indicates that the larvae are able to ingest *Artemia* nauplii (420–480 μm) when they reach 6 mm, after three days of feeding. However, repeated experiments carried out in intensive rearing conditions with large numbers of larvae (10 000–50 000 larvae per tank) showed that about 60–70% of the newly hatched larvae ingested *Artemia* nauplii (small size strain) from the beginning (Kestemont *et al.* 1996). The remaining 30–40% disappeared progressively during the first week of rearing, probably due to exhaustion and to the sibling cannibalism exhibited by the growing larvae which were able to ingest *Artemia* nauplii. Higher survival rates can be achieved by supplying rotifers in combination with *Artemia* nauplii during the first few days, allowing all larvae to capture live prey regardless of mouth gape size. Short-term feeding experiments (10 days) showed that mass-cultured freshwater rotifers (*Brachionus calyciflorus* Pallas) constitute a suitable starting live food for perch larvae, providing all essential nutrients and supporting a rapid growth of fish (Awaïss *et al.* 1992, 1996) (Table 11.3). Based on the technique of Kamler *et al.* (1986) (in which changes in prey numbers are measured in rearing tanks with and without fish), Awaïss *et al.* (1992) determined the maximum daily consumption of rotifers by perch at 20°C as 800 rotifers larva^{-1} day^{-1} at day three (end of yolk sac reabsorption) and up to 2200 rotifers larva^{-1} day^{-1} at day ten. Yellow perch larvae can also ingest *Artemia* nauplii a few days after hatching but the percentage of fish able to ingest them is clearly lower than that of perch and, according to Hinshaw (1985), *Artemia* nauplii are not considered an adequate diet for yellow perch larvae < 13 mm. In this species, supply of adequate size zooplanktonic food, by mass culture of rotifers or by using the green water tank method, is highly recommended during the first days after hatching (NCRAC 1994; Brown *et al.* 1996). This was also suggested by Kolkovski *et al.* (1997) who found, in out-of-season spawned yellow perch, 80% of the larvae with one to four rotifers in the gut at first feeding but detected *Artemia* nauplii only three days after rotifer feeding. Lake zooplankton has also been used to rear Eurasian and yellow perch larvae in aquaria, mainly rotifers *Polyarthra* sp. Ehrenberg and *Conochilus* sp. Hlava and to a lesser degree, copepod nauplii (Hale & Carlson 1972; Siefert 1972; Wang & Eckmann 1994) (Table 11.3).

In contrast to yellow perch, the mouth size of walleye and zander allows early ingestion of relatively large zooplankton organisms, such as copepods and cladocerans of 500–1100 μm total length (Hokanson & Lien 1986; Hilge & Steffens 1996). *Artemia* nauplii, alone or in combination with lake-sieved zooplankton were recommended as starter food for walleye larvae by Colesante *et al.* (1986) and Hokanson & Lien (1986). Colesante (1996) recommended an initial feeding rate of 800–1000 *Artemia* nauplii fish^{-1} day^{-1}, the feeders dispensing nauplii every eight minutes for a three-second interval 24 h day^{-1}. Survival rates averaged 40–60% on day 30 for walleye and 25–30% on day 35 post-hatching for zander

Table 11.3 Survival and growth of percid fishes under different feeding regimens.

Diet	Duration (days)	Initial mass (mg)	Final mass (mg)	Survival (%)	References
Perca fluviatilis					
Live food					
Rotifers (lake sieved)	13	0.8	4.1	41.7	Wang & Eckman 1994
Rotifers (mass cultured)	10	0.8	5.2	83.5	Awaïss *et al.* 1992
Artemia nauplii	18	0.8	20.9	34.3	Kestemont *et al.* 1996
	44	0.8	292	33	Mélard & Kestemont 1994
	22	7.6	135	75.8	Fiogbé *et al.* 1995
	29	12.6	67.7	79.7	Vlavonou *et al.* 1995
Rotifers + *Artemia* nauplii	18	0.8	17.9	50.2	Kestemont *et al.* 1996
Copepodites, daphnids	12	4.1	23.9	65	Wang & Eckmann 1994
Dry food					
	15	0.8	2.9	25	Tamazouzt 1995
	22	8.7	90.2	12	Fiogbé *et al.* 1995
Mixed diet (live + dry food)					
Rotifer + dry food	10	0.8	4.2	42.5	Awaïss *et al.* 1992
Artemia + dry food (50:50)	44	0.8	329	19.8	Mélard & Kestemont 1994
Perca flavescens					
Live food					
Rotifers + *Artemia*	35	0.4	70–90[1]		Kolkovski *et al.* 1997
Dry food	14	0.9[1]	8.0[1]	46.4	Brown *et al.* (pers. comm.)
Freeze dried krill	14	0.9[1]	2.5[1]	12.0	Brown *et al.* (pers. comm.)
Egg solid spray diet	14	0.9[1]	2.4–2.5[1]	7.6–29.6	Brown *et al.* (pers. comm.)
Biokyowa fry feed B	14	0.9[1]	2.1–2.8[1]	8.6–15.2	Brown *et al.* (pers. comm.)
Biomarine larval diet	14	0.9[1]	2.2–2.7[1]	4.2–10.4	Brown *et al.* (pers. comm.)

Stizostedion lucioperca					
Live food					
Sieved-lake zooplankton	19–28[2]	0.3–0.5	70–80[1]	27–44[2]	Schlumpberger & Schmidt 1979
Artemia nauplii	14	3.4	105.8	64.5	Mamcarz *et al.* 1997
Artemia + zooplankton	35	0.5	120–150[1]	26.7	Klein Breteler 1989
Light-attracted zooplankton	7	16	90	77	Jäger *et al.* 1984
Dry food					
	8–18	0.5	3–4[1]	very low	Ruuhijärvi *et al.* 1991
	23	0.3–0.5[1]	120–150[1]	<1	Schlumberger & Proteau 1991
Stizostedion vitreum					
Live food					
Artemia nauplii	30	3–5	180–220[1]	11.6	Colesante *et al.* 1986
Artemia nauplii + zooplankton	30	4.5	180–200[1]	14.6	Colesante *et al.* 1986
Dry food					
	28	4.5	50–100[1]	5.3–6.1	Barrows *et al.* 1988
(In clear water)	21–30	4.5	11–88	0.9–9.4	Bristow & Summerfelt 1994
(In turbid water[3])	21–30	4.5	33–180	5.0–61.2	Bristow & Summerfelt 1994
Mixed diet (live and dry food)					
Artemia nauplii + dry food	65	4.5	300–500[1]	3.9	Colesante *et al.* 1986
Artemia + zooplankton + dry food	65	4.5	300–500[1]	6.0	Colesante *et al.* 1986

[1] Calculated from data in mm (total length). [2] Depending on stocking density. [3] By addition of clay in tank water (20–50 NTU).

(Klein Breteler 1989; Colesante 1996; Steffens *et al.* 1996; Summerfelt 1996) (Table 11.3.).

As frequently reported for other species, development of formulated larval feeds for percid fishes brings several advantages over the use of live feeds. These include savings in labour costs associated with the collection or culture of live feeds and a reduction in the introduction of disease and contaminants. In addition, the cost of formulated larval feeds will probably be less than the costs of *Artemia* cysts (Barrows *et al.* 1988). However, major constraints to feeding dry diets include poor acceptance of feed particles, gradual starvation, cannibalism and reduced water quality with associated diseases. Some experiments have been conducted on feeding both perch and yellow perch larvae with a dry diet only from the time of hatching. Unfortunately, most of them have led to very poor results (Best 1981; Heidinger & Kayes 1986; Dreyer 1987; Awaïss *et al.* 1992). Tamazouzt (1995) investigated the effects of food type on survival and growth of perch. Although perch larvae ingested a dry diet from hatching, maximum survival reached only 26% on day 15 and growth rate was extremely low, about three times less than that usually achieved with *Artemia* nauplii (Table 11.3). Considering the poor growth obtained with a dry diet from hatching onwards, it appears more appropriate to feed the newly hatched larvae with live food or mixed diets (live + dry feeds) up to a stage allowing successful weaning of the larvae. Mélard & Kestemont (1994) compared the growth of perch fed various proportions of *Artemia* nauplii and dry food (*Artemia*:dry food 100:0, 75:25, and 50:50) from hatching up to day 44. Although survival rate was the highest in fish fed *Artemia* only, growth rate was similar in all treatments. Fish fed mixed diets were much more rapidly and successfully trained to accept 100% dry food in comparison with the perch fed 100% *Artemia*. Kestemont *et al.* (1996) demonstrated that the time of weaning had a profound effect on the survival and growth rates of perch larvae, and recommended a progressive transfer of larvae to dry feed from 50 mg body mass when fish are about 21 days old (Fig. 11.4).

In a review of diet and nutrition in walleye, Barrows (1996) reported that the first attempts at culturing walleye fry using starter dry feeds as the sole source were largely unsuccessful, with survival rates typically <10% (Beyerle 1975; Nickum 1978; Colesante *et al.* 1986; Barrows *et al.* 1988) (Table 11.3). In contrast, intensive larval rearing using live feeds produced survival rates as high as 87% at 21 days post hatch (Hokanson & Lien 1986). As recommended for perch larvae, greater survival and growth rates were obtained by starting fry on live food, then converting the fish to formulated diets (Krise & Meade 1986). Apart from the progress made in walleye rearing technology such as improving the rates of swim bladder inflation and reducing cannibalism during early life stages, the development of high quality larval feeds from the late 1980s has allowed a successful replacement of live feeds in walleye larval culture by a formulated diet resulting in rates of survival reaching 28% (Loadman *et al.* 1989) and 47% (Barrows 1990) at 30 days post hatch. Barrows (1996) used a relatively cheap diet based on krill meal, low-temperature herring

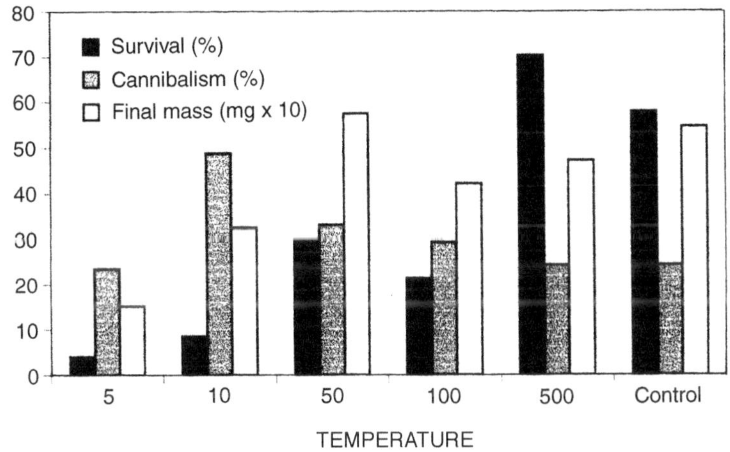

Fig. 11.4 Effects of weaning size on survival, cannibalism and growth of perch larvae. Fish were initially fed *Artemia* nauplii and then progressively conditioned at different body mass, to accept dry food in 5 days. Control fish were fed only *Artemia* nauplii (T = 20°C, duration = 50 days).

meal, spray-dried whole egg, and ground trout fry and eggs (51% protein, 24% lipids). When walleye larvae were fed this diet under appropriate rearing conditions, survival at day 30 from hatching averaged 45% with 95% gas bladder inflation. No significant progress with dry diet has been reported for zander. Comparing three different commercial dry feeds, Ruuhijärvi *et al.* (1991) concluded that zander larvae accept formulated feed from hatching although survival was low and growth very poor in comparison with the results obtained by Klein Breteler (1989) with live food. Very low survival rates (<1%) and poor growth were also reported by Schlumberger & Proteau (1991) in zander larvae fed a starter diet used with success in carp and coregonid larviculture. However, Proteau *et al.* (1993) obtained satisfactory growth when feeding zander with a diet formulated for marine larvae, but sudden high mortalities occurred at the time of intestinal convolution and metamorphosis (10–12 days post hatch). More intensive research on larval and post-larval stages of zander is required to overcome the problems related to diet.

Knowledge of percid larval nutrition is in its infancy and remains one of the greatest constraints to overcome. A series of studies with larval yellow perch and perch have recently been performed. All essential amino acids in whole fish increased similarly as yellow perch grew from 0.4 to 82 mg (Dabrowski *et al.* 1993), without major changes in the amino acid profile (Table 11.4). Similarly, in perch larvae reared under 'reference' conditions (fertilised pond with abundance of zooplankton of appropriate size and composition), the percentage of most essential amino acids remained stable in the whole fish from hatching up to 2 g body mass (Fiogbé 1996). It has been suggested that such stable amino acid profiles can be used as a possible method of predicting essential dietary amino acids in perch larvae (Wilson & Poe 1985; Brown *et al.* 1996). Docosahexaenoic acid (DHA, 22:6n3) was

Table 11.4 Essential amino acid composition (% total amino acids) in pond-reared perch and yellow perch (whole body), during larval development (calculated from Fiogbé 1996 and Dabrowski *et al.* 1993, respectively).

Amino acids (%)	*Perca fluviatilis*			*Perca flavescens*	
	1 mg body mass	75 mg body mass	340 mg body mass	0.4 mg body mass	82 mg body mass
Arginine	7.1	6.5	6.3	6.2	6.9
Histidine	2.9	3.3	3.1	1.8	1.9
Isoleucine	5.2	5.1	5.0	5.9	4.7
Leucine	7.8	7.5	7.3	8.8	9.3
Lysine	8.6	8.0	8.3	9.2	9.2
Methionine	2.5	3.0	3.2	3.9	3.5
Phenylalanine	4.3	4.5	4.6	5.8	4.6
Tyrosine[1]	4.1	4.1	4.0	3.4	3.2
Threonine	5.5	5.3	5.3	4.1	3.3
Valine	5.8	5.7	6.0	7.1	7.0

[1] Non-essential amino acid (sparing effect of phenyalanine).

the predominant fatty acid (9.5%) in phospholipids of pond-reared yellow perch and was found to decrease from eggs to fry (Dabrowski *et al.* 1993). Further, both DHA and eicosapentaenoic acid (EPA, 20:5n3) were higher in juvenile perch than in daphnids, their major food item. These observations and additional data of fatty acid composition in ovaries and eggs from wild and cultured yellow and Eurasian perch (Brown *et al.* 1996; Abi-Ayad *et al.* 1997), indicated that cultured perch larvae and juveniles may require supplements of fatty acids of the ω3 and ω6 families. Enrichment of broodstock diet with DHA, up to 0.9% of the dry feed, appears beneficial for the quality of perch eggs (increased percentage of fertilisation and hatching) and larvae (resistance to osmotic shock and extended starving) (Abi-Ayad *et al.* 1997). No improvements of survival or growth rates were achieved by feeding perch larvae with highly unsaturated fatty acid (HUFA) enriched *Artemia* nauplii (Fiogbé *et al.* 1995). This might be due to the high concentrations of endogenous HUFA at hatching (Dabrowski *et al.* 1993). A 'reserve' provides the larvae with these essential nutritional compounds during early life until bio-conversion pathways become functional (Kestemont *et al.* 1996). In walleye larvae medium levels of ω-3 HUFA source (cod liver oil) in '*Artemia* enrichment' were recommended by Kolkovski *et al.* (1998), and resistance of larvae to a stress test (immersion in 25 ppt salt water) was reduced in groups fed high levels of HUFA (Czesny *et al.* 1998). A supplement of vitamin C to '*Artemia* enrichment' appears to have a positive effect on walleye larvae survival.

Determination of optimal rearing conditions was investigated extensively in the 1990s to reduce mortalities during early life stages. Some of the main areas of investigation were water temperature for optimal growth, tank design and mode of water supply, reduction in the occurrence of non-inflation of the gas bladder (NGB), and lighting conditions to improve the detection of food. The effects of the colour of the tank wall and water turbidity were also investigated in relation to lighting conditions.

Optimal temperature ranges from 17–20°C in perch and walleye (Kestemont *et al.* 1996; Moore 1996; Summerfelt 1996), while higher temperatures (22–26°C) are usually recommended for zander (Hilge & Steffens 1996). Perch larvae (from 16.4 to 118 mg body mass) reared in a recirculating system under graded temperature and feeding levels, had a maximum growth rate (SGR = 22.0% day^{-1}) at 23°C when fed high daily rations (35% body mass) (Fig. 11.5). However, the survival rate was negatively affected above 20°C, mainly due to marked growth heterogeneity and consequent cannibalism and to increased bacterial contamination. Although significantly lower than that obtained at high temperatures, SGR of perch larvae at 14°C reached 10.5% day^{-1}, demonstrating the potential for growth of this species at low temperatures (Kestemont & Mélard 1997; Kestemont *et al.* 1998a).

In walleye, problems related to NGB (e.g. lordosis, slow growth and increased susceptibility to cannibalism) can be reduced by using a surface spray which removes the oil film and cleans the surface of food and debris (Barrows *et al.* 1993; Moore *et al.* 1994; Bristow & Summerfelt 1994). Up to 99.5–100% gas bladder inflation (which occurs between days 7 and 14 at 17–20°C) is now reported in walleye with the proper use of a surface spray (0.5–1 l min^{-1}) in circular tanks. It seems important for the spray to impact the surface with enough pressure to

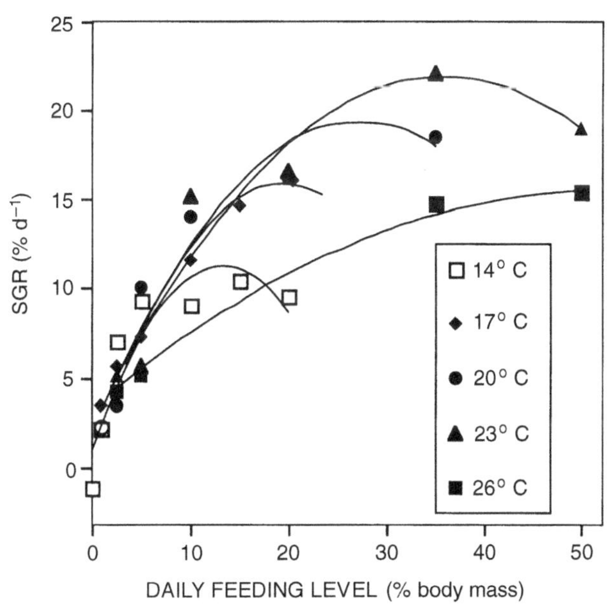

Fig. 11.5 Relationships between temperature, daily feeding level (% body mass) and specific growth rate (% day^{-1}) in perch larvae (initial body mass = 8–16 mg) fed *Artemia* nauplii.

produce a slight depression in the water (Summerfelt 1996). A similar design should be used in zander larval culture as this species is also very sensitive to non-inflation of the gas bladder (Proteau *et al.* 1993).

The effects of light intensity on feeding activity are largely dependent on physical variables such as the colour of the tank walls or water turbidity. The combination of light intensity and tank colour can modify the efficiency of food perception and the feeding behaviour of fish larvae. Several studies (Blaxter 1980; Hinshaw 1986) demonstrated that dark wall tanks facilitated prey detection by improving the contrast between the food and the walls. Corazza & Nickum (1981) and Hinshaw (1985) suggested that for walleye and yellow perch respectively, percid larvae were attracted by the white walls and ceased feeding. As a consequence, they recommended the use of grey walls and a diffuse light source. On the other hand, Tamazouzt (1995) recommended the use of white or light colour walls for the larval culture of perch under a maximum light source of 800 lux at the water surface. Recent results indicate that medium (90 lux) to high (400 lux) light intensity and long photoperiod (24 h of daylight) can significantly reduce mortality due to cannibalism in perch larvae while low intensity (5 and 30 lux) under 16–24 h of day length appears more appropriate for post-larvae (Kestemont *et al.* 1999a, b).

Use of clay slurry to increase water turbidity (up to 50 NTU) in tanks has also led to significant improvement of survival and growth in walleye fry by reducing clinging (i.e. a large proportion of larvae clinging to the walls of the tank, as an innate phototactic response), non-feeding and cannibalism (Fig. 11.6) (Bristow & Summerfelt 1994; Bristow *et al.* 1996; Summerfelt 1996). Turbidity reduces the amount of light reflected from the tank walls and induces dispersion of the fry throughout the water column and improves the ability of larval walleye to distinguish and eat food. This is probably due to the improved contrast between the brownish-orange food and the milky-grey colour of the turbid water. Alternatively, the light scattering effect of the turbidity may improve illumination of food particles (i.e. some light would be reflected on all sides of the food particles as opposed to only the top half in clear water), thereby increasing the ability of the fry to see the food. Similar negative effects of high light intensity during early life stages and improvement of survival and food consumption when water turbidity was increased have been reported for zander (Woynarovich 1960).

11.5 Ongrowing

Investigations of nutrition and animal husbandry of percid fish ongrowing have allowed production performances and the potential of semi-intensive to intensive culture to be evaluated and enhanced. Currently the ongrowing of fry to market size cannot be considered as a bottleneck to percid fish production, but rearing facilities, husbandry conditions and appropriate diet, both in quality and quantity, must be carefully defined in order to ensure significant profitability.

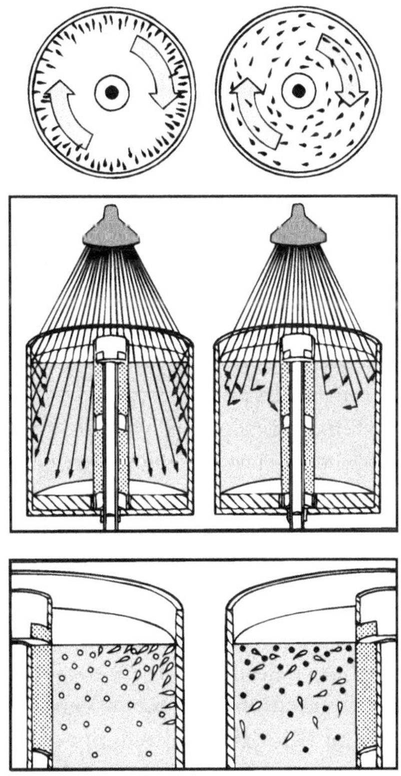

Fig. 11.6 Larval behaviour in clear and turbid water. In tanks with clear water (left), the fish are attracted to light that reflects from the black sidewalls and the meniscus (top and bottom figures). In turbid water (right), the light scattering effect of the suspended clay particles diffuses the light, reducing the reflected light from the sidewalls (middle). Fish distribute across the tanks (bottom, right) and their orientation is into the current (top right). (From Bristow & Summerfelt 1994.)

Feeds and feeding

Once trained to formulated feeds by appropriate procedures, including a mixed diet period (Kestemont *et al.* 1996) or the use of conversion diets containing freeze dried *Artemia* or krill meal (Barrows 1996), all cultured percid species accept dry diet from fry to market size. However, except for walleye which is intensively cultivated commercially and for which specific grower diet formulations have been tested and produced, the development of specific diets for percid fishes is not currently available. To date, very little quantitative research has been published on the nutritional requirements of percid fishes. Protein requirements are generally considered similar to those of salmonids, with diets containing 40–50% crude protein for maximum growth.

Fiogbé *et al.* (1996) investigated the effects of semi-purified diets containing graded levels of crude protein (casein, yeast and cod meal as protein sources) in perch juveniles (3 g initial body mass). They showed significant differences in growth and survival among fish fed diets containing up to 30% crude protein and fish fed diets $\leq 40\%$ crude protein. According to the mathematical model used, these authors determined the optimal crude protein level was between 36.8–43.6%, and the maximum growth rate was obtained between 43.1–56.5%. These results differ markedly from the optimal dietary crude protein concentration of 26% recommended for yellow perch by Ramseyer (1992). Comparisons of commercial diets formulated to meet the requirements of rainbow trout, *Oncorhyncus mykiss*, or channel catfish, *Ictalurus punctatus* (Rafinesque) also suggested that the nutritional needs of yellow perch appear to be closer to the nutritional needs of rainbow trout than those of catfish, with the exception that lower dietary fat levels in conjunction with higher crude protein concentration seems beneficial (Brown *et al.* 1996). In a series of studies, Brown *et al.* (1996) identified some potential experimental diets (with casein/arginine or crystalline amino acids as sources of protein) in nutritional studies for yellow perch juveniles. Based on such purified diets, some of the amino acid requirements for yellow perch were evaluated. The total sulphur amino acid (TSAA) requisite was estimated between 1.01–1.02% dry diet, with a cysteine replacement value of 51% of the TSAA requirement (Brown *et al.* 1998). The dietary lysine needed was estimated between 1.09–1.12% dry diet, but excessive dietary arginine concentrations increased the requirement for lysine in diets for yellow perch (Twibell *et al.* 1998).

In walleye, several grower diets were developed by the US Fish and Wildlife Service (USFWS), progressively leading to a significant improvement of growth and survival of juveniles and adults. The optimal composition of walleye grower diet contains 48% protein (mainly anchovy, blood and soybean meals), 10% fish oil (menhaden or cod liver oil) and 3500 kcal ME kg^{-1} (Barrows 1996). Knowledge of the nutritional requirements of zander is limited. Hilge (1990) observed a rapid growth of zander juveniles, from 2 to more than 300 g, in about one year under constant warm water conditions (22–24°C) with different commercial trout feeds containing 42–47% crude proteins and 8–14% of lipids.

In terms of protein content and amino acid profile, salmonid feeds are relatively appropriate for percids, however the lipid requirements of percids are probably much lower than those of salmonids. As reported for other coolwater species (e.g. hybrid striped bass, *Morone saxatilis* (Walbaum) × *M. chrysops*), percids fed diets containing high lipid levels (up to 25–33% in some salmonid diets) often have pale and fatty livers (hepatic steatosis) with reduced growth and survival rates. Given the importance of hepatic metabolism in the overall health of fish, impaired liver function could be a contributing factor in unexplained mortality and needs further research (Kestemont *et al.* 1997). Experiments conducted by the authors and colleagues, but unpublished at the time of writing, showed the importance of dietary lipid levels on growth and nutritional status of perch juveniles. Growth of fish fed 6% dietary lipid was slower than that of fish fed 12 or 18% lipid, but no significant differences in growth rate were observed between these two latter groups. On the other hand, accumulation of lipid was increased at the highest lipid level, both in number and size of lipid droplets inside the hepatocytes. Perch appear to be extremely sensitive to oxidative rancidity, unstabilised dietary lipid inducing rapid liver degeneration and high mortality (Kestemont *et al.* 1997). Several levels of dietary lipid (6, 12 and 18%), from five sources (menhaden oil, soybean oil, coconut oil, tallow, 1:1 mixture of menhaden and soybean oils) were also investigated in yellow perch juveniles by Cartwright *et al.* (1998). With all diets, fish exhibited improved growth at the lower lipid levels. Diets consisting of high saturated fats, such as coconut oil and tallow, induced the lowest increases in mass.

Effects of daily feeding levels on growth rate have been investigated in perch and walleye. The maintenance (R_{maint}), optimum (R_{opt}) and maximum (R_{max}) daily food rations of perch ranging from 3 to 300 g body mass reared at 23°C were modelled by Mélard *et al.* (1996b) as follows:

$$R_{maint} = 1.09\,w^{-0.23} \qquad R_{opt} = 3.30\,w^{-0.24} \qquad R_{max} = 7.60\,w^{-0.31}$$

where w = fish body mass (g) and R = daily ration (% body mass day^{-1}).

In addition to growth, the feeding level also affects mass heterogeneity in perch. Fontaine *et al.* (1997a) showed that after 84 days the coefficient of variation ($cv = sd/$ mean) for mass in a perch population reared in a tank decreased from 45–50 to 30–35% when the fish were fed one or two per cent body mass per day feeding level, while the *cv* remained constant at three per cent. The final autopsy of perch revealed that a three per cent feeding level favoured a sexual growth dimorphism (in favour of females) while no significant mass differences between females and males were recorded at the lower feeding levels.

The effect of feeding frequency and mode of feeding (hand, automatic or on-demand) has been investigated in several percid species. In perch, highest growth rates were recorded when 30 g fish were fed at the maximum daily food ration three times a day, while a feeding frequency of six meals per day supported the highest growth rate at a restricted ration, but increased the size heterogeneity of the population (Kestemont & Mélard 1996; Mélard *et al.* 1996b). Automatic feeding provides a satisfactory food conversion ratio (*FCR* = mass of food supply/fish mass gain), usually 1.1–2.0 in perch ranging in size from 3–300 g. Significant improvements of growth and feed efficiency were recently reported by Fontaine *et al.* (1997b) and Anthouard & Fontaine (1998) with the use of on-demand feeders. The importance of feeding strategy on walleye growth, health and production economics was discussed by Summerfelt & Summerfelt (1996). Since walleye do not feed at the surface but consume feed while it is sinking, it is difficult to observe feeding in this species and to evaluate the satiation level. The authors suggest that the amount of unused food can be determined from feed pellets in the effluent stand-pipe or solids trap, or the use of an ultrasonic waste feed controller (Juell 1991; Summerfelt *et al.* 1995).

Animal husbandry aspects of percid ongrowing

All cultured percid species can be produced up to a market size in various culture systems (ponds, cages and tanks), but rearing at ambient temperatures is impeded by the relatively slow growth rate of percid fish at temperatures < 20°C and the long turnover time from fry to market size. Percid fish are thermophilic species and their optimum temperature for growth ranges from 22–24°C in perch and yellow perch (Brown *et al.* 1996; Mélard *et al.* 1996b), from 22–26°C in walleye (Summerfelt & Summerfelt 1996) and from 24–30°C in zander (Hilge and Steffens 1996). Good growth of percids in ponds and cages can only be achieved in regions with long warm periods of weather which do not exist over the entire range of these species.

In perch, minimum marketable size (80–100 g, Tamazouzt *et al.* 1993) can be reached by the end of the second summer in lake cages under ambient temperatures, only if one year old perch (50–60 g) have been reared in intensive conditions under constant heated water (20–23°C) during their first year (Kestemont *et al.* 1998b). Based on data obtained during a three year study of perch ongrowing in cages, a simulation model of production indicated that under ambient temperatures ranging from 14–20°C during the growing season (May to October), production of perch varied from 15 to 160 $g\,m^{-3}\,day^{-1}$ when fish were reared at relatively high stocking density (30 $kg\,m^{-3}$ at the end of the production cycle) (Kestemont *et al.* 1998b). Cage ongrowing was also recommended by Fontaine *et al.* (1996), for one year old perch (10–20 g) harvested from ponds in spring, but the proportion of fish reaching market size by the end of the second summer was relatively low. Two to three years are also required to grow marketable sized yellow perch under ambient temperatures in ponds (Heidinger & Kayes 1986). Summerfelt & Summerfelt (1996) considered the production of walleye to market size in ponds was not practical. Culture included the following steps: fingerlings (32–64 mm) harvested from ponds and trained to eat dry feed in tanks or grow-out ponds; grown up to 127–203 mm TL by the end of the first summer; over-wintered with low mortality if ponds are adequately aerated but a decline in fish condition; cultured during the second summer; and reaching a market size (*c.* 250 mm and 560 g) between the end of the second summer and the middle to later part of third summer, or even later depending on geographical location and climate conditions. The cage production of food size walleye requires from two to two and a half growing seasons (Stevens 1996). The ongrowing of zander in ponds also varies largely according to the area of production. By the end of the first summer, zander juveniles reach a size ranging from 50–80 mm in Finland (Ruuhijärvi & Hyvärinen 1996), 100–120 mm in Belgium (Kestemont 1995), 130 mm (50 g) in Hungary (Hilge & Steffens 1996) and to 150–180 mm in the south of France (Schlumberger & Proteau 1991). Under very favourable temperature conditions (southern regions of Russia), zander may reach 270–300 mm (300 g) at the end of the first summer. In most regions, a minimum market size of 350–380 mm (400–500 g) requires three to five summers.

Intensive ongrowing under a controlled temperature of 22–24°C (slightly higher for zander) is considered as the most suitable alternative for the production of percid fishes, even if not extensively applied to date at a commercial level. Such high temperatures maintained during the whole ongrowing phase allow market size fish to be obtained in about 14 months for perch (Mélard *et al.* 1996b) and zander (Hilge 1990), 16–18 months for yellow perch (Malison personal communication) and 16 months for walleye (Summerfelt & Summerfelt 1996) (Fig. 11.7).

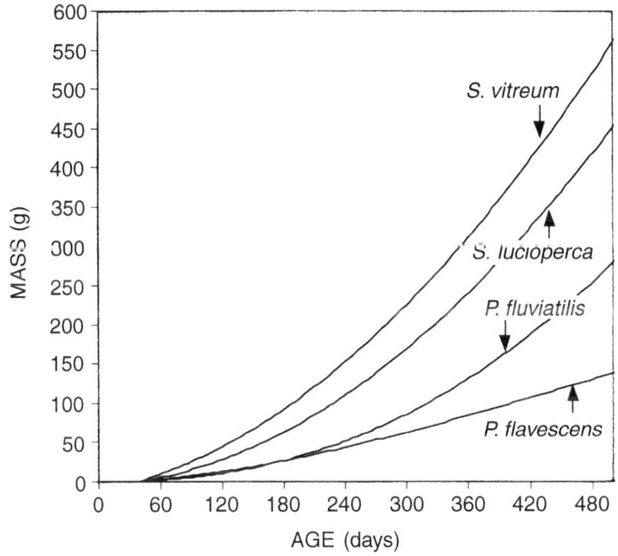

Fig. 11.7 Typical growth curves of *P. fluviatilis*, *P. flavescens*, *S. lucioperca* and *S. vitreum* reared under intensive conditions and optimal temperature. (Data from Mélard *et al.* 1996b; Malison personal data; Hilge & Steffens 1996; and Summerfelt & Summerfelt 1996.)

Percid fish are tolerant to crowding and to rearing at high densities (50–120 kg m^{-3}) in recirculating systems, as long as oxygen levels are maintained at > 4–5 mg l^{-1}. In perch juveniles (from 1–20 g body mass), increased stocking density from 400–10 000 fish m^{-3} resulted in a 67% increase in growth rate, probably because of a density-dependent inhibition of territorial and agnostic tendencies that potentially limit access to food. In larger fish, growth rate decreased with increasing stocking densities, mainly due to increase in waste. According to Mélard & Kestemont (1994) the maximum production rate (0.8 kg m^{-3} day^{-1}) was achieved at 23°C at a stocking biomass of 60 kg m^{-3}. A similar stock biomass, at the end of the growing period, was achieved in walleye reared in a recycle culture system with ambient oxygen (Summerfelt & Summerfelt 1996).

Growth heterogeneity is a major concern in percid species and must be considered in stocking and harvesting strategies. As an example the mass of seven month old perch (average body mass = 26 g) ranged from 7–89 g (Mélard *et al.* 1996b). Moreover, once sorted in three non-overlapped mass groups, the specific growth rate of each fish group was similar and fast-growing fish emerged within each group. This suggests that growth heterogeneity is not exclusively genetic but could also be a consequence of social interactions (Mélard *et al.* 1995). The technique of regular size-sorting during ongrowing remains questionable and preliminary results (Mélard *et al.* 1996b) tend to demonstrate that the size-sorting may cause a slight deficit (5–6%) in growth and production of perch, compared to non-sorted fish. Another common feature of percid growth is sexual dimorphism with faster growth and larger ultimate size of females (Scott & Crossman 1973; Schott *et al.* 1978; Colby *et al.* 1979; Guti 1993; Chapter 3). In wild walleye differences in growth between the sexes begin when fish reach 1–3 years of age, while in intensive culture conditions, female walleyes begin to outgrow males at 200–250 mm TL and 100–150 g total mass (Malison *et al.* 1998b). Sexual dimorphism of perch occurs at a much smaller size, 80–100 mm TL and 5–10 g total mass, corresponding to the onset of physiological events such as vitellogenesis and spermatogenesis (Malison *et al.* 1986). Malison *et al.* (1985) demonstrated that oestradiol-17β (E_2) promotes yellow perch growth after a certain size-related maturity stage is reached. Although the onset of sexual growth dimorphism has not yet been clearly determined in perch, similar effects of oestrogens on food consumption and growth promotion in both sexes have been recently demonstrated (Kestemont unpublished data).

11.6 Genetic improvements

The faster growth of females and the negative impact of sexual development (which in perch begins before fish reach a marketable size) on somatic growth have led to the development of methods for producing monosex female populations and for inducing sterility. These methods have been mainly applied to yellow perch and walleye. They are under investigation in perch but have not been recorded for zander. This section will thus be limited to the recent development in sex control and ploidy manipulations in the North American species. Finally, survival and growth performances will be compared between pure-bred and hybrid percid fish.

Production of monosex female populations

Since yellow perch and walleye females are the homogametic sex (Best 1981), three methods of producing monosex females can be potentially applied: (a) the direct oestrogen treatment of juveniles, (b) gynogenesis and (c) the indirect use of hormones, in which juveniles are treated with androgens to induce phenotypic sex reversal of genetic females (xx) and the sperm from these 'masculinised' females is used to fertilise normal eggs (Malison & Garcia-Abiado 1996). The last method is recommended for the production of monosex females in both yellow perch and walleye (Malison personal communication).

According to Malison *et al.* (1986) the first sign of sexual differentiation in yellow perch was apparent in fish of 16–18 mm TL when female gonads had fused. In walleye, where the ovaries remain paired, the onset of sexual differentiation occurs at a larger size (Malison *et al.* 1990). The sexes can be distinguished morphologically at 100 mm TL based upon the greater cross-sectional area of ovaries compared to testes. In yellow perch of 20–35 mm TL, the administration of 17α-methyl-testosterone (MT) at 60 $\mu g\ g^{-1}$ diet for 84 days at 21°C induced spermatogenesis and the formation of ova-testis in females (Malison *et al.* 1986). In perch, MT treatment must be applied when fish reach 40–70 mg body weight. The strategy for producing monosex female populations of walleye is similar to that of yellow perch, but young fish are fed a diet containing 17α-methyl-testosterone at 15 μg g^{-1} of food for 60 consecutive days when they reach a mean size of 50 mm TL (Malison *et al.* 1998b). Hormonal treatments applied to walleye of 70–90 mm TL did not alter germ cell differentiation in females, probably because, at that size, gonad differentiation in females had proceeded to a sufficient degree such that phenotypic sex could no longer be redirected by MT treatment (Malison *et al.* 1990). At two years, the masculinised females can be identified by the presence of both ovarian and testicular tissue. For perch, yellow perch and walleye, the induction of partial rather than complete sex reversal can be accomplished by timing the initiation of MT treatment to approximately coincide with the onset of oogenesis. The main advantage of partial sex inversion is the ease with which, at sexual maturity, genetic males can be distinguished from genetic females. In both species milt collected from masculinised females was used to fertilise normal eggs, and in all instances the resultant offspring were 100% females (Malison & Garcia-Abiado 1996; Malison *et al.* 1998b). The direct oestrogen treatment consists of administrating E_2 at 15–120 μg g^{-1} diet for 84 days in yellow perch of 20–35 mm TL. Meiotic gynogenesis of yellow perch and walleye should include the production of 100% haploid embryos by fertilising normal eggs with genetically inactivated spermatozoa (by exposure of yellow perch or walleye spermatozoa to ultraviolet irradiation at doses of 3240–6480 ergs mm^{-2}), and subsequently exposing haploid zygotes to either physical (heat or hydrostatic pressure shocks) or chemical treatments (colchicine, cytochalasin B or nitrous oxide) to disrupt either the second meiotic division or the first mitotic division and thereby restore the normal diploid state (Malison & Garcia-Abiado 1996). However, according to these authors, production of gynogens has not yet been achieved in percid species.

Induction of triploidy and tetraploidy

Triploidy can be induced in fishes by shocking eggs shortly after fertilisation to inhibit the extrusion of the second polar body (Purdom 1983; Thorgaard 1983; Ihssen *et al.* 1990). In both yellow perch and walleye, initial efforts at inducing sterility focused on the direct use of either heat or hydrostatic pressure shocks to produce triploid fish (Ewing 1989; Malison *et al.* 1993a). In yellow perch Malison *et al.* (1993a) recommended heat shocks from 11 to 28–30°C applied at a time of initiation (TI, minutes after fertilisation) of five minutes for 10–25 min, and hydrostatic pressure shocks of 627 or 766 kg cm^{-2} (9000–11 000 psi) applied at a TI of five minutes for 12 min. In walleye, the most efficient methods are heat shocks applied at a TI of one to five minutes at 28–31°C for 25 min, and hydrostatic pressure shocks of 557 kg cm^{-2} (8000 psi) applied at TI of four minutes for 30 min (Ewing 1989; Malison & Garcia-Abiado 1996). Similar heat shock treatments have been successfully applied to perch resulting in 100% triploid fish. Heat shock is preferred for induction of triploidy in yellow perch while hydrostatic pressure shock is better in walleye. The gonadal development of triploid yellow perch and walleye of both sexes is retarded compared to that of diploids, and triploid yellow perch can have higher fillet yields than diploids (Malison *et al.* 1993b; Malison & Garcia-Abiado 1996). In order to avoid the negative effects of heat or pressure shocks on growth, methods of producing triploids by crossing fertile tetraploids with diploids are still under investigation. Malison *et al.* (1993a) reported, in yellow perch, a rate of 89–100% of tetraploids with post-hatch survival of up to 80% by applying hydrostatic pressure shocks of 9000 psi at a TI of 192 min for 16–24 min. Less successful results (40% tetraploids and 10% post-hatch survival) were obtained in walleye using pressure shocks of 9000 psi at TI of 192 min for eight minutes (Malison & Garcia-Abiado 1996).

Performances of hybrid percids

With certain species, inter-specific cross-breeding produces hybrids with growth performances better suited for intensive culture than are those of pure-bred species. This is the case for the hybrids of muskellunge (*Esox masquinongy*) × pike (*E. lucius*) or striped bass (*Morone saxatilis*) × white bass (*M. chrysops*). In percid fishes, natural hybridisation between walleye and sauger was reported by Colby *et al.* (1979) and both reciprocal hybrids were artificially propagated under laboratory conditions (Nelson *et al.* 1965). Malison *et al.* (1990) compared the survival, growth and reproductive development of juvenile walleye, sauger and their hybrids reared under intensive culture conditions. After an 84 day experiment with juveniles ranging from 9–18 g body mass, saugeye had significantly greater mass and length gains and higher condition factors than either pure-bred walleye or S♀ × W♂ hybrids, whereas pure-bred sauger grew slower than the other fishes. Malison *et al.* (1990) observed that saugeye were less aggressive and more tolerant to routine disturbances and handling. Gonadal development of juvenile walleyes and sauger was similar.

Kestemont & Malison (unpublished data) have succeeded in the hybridisation of perch and yellow perch. Preliminary results indicate that both hybrids grow faster than pure-bred fish, with a mass gain about 30% higher in the hybrid after one year of rearing under intensive conditions. No growth differences have been observed between the progeny of the cross between perch ♀ × yellow perch ♂ and those of the reciprocal cross. Experiments are currently in progress to evaluate the reproductive development of the hybrids.

11.7 Pathologies in aquaculture environment

Percids are cultivated in a variety of environments (e.g. ponds, cages and tanks with flow-through or recirculating systems) and are exposed to a wide range of parasites and diseases with all major groups being represented (Chapter 8). A preliminary overview of parasites and diseases of perch in intensive culture systems was reported by Grignard *et al.* (1996a). The need to pay particular attention to viruses associated with diseases of walleye, sauger and their hybrids has been discussed by McAllister (1996).

Protozoa are the most frequent parasites in cultured perch. They do not normally have any significant impact unless they proliferate (Grignard *et al.* 1996a). *Ichthyobodo necator* and *Trichodina* sp. have been found in skin smears and gills of cultured perch, both in flow-through tanks and lake-based cages. Clinical signs of infection of perch and walleye fry and juveniles by *Trichodina* sp. are failure to feed, tail chasing, emaciation, pinheads, erosion of fins and excess mucus production (Grignard *et al.* 1996a; Marcino 1996). These protozoa may cause massive mortality in larvae and juveniles although adult fish seem unaffected. *Ambyphria* sp. was also found on fry less than 1 g, affecting respiration under severe infestation, but was not recorded in juveniles and adults. The joint infestation by *Heteropolaria* sp. (Epistylididae) and *Aeromonas* sp. represents a symbiotic pathogenic complex, well known as 'red sore disease' (Lom & Vàvra 1961). The colonial ciliate is most frequently detected at temperatures >17°C, when it colonises the spiny dorsal fin of young perch and spreads all over the body in older individuals (Grignard *et al.* 1996b). It can affect survival and growth in perch cultured in flow-through systems at high rearing density, the infestation frequency varying from 10, 50 and 91% at stocking densities of *c.* 300, 600 and 2000 40 g-fish m^{-3} respectively. Infestation and impact of *Heteropolaria* sp. can be greatly reduced by improvement of water quality in recirculating systems or by reduction of fish handling and sorting in flow-through systems, combined with regular preventive treatment of formalin + salt (Mélard & Kestemont 1994). Among the other ectoparasites, *Gyrodactylus* sp. and *Dactylogyrus* sp. were rarely reported in perch reared in tanks but were observed in perch stocked in cages in a eutrophic lake (Kestemont *et al.* unpublished data). The large ciliated protozoan, *Ichthyophthyrius multifiliis*, is a common parasite of walleye (Marcino 1996), even in recirculating systems. It can cause mass mortality although it apparently does not produce symptoms of disease in cultured perch (Grignard *et al.* 1996a).

Viruses, bacteria and fungi may induce severe diseases often resulting in high mortality in intensive or cage culture systems. Virus-associated disease in adult walleye and sauger is well documented, while the viral infections appear much less important in perch. According to McAllister (1996), the four viral diseases of walleye are lymphocystis, dermal sarcoma, diffuse epidermal hyperplasia and discrete epidermal hyperplasia. Infectious pancreatic necrosis has also been isolated from walleye, but no signs of disease were evident. Only one virus-associated disease, lymphocystis, is documented for sauger and no virus associated disease has been found in saugeye. While bacteria were rarely identified as the primary cause of disease outbreak in wild percids (Chapter 8), it appears that these organisms are involved in mass mortality in cultured perch. The main species identified by Grignard *et al.* (1996a) in intensive systems were *Aeromonas* sp., *A veronii*, *A. hydrophila*, *Streptococcus* sp., *Staphylococcus* sp., *Vibrio fluviatilis* and *Enterobacter agglomerans*. In walleye, only two types of bacteria were observed from nine batches of fish from private hatcheries (Marcino 1996): *Yersinia ruckeri* (enteric redmouth disease) and *Cytophaga* (*Flexibacter*) *columnaris*, the latter being considered as the primary bacterium of concern in walleye culture (Nickum 1978).

Good animal husbandry is important in the culture of percids to avoid or reduce the effects of pathogens. Reduction in handling and sorting of perch reduces disease outbreak. Perch are particularly sensitive to manipulation and transport at low temperature, with disease and mortality appearing within a few of weeks after transport. This is probably due to loss of mucus during handling and very slow mucus production below 15°C. Such sensitivity to handling has not been reported in yellow perch. Zander is also considered to be highly sensitive to transport but mortalities usually occur during the transport itself, probably due to acute stress. For all percids addition of salt at a concentration of 3 g $NaCl\,l^{-1}$ is usually recommended to alleviate stress during handling and transport.

11.8 Conclusions

The culture of percid fishes has significantly progressed during recent decades, with many improvements in husbandry. Among the different techniques, the most significant are the control of out-of-season spawning in yellow perch and walleye, and progress in the control of the perch reproductive cycle. Although mainly based on live food during the early life stages, larviculture of percid fishes has also been improved recently. It does remain dependent on the controlled production or supply of live food during the first weeks after hatching. Research efforts have been directed to larval nutrition but techniques are still needed to develop a suitable artificial dry diet, which might be used from hatching onwards. Recent research has also been focused on practical diets for grow-out, production of monosex (female) populations and increasing growth rates. In this last respect, the relatively poor growth of both perch and yellow perch probably constitutes the major bottleneck for their intensive large scale production. However, genetic improvements, including the selection of suitable strains, combined with an increasing knowledge of nutritional requirements, should support the development a profitable percid fish industry in the near future.

Acknowledgements

The authors are grateful to several colleagues who contributed substantially to the preparation of this chapter by providing unpublished data, reports and figures. Special thanks to Jeff Malison and Jim Held (Wisconsin–Madison University), Robert Summerfelt (Iowa State University), Paul Brown (Purdue University), Olivier Schlumberger (CEMAGREF–Montpellier, France), Jacques Rinchard, Sergiuz Czesny and Konrad Dabrowski (Ohio State University). Thanks also to John and Hilary Craig for editing.

Chapter 12
Etheostomatinae

Lawrence M. Page
Illinois Natural History Survey, Champaign, Illinois 61820 USA

12.1 Introduction

Darters are small percids (43–200 mm TL) restricted to North America. They are found in the Mississippi River basin and drainage of the Great Lakes, Hudson Bay, Atlantic Coast, Gulf of Mexico and Pacific Coast of Mexico. With 173 recognised species, darters are second only to minnows (Family Cyprinidae) in diversity among freshwater fishes in North America. Darters often occur in large populations and have a considerable impact on aquatic ecosystems.

The largest diversity of darters is found in the Tennessee and Cumberland River systems in central and eastern Tennessee and northern Alabama. A second centre of diversity is in streams in the Ozark and Ouachita Mountains in western Arkansas and eastern Oklahoma. These centres of darter diversity are characterised by a large amount of topographic relief and are located near the southern edge of the area of eastern North America that lies between the coastal plain and the maximum extent of Pleistocene glaciation. Significant characteristics of these centres of abundance are their geological antiquity (providing ample time for species diversification), spatial heterogeneity (providing opportunities for isolation and diversification) and, because of their southern positions, relatively stable climates (perhaps resulting in little extinction).

All percids have ctenoid scales, two dorsal fins that are separate or narrowly joined to one another (the first with spines, the second with soft rays), one or two anal spines, pelvic fins that are thoracic in position and have one spine and five soft rays and five to seven branchiostegal rays. In contrast to other percids, the supraoccipital bone of darters is greatly reduced (Collette & Bănărescu 1977) and they have an enlarged coronomeckelian, simple (rather than enlarged and spatulate) proximal ends on branchiostegals V-VII and four or fewer sensory pore foramina on the dentary (Wiley 1992). The four genera of darters (Plate 5), *Percina*, *Crystallaria*, *Ammocrypta* and *Etheostoma* constitute the Subfamily Etheostomatinae of the family Percidae (Song *et al.* 1998; Chapter 1).

The generic and sub-generic classification (Table 12a) followed in this chapter is that originally proposed by Bailey (Bailey 1951; Bailey *et al.* 1954; Bailey & Gosline, 1955) with modifications by Collette and Yerger (1962), Cole (1967), Howell (1968), Page (1974a, 1981), Page and Whitt (1973a), Simons (1991) and Stephens *et al.* (pers. comm.). Simons (1992), using osteological data, suggested that *Ammocrypta* is closely related to the subgenera *Boleosoma* and *Ioa* of *Etheostoma* and that *Ammocrypta* should be a subgenus of *Etheostoma*; however, subsequent studies using molecular data indicate that *Ammocrypta* is not closely related to *Boleosoma* and *Ioa* (Wood & Mayden 1997; Song *et al.* 1998) and is recognised here as a genus.

The genus *Percina* contains nine subgenera and 40 species (Table 12a). Males of all species of *Percina* have modified (enlarged and strongly toothed) scales (Plate 4) on the breast. Modified scales are also on the breasts of most females of *Percina* and in a row along the midline of the belly in males (and occasionally in females) in all subgenera of *Percina* except *Odontopholis* (Page 1976). Darters in genera other than *Percina* do not have modified scales. The monophyly of *Percina* also is supported by a unique lactate dehydrogenase (LDH) allele that is apparently fixed in *Percina* and absent in *Ammocrypta, Crystallaria* and *Etheostoma* (Page & Whitt 1973a,b; Dimmick & Page 1992).

Other characteristics of *Percina* are a complete and nearly straight lateral line, uninterrupted head canals, modally 10 preoperculomandibular pores and eight infraorbital pores, two anal spines, opaque flesh (i.e. not strongly translucent when alive) and a moderate to deep body (mean body depth/standard length (SL) = 0.14–0.20).

Ammocrypta includes six species of highly distinctive darters that inhabit sandy runs of streams. Unlike species of *Percina* and all but one species of *Etheostoma*, *E. vitreum* (Cope), species of *Ammocrypta* are translucent when alive. The body is long and slender; the body depth goes seven to nine times into the SL (mean body depth/SL = 0.11–0.13). Only one anal spine is present, head canals are uninterrupted, the premaxillary frenum is absent, the lateral line is complete and straight anteriorly (deflected downward posteriorly in some species), the preoperculomandibular canal modally has eight to 10 pores and the infraorbital canal usually has eight pores.

The genus *Crystallaria*, containing only *C. asprella* (Jordan), was until recently (Simons 1991) considered a subgenus of *Ammocrypta*. It shares with *Ammocrypta* the long slender body, translucent flesh, one anal spine, uninterrupted head canals, lack of modified scales and lack of the *Percina* LDH allele. It is distinguished from *Ammocrypta* by the presence of a narrow premaxillary frenum, deeply forked caudal fin, four large dark dorsal saddles, 12–15 anal rays and modally 12–15 dorsal spines and 13–14 dorsal rays (*Ammocrypta* species have seven to 12 anal rays and modally nine to 12 dorsal spines and 10–11 dorsal rays). It differs from all other darters in possessing a bifurcate supraoccipital crest (Simons 1991). The lateral line is complete and straight; the preoperculomandibular canal and infraorbital canal modally have 10 and eight pores, respectively.

Etheostoma, the largest genus of darters (and the largest genus of North

American freshwater fishes), includes 126 recognised species in 17 subgenera (Table 12a). *Etheostoma* is an assemblage of darters not related to *Percina*, *Crystallaria*, or *Ammocrypta*, has no derived characteristic shared by all species and may not represent a monophyletic group (Song *et al.* 1998). All species of *Etheostoma* lack the modified scales and LDH allele of *Percina* and all species, except *E. vitreum*, lack the translucent flesh of *Ammocrypta* and *Crystallaria* (*E. vitreum* often has two anal spines, has more pigment and is somewhat less slender than species of *Ammocrypta* and *Crystallaria*). As suggested by the large number of species in the genus, other morphological traits are highly variable. The lateral line is complete or incomplete and straight or arched upward anteriorly, one or two anal spines are present, head canals are uninterrupted or interrupted, preoperculomandibular pores modally number six to ten and infraorbital pores number three to eight.

12.2 Morphology

The strong correlations between certain morphological traits and habitat suggest that physical constraints imposed by the habitat have been of major influence in the evolution of darter morphology. Among these traits (some of which are discussed for fishes in general e.g. by Breder 1926, Hubbs 1941, Alexander 1967, Aleev 1969 and Webb 1975 and for darters in particular by Page & Swofford 1984) are the degree of fusiformity, snout and body shape, premaxillary protrusibility, size of the gas bladder and features of pigmentation.

Some species are less restricted than others to particular habitats and many undergo ontogenetic shifts in habitat preference. The following species are among the most representative, as adults, of particular habitats:

Gravel run, mid-water – *Percina sciera* (Swain)
Pool with current, mid-water – *P. maculata* (Girard)
Sand run, benthic – *Ammocrypta* spp., *Etheostoma vitreum*
Rock riffle, benthic – *P. roanoka* (Jordan & Jenkins), *E. blennioides* Rafinesque, *E. caeruleum* Storer
Quiet pool, benthic – *E. chlorosomum* (Hay), *E. gracile* (Girard).

Mid-water species of darters are relatively fusiform (i.e. the body is more or less cylindrical in cross-section, broadest in the middle and tapering at both ends); benthic species have reduced fusiformity resulting from varying combinations of a more flattened venter, more compressed sides, an anterior shift of the greatest body depth (riffle species) and a blunter snout (pool and some riffle species). A flattened venter reduces exposure to current by shifting the centre of gravity of the darter closer to the stream bed where the velocity of the water is less than that in mid-water (Hynes 1970). Although living in fast-flowing water, riffle darters may further avoid current by selecting shelters among components of the substrate (Harding *et al.* 1998).

Riffle species are more robust than other forms, as shown by proportional measures of body depth and head width (Page 1981, 1983). Riffle species have proportionally deep caudal peduncles; species living in sandy runs have narrow peduncles. The pectoral fins of mid-water darters are proportionally shorter than those of benthic forms. The second dorsal fins of riffle darters are longer than those of pool species. The snouts of sand-run species are usually proportionally longer than those of riffle and pool species.

The efficiency of securing food is probably the most important factor influencing the morphology of the snout. Almost all darters studied have been found to feed mainly on immature aquatic insects and variation in snout shape among darters is determined by where darters feed more than what kinds of organisms they feed on. Mid-water species feed off the bottom at an angle, grope among the substrate materials for food items and have pointed snouts and terminal mouths. Attenuation of the snout of some species of *Percina*, e.g., *P. nasuta* (Bailey), *P. squamata* (Gilbert & Swain) and *P. macrocephala* (Cope), is extreme. Benthic species often capture food organisms while remaining nearly parallel to the substrate or to the rock on which they are foraging; for these species a blunt snout puts the sub-terminal mouth in the most proximal position to the food. In investigating five species of *Etheostoma*, Wehnes (1973) concluded that sub-terminal mouths have developed for feeding on insects inhabiting the tops of rocks and that terminal mouths facilitate feeding on the sides of rocks.

Species living in riffles or gravel runs have a well developed frenum connecting the premaxillary bones to the snout. Species living in quiet pools (e.g. *E. nigrum* Rafinesque) or sand runs (e.g. *Ammocrypta* spp. and *E. vitreum*) often have no, or a much reduced, frenum. A frenum may be selected for or against depending on characteristics of the substrate where the darter is living. Darters on gravel and rocks feed by darting and grabbing their prey, whereas species on sand or mud have to remove their food more carefully from the less stable substrate. Greenberg (1991) observed *E. jessiae* (Jordan & Brayton), a species lacking a frenum, to pick food from sand. The absence of a frenum may provide more flexibility to feeding manoeuvres of pool and sand run species.

Gas bladders are well developed in large percids such as the yellow perch, sauger and walleye. They remain present in mid-water darters but have been lost in benthic darters. In 17 species of darters examined (Page 1983), large gas bladders were found only in the highly mid-water *P. cymatotaenia* (Gilbert & Meek) (the bladders in three specimens examined averaged 26% SL) and moderate gas bladders were found in *P. sciera*, *P. phoxocephala* (Nelson) and *P. maculata* (mean bladder length for each was about 15–16% SL). A small gas bladder (8% SL) was found in one of three adult *P. evides* (Jordan & Copeland) (a moderately benthic species) examined; the other two had no gas bladders. Gas bladders were absent in *A. pellucida* (Putnam), *E. caeruleum*, *E. nianguae* (Gilbert & Meek), *E. nigrum*, *E. vitreum*, *E. zonale* (Cope), *P. copelandi* (Jordan), *P. palmaris* (Bailey), *P. roanoka*, *P. shumardi* (Girard), *P. ouachitae* (Jordan & Gilbert) and *P. antesella* Williams & Etnier, all of which live on the bottoms of streams. As expected, mid-water species move around much more than do benthic species (Greenberg 1991).

Except for the bright colours of breeding males, the pigments and body patterns of darters probably enhance the cryptic nature of these fishes. Most mid-water darters lack bright colours. Mid-water species of *Percina* characteristically have a dark horizontal mid-lateral stripe; the stripe reaches its extreme development in the subgenera *Hadropterus*, *Alvordius* and *Odontopholis*. The stripe is black, dorsal to a white venter and ventral to a pattern of alternating black and white blotches or vermiculations. The mid-lateral stripe serves to disrupt the body outline of mid-water darters viewed laterally by potential predators and in doing so makes the fish less vulnerable. Such disruption of the body outline is a common form of camouflage among animals. Camouflage is further enhanced by the forward extension of the stripe through the eye, absorbing the otherwise unmitigated black pupil (Barlow 1972).

Most quiet-pool species are drably coloured, speckled with black or brown pigment and blend in well with the small-grained substrate characteristic of quiet pools. Exceptions are species such as *E. gracile* that live in heavily vegetated habitats and have green bars and bands in the fins to conceal them when viewed among green plants. Species inhabiting sand runs are translucent, without bright colours (except yellow) and are nearly invisible on sand.

Darters in gravel riffles are often brightly coloured and vertically barred. The bright colours may help to conceal the fishes. Many species living among the coloured stones of stream riffles (e.g. species of the subgenera *Nothonotus* and *Oligocephalus* of *Etheostoma* and *Ericosma* of *Percina*) have hues of red, orange and yellow and, those showing an affinity for plants (e.g. *E. blennioides* and *E. zonale*), are green. It may also be that living in the shallow water of riffles reduces the danger of predation to a level at which selection for crypticity is no longer a dominating influence on pigmentation and the epigamic advantages of bright pigmentation prevail. Males of most riffle-inhabiting species are coloured more brightly than females.

Whereas mid-water species are probably viewed laterally by potential predators and often have a mid-lateral stripe, benthic species are seen dorsally or dorsolaterally and often have vertical bars disrupting their body outlines. A common pattern in darters and other North American fishes living on rocky substrates is four dark saddles on a light background. The light spaces between the saddles mimic rocks and the dark saddles appear as shadows or gaps between rocks. The saddles are spaced unevenly on the body as rocks in streams are a mixture of sizes and a fish that mimics rocks of different sizes is less conspicuous than one that mimics a series of rocks of similar sizes (Armbruster & Page 1996).

The relationship of pattern to the habitat occupied is demonstrated by the ability of some species to change their colour pattern as they change habitat. For example an adult *P. palmaris* captured in deep riffles is strongly vertically barred; on an adult captured elsewhere, the vertical bars are nearly absent and the mid-lateral dark stripe, characteristic of mid-water species of *Percina*, is present.

A prominent black basicaudal spot (or spots) is common among darters living in gravel runs and pools with a current. Black basicaudal spots are even present on mostly brown forms such as *P. phoxocephala* and *E. nianguae.* The spots probably serve as false eyespots and thwart predators by directing their strikes toward the rear of the body. Riffle-inhabiting species do not have well-defined black basicaudal spots, again suggesting that predation pressure in the shallow water of riffles may be considerably less than elsewhere in the stream environment. Rather than a medial black spot, riffle forms, such as *E. caeruleum*, *E. camurum* (Cope), *E. zonale*, *P. palmaris* and *P. roanoka*, have two large, vertically aligned, white or yellow blotches on the basicaudal region. The function of the white or yellow blotches is unknown.

Almost all darters have black preorbital bars extending from the eyes to the tip of the snout. The bars may function as sights to facilitate striking at prey or in camouflaging the otherwise conspicuous eyes (Barlow 1972), or both. Suborbital and postorbital bars are present on most species and probably disrupt the outline of the eye. Suborbital bars are generally characteristic of deep-bodied fishes; pre- and postorbital bars are most characteristic of slender fishes (Barlow 1972).

The largest species of darters are members of *Percina* and the smallest are advanced species of *Etheostoma.* At its maximum recorded length of 200 mm TL (Douglas 1968), *P. lenticula* Richards & Knapp, the largest darter, is almost five times as long as the smallest darter, *E. fonticola* (Jordan & Gilbert), which reaches a maximum length of only 43 mm TL (Page & Burr 1979). *Percina lenticula,* at a maximum mass of about 70 g, exceeds *E. fonticola,* at a maximum of about 0.6 g, in mass by a factor of over 100.

Darters are much smaller than their percid relatives. The yellow perch reaches 400 mm, the sauger 760 mm and the walleye 910 mm. Reduction in size, coupled with their benthic habits, may be the most important factor in explaining the considerable evolutionary success (both in numbers of species and of individuals) of darters. The large numbers of very small benthic organisms (<5 mm long) available in a stream as food promote the evolutionary diminution of consumers, especially when the consumers are forced to forage in a complex substrate (e.g. a rubble riffle) (Miller 1979). A predator's body size should relate to its prey size in a way to maximise net energy gain per unit time spent foraging (Schoener 1971); in fishes, growth efficiency increases as predator size decreases relative to that of the prey (Kerr 1971).

Smaller size increases exposure to a greater number of predators. However, diminution in benthic darters also enhances their concealment among the components of the substrate. Sand-dwelling darters (species of *Ammocrypta* and *E. vitreum*), benthic and relatively small, occupy a simple substrate but reduce exposure to predators through cryptic colouration (translucency) and by burrowing into the sand (Williams 1975). Mid-water darters, more exposed than benthic darters to predators, are relatively large.

In addition to a reduction in body size, there are, in more advanced darters, reductions in the lateralis system and in meristic (scale, fin-ray and vertebral) counts. Basal species (e.g. *Crystallaria asprella*) have complete lateral lines and uninterrupted head canals and they usually have high meristic counts (Page 1981, 1983). Derived traits include incomplete lateral lines, interrupted head canals (Page 1977) and low meristic counts. *Etheostoma microperca* Jordan & Gilbert, possibly the most derived darter (Bailey & Gosline 1955; Burr 1978), has extremely reduced meristic values (Page 1981, 1983).

Suites of traits are characteristic of darters found in particular habitats. Darters in gravel runs and flowing pools are mid-water, large, fusiform, have pointed snouts, a large interorbital width, short pectoral fins, non-protusibile premaxillaries, one or more prominent black basicaudal spots, usually are without bright colours, often have a black mid-lateral horizontal stripe and, seasonally, moderate to weak sexual dimorphism. Most have a relatively large gas bladder.

Riffle darters are benthic, usually moderate in size, robust, without gas bladders, have reduced fusiform shape, deep caudal peduncles, long second dorsal fins, non-protusibile premaxillaries, a narrow interorbital width, are usually brightly pigmented, often have vertical bars and pronounced sexual dimorphism and are without a prominent black basicaudal spot.

Sand run darters are benthic, semi-translucent, have pointed snouts, protrusibile premaxillaries, slender bodies and narrow caudal peduncles, are without gas bladders, moderate in size, have a weak black basicaudal spot and have weak sexual dichromatism. The slender shape of these species enables them to bury in the sand and conserve energy in a homogeneous environment where they otherwise are constantly exposed to current (Williams 1975; Daniels 1989; Simon 1991).

Quiet pool darters are small, benthic, without gas bladders, have reduced fusiform shape, blunt snouts, short second dorsal fins, protrusibile premaxillaries and a weak black basicaudal spot. Some species (e.g. *P. macrolepida* Stevenson and *E. chlorosomum*) are dully pigmented with only moderate development of sexually dimorphic colours. However, species living in pools with vegetation or other forms of cover, are often brightly coloured and strongly sexually dichromatic, e.g. *E. gracile* and *E. exile* (Girard).

In many species of *Etheostoma*, males are brilliantly coloured during the spawning season and contrast sharply with the cryptic females. Among the most brightly coloured species are those of the subgenera *Nothonotus*, *Oligocephalus* and *Etheostoma*. Conversely, most species of *Percina* exhibit relatively minor sexual dichromatism; during the spawning season the male characteristically assumes an overall duskiness and the female retains virtually the same colour as throughout the rest of the year. Notable exceptions among species of *Percina* are *P. roanoka*, *P. aurantiaca* (Cope), *P. evides* and *P. palmaris*, in which the males become brightly coloured during the breeding season and contrast markedly with the females.

The dorsal fins of breeding males of many species of darters are brightly banded and apparently serve as flags to facilitate species and sexual recognition. Although extremely variable interspecifically, patterns consisting of red or orange bands outlined by contrasting blues and greens are the most common. Males characteristically have larger fins than females. Large fins may enhance sexual recognition (particularly the colourful dorsal fins of breeding males) or increase swimming ability (particularly the paired fins) of the breeding male during spawning when increased activity is required. Extreme adaptations are the large pelvic fins in adult males of the *E. microperca* species group (including *E. proeliare* (Hay) and *E. fonticola*) and the long anal fins in adult males of the subgenus *Imostoma* of *Percina*. The large pelvic fins in males of *E. microperca* and closely related species aid in clasping the back of the female during spawning (Petravicz 1936). The long and highly tuberculate anal fins in males of *Imostoma* may aid in stimulating females during spawning, or may facilitate the transfer of milt from the male's genital opening to ova buried in the substrate. The semicircular keel on the caudal peduncle of breeding males of *Percina cymatotaenia*, *P. stictogaster* Burr & Page and *P. antesella* appears to function as a ploughshare which, in conjunction with the long anal fin, delivers spermatozoa to buried ova (Page & Sabaj 1994).

Breeding tubercles are keratinised epidermal protuberances found in about half of the species of darters. Generally the presence of tubercles is restricted to males (Collette 1965). However, tubercles are found characteristically on both males and females in the subgenus *Poecilichthys* (described for the '*E. variatum* Kirtland group' by Collette 1965) and in *E.* (*Ioa*) *vitreum* (Jenkins 1971) and are found exceptionally on females in other species.

Collette (1965) recognised 23 species of *Percina*, five of *Ammocrypta* (then including *Crystallaria asprella*) and 84 of *Etheostoma*; of these, five species of *Percina* (21%), all *Ammocrypta* and *C. asprella* and 36 of *Etheostoma* (43%) were known to develop tubercles. Since 1965 additional species have been found to be tuberculate (Jenkins 1971; Denoncourt 1976; Burr 1979) and more may be found as additional collections of breeding specimens become available.

Tubercles reach full development during the breeding season and function as contact organs during spawning, probably aiding in tactile stimulation of the female to lay ova. Linear ridges of keratinised epidermal material (tubercular ridges) are present, usually along fin rays, in some species which also develop individualised tubercles and in some which do not (Wiley & Collette 1970; Mayden 1985). Tubercles are thought to be specialisations derived from tubercular ridges (Wiley & Collette 1970). The erratic occurrence (some males in a species may have tubercles whereas others do not) and somatic distribution (some males have tubercles on parts of the body and fins where other males of the same species do not) of tubercles and tubercular ridges reduce their taxonomic and phylogenetic utility.

Modified (enlarged and more strongly toothed) scales (Plate 4) are present on the breast of most individuals of both sexes of all species of *Percina* and in a row along the belly midline of males of most species of *Percina* but are absent in other genera. The characteristic absence of modified scales on the bellies of females (rarely a mid-belly row is found on a female) and changes which occur in the scales on males during the spawning season suggest that the scales are used to induce tactile stimulation of the females during spawning (New 1966; Page 1976).

The genital papilla is a fleshy protuberance surrounding an aperture through which ova or spermatozoa are discharged. Specialisations in reproductive behaviour have been accompanied by modifications of the papilla of the female from the generalised short cone found in basal species into more complex structures (Page 1983). Often it is possible to relate the structure of the papilla to the reproductive behaviour, or more specifically to the site of ova deposition. The papillae of egg-buriers usually are conical tubes which inject the ova into the substrate. In egg-clumping and egg-clustering species the female has a wide flat papilla which enables her to push an ova, as it is released from the papilla, against the flat underside of a stone. In egg-attaching species a diverse array of papillae exists perhaps reflecting the many independent origins of egg-attachment (Page 1985), although many are conical or tubular. The genital papillae of males generally have remained simple conical structures.

12.3 Ecology

Habitat

Most darters have strict habitat requirements and generally a species can be found only by sampling a particular habitat in a stream of a certain size within the range of the species. For example, *P. shumardi* is found in rocky riffles of large rivers, *P. maculata* and *E. nigrum* are found in pools of creeks and *E. spectabile* (Agassiz) is found in gravelly headwaters. Among the most extreme species are those restricted to springs and spring-fed runs (including *E. nuchale* Howell & Caldwell, *E. ditrema* Ramsey & Suttkus and *E. fonticola*). Among the least habitat-specific is *P. caprodes* (Rafinesque), a species found in water bodies as diverse as muddy oxbow lakes and boulder riffles of swift streams.

In general, syntopically occurring darters and other benthic fishes segregate from one another by water depth (Fisher & Pearson 1987; Greenberg 1991; Chipps *et al.* 1994), current velocity (Fisher & Pearson 1987; Greenberg 1991; Kessler & Thorp 1993; Chipps *et al.* 1994) and substrate composition (Hlohowskyj & Wissing 1986; Greenberg 1991; Kessler & Thorp 1993). For most darters, composition of the substrate appears to be the most distributionally restrictive habitat characteristic. However, it is difficult to distinguish among substrate composition, water velocity and water depth because in streams all three usually vary concomitantly.

Welsh and Perry (1998) found that *Percina* (*P. caprodes*, *P. macrocephala* and *P. sciera* were the species studied) typically stayed in the water column, *Ammocrypta* (*A. pellucida*) buried in sand and species of *Etheostoma* were benthic and segregated from one another by occurring under (*E. variatum* and *E. camurum*), between (*E. caeruleum* and *E. zonale*) and on top of rocks (*E. blennioides*).

Etheostoma maculatum Kirtland and *E. bellum* Zorach, two closely related species that sometimes occur together in rocky riffles, differ in microhabitat preferences; *E. maculatum* lives mostly under large rocks in deep parts of riffles whereas *E. bellum* spends much of its time among smaller rocks in shallower areas of riffles (Kessler & Thorp 1993). Microhabitat use and spatial relations among species vary seasonally as current velocity and water depth change (Kessler *et al.* 1995).

Larvae occupy quiet water regardless of the habitat of the adults and juveniles occupy a habitat transitional between that of the larvae and that of the adults (Matthews 1985). In Little River, Tennessee, young-of-the-year darters were found most often in deep, slow-flowing water with silt-free coarse substrates (Greenberg & Stiles 1993). Usually the larger the individual, the more extreme the habitat occupied; e.g. the largest individuals of *E. sanguifluum* (Cope) are found in the fastest and deepest riffles; smaller individuals are found in slower, shallower riffles.

The vast majority of darters inhabit only flowing water habitats. A few species, especially those of the subgenus *Boleichthys*, inhabit swamps and other standing water bodies but also commonly are found in backwaters and quiet pools of streams. Because most darters require flowing water, dams, which transform flowing streams into standing bodies of water, eliminate all or almost all populations of darters originally present in the impounded streams.

Reproduction

Provided they reach a minimum size by the spawning season, individuals of most and perhaps all, species of darters reach sexual maturity at one year of age (Page 1983). However, individuals of some northern populations fail to reach maturity at one year even though more southern populations of the same species do. For example, Lutterbie (1979) found no mature one-year-olds among specimens of *E. zonale* from Wisconsin, but Lachner *et al.* (1950) found *E. zonale* to reach sexual maturity at one year in Pennsylvania. A possible exception to maturity at one year are females of *P. aurantiaca* which may not mature until their second year (Howell 1971). Raney & Suttkus (1966) thought that in *E. rubrum* Raney & Suttkus neither sex matured until two years of age. However, in a study on age and growth in *E. rubrum*, Knight & Ross (1992) documented sexual maturity in one-year-olds. In some species, e.g. *E. squamiceps* (Jordan), both sexes may reach maturity at one year, but competition from older males for superior spawning sites may prevent first-year males from spawning (Page 1974b).

Among species studied, the minimum SL required for females to be sexually mature varies from 24 mm in *E. ditrema* to about 40 mm in *P. sciera* and *P. phoxocephala* and that for males varies from 24 mm in *E. ditrema* to about 50 mm in *P. sciera* (Ramsey & Suttkus 1965; Page & Smith 1970, 1971).

In a typical reproductive cycle of a darter, small white ova are discernible in dissected ovaries as early as late summer or early autumn. Intermediate yellow ova appear between late autumn and winter in most species and later in the year develop into large, orange ova. White and yellow ova continue to appear and during late winter–early spring three distinct classes of ova are present (Fig. 12.1). Distinct size-groups of ova occur in other fishes as well and Johnson (1971) suggested that these size groups develop in response to a hormonal cycle functioning to clump ova in preparation for spawning. As females ripen just prior to spawning, the largest ova enlarge even more and become transparent except for orange oil droplets.

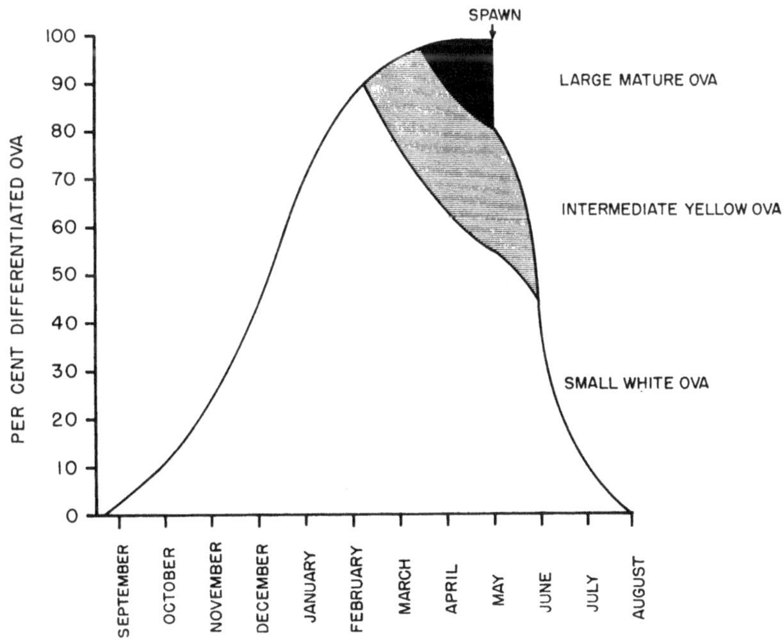

Fig. 12.1 Annual cycle in the differentiation of ova in darters. (After Page & Smith 1970.)

Fully mature ova of darters are adhesive, demersal, translucent except for one or more oil droplets, usually spherical and vary interspecifically in mean diameter from 0.7 mm in *E. proeliare* (Burr & Page 1978) to 2.9 mm in certain populations of *E. flabellare* Rafinesque (Cooper 1979). The mean diameter of the egg is not correlated with the body size of the species but is related to the site of egg deposition, as discussed below.

Estimates of fecundity from ovarian counts are available for several species of darters (Page 1983). However, estimates were made using various assumptions and, consequently, have given widely differing results (Page 1983; Heins & Baker 1989; Weddle & Burr 1991). As noted by Heins *et al.* (1992) and Heins and Baker (1993), standardising methods for estimating fecundity is necessary before interspecific comparisons can be meaningful.

The best estimates of fecundity are probably those of Gale and Deutsch (1985) and Weddle and Burr (1991), who counted eggs laid by individual females in in-stream cages and that of Fuller (1998b) who counted eggs laid in aquaria. Gale & Deutsch (1985) found that female *E. olmstedi* Storer laid 97–1435 (mean = 727) eggs during the spawning season, Weddle & Burr (1991) found female *E. rafinesquei* Burr & Page to lay *c.* 500–1000 eggs and Fuller (1998b) found *E. caeruleum* to lay an average of 309 eggs per season.

By holding a female *E. lepidum* (Baird & Girard) at an aquarium temperature of 20–23°C, Hubbs and Strawn (1957) were able to keep her producing ova for 251 days, which suggests that immature ova continue to mature during the spawning season. Starnes (1977) found that in *P. tanasi* Etnier, which spawns from February to April, the number of mature ova per female averaged over 600 in December-collected females, 427 in February-collected females and about 100 in April-collected females, indicating that females collected late in the season already had laid a portion of their ova. Strawn & Hubbs (1956), studying *E. grahami* (Girard) in Texas, found that within an acceptable temperature range a female would spawn every 4–10 days.

Within a species, larger females produce more ova. In *E. spectabile* in central Texas, the number of ova produced was found to be related exponentially to the length and linearly to the mass of the female (Hubbs *et al.* 1968). Egg size may vary intraspecifically among populations, among females collected at the same time and place and within populations through the course of the breeding season (Hubbs 1967; Hubbs *et al.* 1968; Marsh 1984, 1986). Ova tend to be largest in the middle and smaller at the beginning and end of the breeding season (Hubbs *et al.* 1968; Hubbs & Armstrong 1962; Pfleiger 1975; Marsh 1984). Several studies have shown that larger females do not produce significantly larger ova; i.e. no correlation existed between the average size of ova produced and the length or mass of the female (Hubbs *et al.* 1968; Mathur 1973b; Marsh 1984; Heins 1985; Heins & Baker 1993). In contrast, Fuller (1998a) found larger *E. caeruleum* produced larger ova. Larger larvae develop from larger ova which may have a survival advantage over smaller larvae (Marsh 1986; Fuller 1998a). Hubbs and Johnson (1961) discovered that the number and size of the ova of *P. sciera* varied geographically. Hubbs (1958) found the ova of *E. spectabile* to vary geographically in size but not in number. Darters with high fecundity and small ova exhibit high gene flow; those with low fecundity and large ova exhibit low levels of gene flow (Turner & Trexler 1998). Levels of gene flow influence patterns of geographic variation and the propensity to speciate.

Between the postspawning period (summer) and the following spring, ovaries increase markedly in size in proportion to the body of the female. In *E. barbouri* Kuehne & Small, the mass of the ovaries, expressed as a percentage of total body mass (GSI), ranges from an average of 0.4% in August to 11.5% in April (Flynn & Hoyt 1979). In *E. flabellare* the GSI reaches a maximum in April (*c.* 15.5%) (Baker 1978). Maximum values for an index calculated as the mass of the ovaries divided by the 'adjusted body mass' (= total mass minus the mass of the ovaries and digestive system) range from 15.2% in *E. kennicotti* (Putnam) (Page 1975) to 23.0% in *E. microperca* (Burr & Page 1979).

Most darters spawn only in spring. At southern latitudes darters generally spawn from February to April and at northern latitudes from March to June. Exceptions are species of *Nothonotus* which spawn from May to August (Stiles 1972; Bryant 1979; Knight & Ross 1992) and species of *Ammocrypta* which spawn in late spring and summer (Williams 1975; Heins & Rooks 1984; Heins 1985; Johnston 1990). Other exceptions to spring-only spawning are found in southern states where seasonal variation in climate is less extreme than in northern latitudes. In Texas, *E. lepidum* and *E. spectabile* may begin spawning as early as October and November respectively (Hubbs 1961a; Hubbs *et al.* 1968). *Etheostoma edwini* (Hubbs & Cannon), a species restricted to Alabama, Georgia and Florida, may spawn in the autumn and in spring (Collette & Yerger 1962). *Etheostoma ditrema*, an inhabitant of constant-temperature springs in Alabama, Georgia and south-eastern Tennessee, spawns from March to September (Seesock *et al.* 1978). *Etheostoma fonticola*, a spring-inhabiting species in Texas, spawns year-round in water varying only slightly from 21°C (Schenck & Whiteside 1977b).

Hubbs & Strawn (1957), studying *E. lepidum*, concluded that temperature and condition of the fish control egg production and that photoperiod and light intensity do not affect fecundity. Furthermore, aquarists have concluded that an increasing temperature, but not a change in photoperiod, stimulates darters to spawn. Darters held in aquaria for long periods will seldom spawn unless they have gone through a cold period and are subsequently exposed to warming temperatures. Other factors stimulating darters to spawn include frequent water changes, abundant food and bright sunlight (Katula, personal communication).

Observations on spawning are available for 103 species of darters (Table 12.1). Among these species, four spawning modes have been recognised: burying, attaching, clumping and clustering (Page 1985). In egg-burying, a female that is receptive to the advances of a courting male buries her body partially below the surface of the substrate. A male mounts her, the two vibrate and release ova and spermatozoa. The fertilised eggs are buried in the substrate and, consequently, hidden from many potential predators. After spawning at one site, the female moves to another and the spawning sequence is repeated with the same or another male. The result is that ova are laid over a rather large area and are abandoned by both parents. No direct parental care is given to the eggs.

In egg-burying species, supernumerary males ('sneakers') may position themselves next to the spawning pair (e.g. Reeves 1907; New 1966; Fisher 1990; Jenkins & Burkhead 1994) and fertilise some of the ova. To increase the likelihood of successful fertilisation, males of *E. caeruleum* release more milt when spawning in the presence of competing males than when spawning in the absence of competing males (Fuller 1998a). Large males may forego spawning opportunities in the presence of many other males, presumably to wait for opportunities when sperm competition is less.

In egg-attaching species, eggs are attached to an object, usually a plant or large rock, above the substrate rather than being buried below the surface of the

Table 12.1 The four spawning modes of darters (Page 1985), species observed in each mode, and source of information.

Species	Information source
Buriers (N = 39)	
Percina (*Odontopholis*) *cymatotaenia*	Pfleiger 1984
P. (*Hypohomus*) *aurantiaca*	Howell 1971
P. (*Hadropterus*) *sciera*	Evans & Page, personal observation
P. (*Swainia*) *phoxocephala*	Page & Simon 1988
P. (*Alvordius*) *maculata*	Petravicz 1938
P. (*A.*) *notogramma*	Loos & Woolcott 1969
P. (*A.*) *pantherina*	James & Maughan 1989
P. (*A.*) *peltata*	New 1966
P. (*A.*) *roanoka*	Jenkins & Burkhead 1994
P. (*Ericosma*) *evides*	Page *et al.* 1982
P. (*Imostoma*) *ouachitae*	J.D. Evans, personal communication
P. (*Cottogaster*) *copelandi*	Winn 1953
P. (*Percina*) *caprodes*	Winn 1958a,b
P. (*P.*) *rex*	Jenkins & Burkhead 1994
Crystallaria asprella	R.S. Katula, personal communication
Ammocrypta pellucida	Johnston 1990
Etheostoma (*Nothonotus*) *acuticeps*	W.D. Voiers, personal communication
E. (*N.*) *bellum*	Fisher, 1990; Voiers, personal communication
E. (*N.*) *camurum*	Mount 1959
E. (*N.*) *jordani*	Page & Simon 1988; Orr & Ramsey 1990
E. (*N.*) *juliae*	James & Taber 1986
E. (*N.*) *rubrum*	Ross & Wilkins 1993
E. (*N.*) *rufilineatum*	Stiles 1972
E. (*N.*) *tippecanoe*	Trautman 1957; Warren *et al.* 1986
Etheostoma (*Litocara*) *nianguae*	Pflieger 1978
E. (*L.*) *sagitta*	Page & Simon 1988
E. (*Poecilichthys*) *tetrazonum*	Pflieger 1978
E. (*P.*) *variatum*	May 1969
E. (*Etheostoma*) *swannanoa*	Jenkins & Burkhead 1994
E. (*Doration*) *stigmaeum*	Winn 1958a,b
E. (*Belophlox*) *fricksium*	Layman 1993
E. (*Ozarka*) *cragini*	Distler 1972
E. (*O.*) *punctulatum*	Page & Simon 1988; Simon & Garcia 1990
E. (*O.*) *tuscumbia*	Koch 1978
E. (*Oligocephalus*) *caeruleum*	Winn 1958a,b
E. (*O.*) *collettei*	Katula 1991
E. (*O.*) *radiosum*	Scalet 1973
E. (*O.*) *spectabile*	Winn 1958a,b
E. (*O.*) *swaini*	Ruple *et al.* 1984
Attachers (N = 37)	
Etheostoma (*Allohistium*) *cinereum*	Jenkins & Burkhead 1994
E. (*Etheostoma*) *baileyi*	Page 1985; Porterfield 1998
E. (*E.*) *barrenense*	Winn 1958a,b
E. (*E.*) *blennioides*	Fahy 1954
E. (*E.*) *brevirostrum*	Johnston & Shute 1997

Cont.

Table 12.1 Contd.

Species	Information source
E. (*E.*) *coosae*	O'Neil 1981
E. (*E.*) *duryi*	Page *et al.* 1982
E. (*E.*) *etnieri*	Porterfield 1998
E. (*E.*) *flavum*	Keevin *et al.* 1989; Porterfield 1997
E. (*E.*) *histrio*	Steinberg *et al.* in press
E. (*E.*) *pyrrhogaster*	Carney & Burr 1989
E. (*E.*) *rafinesquei*	Winn 1958a,b
E. (*E.*) *raneyi*	Johnston & Haag 1996
E. (*E.*) *scotti*	Bauer *et al.* 1995
E. (*E.*) *simoterum*	Page & Mayden 1981; Porterfield 1997
E. (*E.*) *zonale*	Winn 1958a,b; Walters 1994
E. (*E.*) *zonistium*	Carney & Burr 1989
E. (*Fuscatelum*) *parvipinne*	Johnston 1994
E. (*Ioa*) *vitreum*	Winn & Picciolo 1960
E. (*Vaillantia*) *chlorosomum*	Page *et al.* 1982; Bart 1992
E. (*V.*) *davisoni*	Bart 1992
E. (*Belophlox*) *okaloosae*	Collette & Yerger 1962
E. (*Villora*) *edwini*	Williams 1976
E. (*Ozarka*) *boschungi*	Boschung 1979
E. (*O.*) *pallididorsum*	Johnston 1995
E. (*O.*) *trisella*	Ryon 1986
E. (*Oligocephalus*) *asprigene*	Page *et al.* 1982
E. (*O.*) *ditrema*	Seesock *et al.* 1978
E. (*O.*) *grahami*	Strawn 1956
E. (*O.*) *lepidum*	Strawn 1956
E. (*Boleichthys*) *collis*	Jenkins & Burkhead 1994
E. (*B.*) *exile*	Winn 1958a,b
E. (*B.*) *fonticola*	Schenck & Whiteside 1977b
E. (*B.*) *fusiforme*	Fletcher 1957
E. (*B.*) *gracile*	Braasch & Smith 1967
E. (*B.*) *microperca*	Burr & Page 1979
E. (*B.*) *proeliare*	Burr & Page 1978
Clumpers (N = 5)	
E. (*Nothonotus*) *aquali*	Page *et al.* 1982
E. (*N.*) *maculatum*	Raney & Lachner 1939
E. (*N.*) *microlepidum*	Page *et al.* 1982
E. (*N.*) *vulneratum*	Stiles 1972
E. (*N.*) *wapiti*	Burkhead & Williams 1992
Clusterers (N = 22)	
Group A. Female and male briefly invert	
E. (*Catonotus*) *chienense*	Warren *et al.* 1994; Piller & Burr 1999
E. (*C.*) *corona*	Page, personal observation
E. (*C.*) *crossopterum*	Page 1974b; Braasch & Mayden 1985
E. (*C.*) *forbesi*	Page, personal observation
E. (*C.*) *neopterum*	Page, personal observation
E. (*C.*) *nigripinne*	Braasch & Mayden 1985

Contd.

Table 12.1 Contd.

Species	Information source
E. (*C.*) *olivaceum*	Page 1980
E. (*C.*) *oophylax*	Page & Mayden 1979
E. (*C.*) *pseudovulatum*	Page, personal observation
E. (*C.*) *squamiceps*	Page 1974b
Group B. Female only has prolonged inversion	
E. (*C.*) *barbouri*	Page *et al.* 1982
E. (*C.*) *flabellare*	Lake 1936; Winn 1958a,b
E. (*C.*) *kennicotti*	Page 1975
E. (*C.*) *obeyense*	Page *et al.* 1981
E. (*C.*) *percnurum*	Layman 1984
E. (*C.*) *smithi*	Page & Burr 1976
E. (*C.*) *striatulum*	Page 1980
E. (*C.*) *virgatum*	Kornman 1980
Group C. Female and male have prolonged inversions	
E. (*Boleosoma*) *nigrum*	Winn 1958a,b
E. (*B.*) *longimanum*	Page *et al.* 1981
E. (*B.*) *olmstedi*	Atz 1940; Lindquist *et al.* 1981
E. (*B.*) *podostemone*	Jenkins 1980

substrate. As in egg-buriers, the female selects the site of egg deposition. With the male following, she swims to the site where she will attach the egg. As she reaches the site, the male mounts her, the two vibrate and ova and spermatozoa are released. One to three ova are released and pushed by the female onto the plant or rock. Then, as in egg buriers, the female goes to another site and spawns with the same or another male. The eggs are abandoned by both parents. No direct parental care is given to the eggs.

In egg-clumping, the male selects and defends a cavity under a large rock and displays to passing females. A responsive female will swim into the cavity being guarded by the male and wedge herself into the interface between the large rock and the substrate beneath the rock. The male aligns himself next to the female and releases milt as she releases ova. After spawning, the female leaves but, unlike in burying and attaching species, the male remains to guard the eggs. Other females may add ova to the clump laid by the first female.

As in egg-clumping species, the male of egg-clustering species establishes a territory under a large rock. A female enters the territory and, if she decides to spawn after being courted by the male, she turns upside down and rises to the underside of the rock. As she does so, the male also turns upside down and positions himself next to her. She attaches the adhesive ova to the roof of the cavity (underside of the rock) and he releases milt. A few ova, usually three to five, are released during each spawning and attached to the rock in a short, straight row as the female moves forward slightly. This process is repeated until the female has laid her ripe ova and she leaves the nest. The male stays to guard the eggs and other females may add eggs to the nest.

In egg-clustering, ova are laid in a single layer on the underside of a stone. By being in a single layer, all eggs are exposed to care by the guarding male. He fans the eggs to reduce the likelihood of infection and removes organisms, such as snails or crayfish, that might move onto the nest to eat the eggs. Egg-clustering is characteristic of some of the most common species of darters (e.g. *E. nigrum*, *E. olmstedi* and *E. flabellare*) and probably is a major factor in the success of these species.

Females of egg-clustering species of darters prefer to add ova to an existing nest (i.e. to lay ova on a nest-stone with eggs already on it) rather than begin a new nest (Knapp & Sargent 1989). This preference arose because of benefits associated with nests with eggs (e.g. reduced egg predation because of the nest location or high egg density; by choosing to add eggs to those of other females, a female is imitating the choice of male made by other females and thereby will have male offspring that are likely to enjoy high reproductive success) and has led to the evolution of some particularly interesting behavioural and morphological characteristics. For example some egg-clustering species have evolved egg mimics (Page & Swofford 1984; Page & Bart 1989; Bart & Page 1991). As the male displays the mimics close to or even touching the stone where ova are to be laid, the female reacts as if eggs are present even on an empty nest-stone and is attracted to the male's territory.

In the fantail darter and closely related species, large yellow mimics are on the dorsal spines (Plate 6). In a second group, the spottail darters, egg mimics develop on the tips of soft rays of the second dorsal fin. The species in this group show a progression in the mimics from small white knobs to large, egg-yellow knobs held well above the fin membrane to enhance visibility (Plate 6). As a male with mimics on either dorsal fin displays to a female, mimics are pressed against the nest-stone where real eggs would be located.

Species in the third group of mimic-bearing darters are called barcheek darters and have two dark rectangles on each cheek. In each rectangle is a spot which becomes more brightly coloured as the fish grows. In a large individual, the upper spot is red, the lower one is white and each is surrounded by an iridescent yellow circle. On a large breeding male, the upper spot changes from red to bright yellow-gold and is surrounded by a dark ring of melanophores that give the yellow spot a spherical egg-like appearance (Plate 6).

All of the above species are in the subgenus *Catonotus*. The fourth group in which egg mimics develop is in *Boleosoma*. Spawning behaviour in *Boleosoma* is similar to that in *Catonotus* except that the male displays to a female upside down under his nest-stone (Page & Bart 1989). It is the pectoral and pelvic fins that are against the rock as the male is displaying upside down and egg mimics have evolved on these fins rather than on the dorsal fins as in *Catonotus*. Egg mimics in *Boleosoma* are large white knobs that develop on the tips of the rays of the otherwise black fins (Plate 6).

From field and aquarium observations, Constantz (1979) noted that males of *E. olmstedi* clean and guard eggs that have been fertilised by conspecific males. This behaviour seems to result from a preference for nest stones with a large area available for egg deposition (and thus ample opportunity for spawning). Leaving a nest of eggs (the presence of which reduces the area on the nest stone still available for egg deposition), to find another and better stone in an area with an excess of males waiting to secure nest stones of their own, adequately insures that the original nest will be reoccupied.

Egg-clustering species produce the largest ova and egg-attaching species produce the smallest eggs (Page 1983; Bart & Page 1992). Clustered eggs are guarded by males and presumably are more likely than those of non-clustering species to survive until hatching. Hence, it probably is beneficial for females of egg-clustering species to invest more in each egg. Ova of egg-attaching species, although presumably benefiting from being distributed over a large area, are not buried and are not guarded. Hence, they are exposed to predation but their small size may lessen the likelihood of their being found and eaten.

Growth

Data on 27 species of darters (Table 12.2; Fig. 12.2) suggest that the incubation period (time from fertilisation to hatching) is strongly temperature-dependent and that differences associated with phylogeny, spawning mode or other variables are minor. Darter eggs hatch in about 27 days at 10°C, nine days at 20° and five days at 28°C. Ten and 30°C appear close to the limits of tolerance for embryonic development in darters (Hubbs 1961a,b; Burr & Page 1978, 1979).

Recent studies by Simon (e.g. Simon & Faber 1987; Simon & Garcia 1990; Simon 1992) provide a large amount of information on larval development in darters. Hatchlings (prolarvae) vary from about 3 to 7.5 mm in length, are scaleless and have a large yolk sac, fin-buds (but with rays absent or developed in the pectorals only), a functional (movable) operculum, a functional lower jaw and well-developed eyes. Available data suggest that hatchlings of *Etheostoma* are precocious relative to those of *Percina* (Fahy 1954; Braasch & Smith 1967; May 1969; Page & Smith 1970; Page 1974b; Cooper 1978, 1979; Simon & Faber 1987; Simon & Garcia 1990; Simon 1992).

Darters have indeterminate growth but the rate of increase in length rapidly diminishes with age (Fig. 12.3). Individuals at one year usually average 60–70% of the average length of the oldest age-class. In species living three or more years, individuals at the end of their second year characteristically average 80–90% of the average length of the oldest age class (Page 1983). Large species of darters grow faster, mature at a larger size, produce bigger clutches of eggs and have longer reproductive periods than do small species (Paine 1990). Looking at life-history characteristics of all percids, Bart & Page (1992) found a general pattern grouping large, long-lived, late-maturing, multiple spawners with high fecundity at one

Table 12.2 Hatching periods of darters.

Species	Temperature °C	Hatching period Days	Reference
E. lepidum	9	40	Hubbs 1961b
P. macrocephala	10	27	Etnier & Starnes 1993
E. nigrum	12	16	Speare 1965
E. trisella	12	30	Ryon 1986
P. tanasi	12.5	15	Starnes 1977
E. smithi	13	30	Page & Burr 1976
E. blennioides	14	18	Fahy 1954
E. proeliare	15	12–13	Burr & Page 1978
P. copelandi	15	13	Hubbs & Bryan 1975
E. exile	13–16	18–26	Winn 1958b
E. microperca	15.5	11	Burr & Page 1979
P. copelandi	16	10	Hubbs & Bryan 1975
P. caprodes	16.5	9	Cooper 1978
E. spectabile	16.5–18.5	9–10	Winn 1958b
E. caeruleum	17–18.5	10–11.5	Winn 1958b
E. tuscumbia	17	5	Koch 1978
P. copelandi	17	9	Hubbs & Bryan 1975
P. copelandi	18	8	Hubbs & Bryan 1975
E. flabellare	17–20	30–35 (?)	Lake 1936
E. stigmaeum	17–20	9–10	Winn 1958b
E. microperca	18–20	6	Burr & Page 1979
E. microperca	20	7.5	Burr & Page 1979
E. nigrum	20	10	Speare 1965
E. proeliare	20	9–10	Burr & Page 1978
E. zonistium	20	7	Carney & Burr 1989
P. copelandi	20	6	Hubbs & Bryan 1975
E. squamiceps	18–22	12	Page 1974b
E. camurum	19–23	7–10	Mount 1959
E. kennicotti	19–23	6	Page 1975
E. smithi	21	13	Page & Burr 1976
E. nigrum	21	6	Speare 1965
E. flabellare	21–22	21 (?)	Lake 1936
E. fricksium	22	6–11	Layman 1993
P. copelandi	22	5	Hubbs & Bryan 1975
P. caprodes	21–23	5–7	Grizzle & Curd 1978
E. microperca	22–23	6	Burr & Page 1979
E. proeliare	22–23	5–6	Burr & Page 1978
E. nigrum	22–24	5.5–8	Speare 1965
E. gracile	23	5	Braasch & Smith 1967
E. bellum	23	7–9	Fisher 1990
E. flabellare	23.5	14–16 (?)	Lake 1936
E. tippecanoe	22–25	9	Warren *et al.* 1986
E. squamiceps	22–26	5	Page 1974b
P. sciera	25	4	Page & Smith 1970
E. pyrrhogaster	25	6–8	Carney & Burr 1989
E. lepidum	28	4–5	Hubbs 1961b

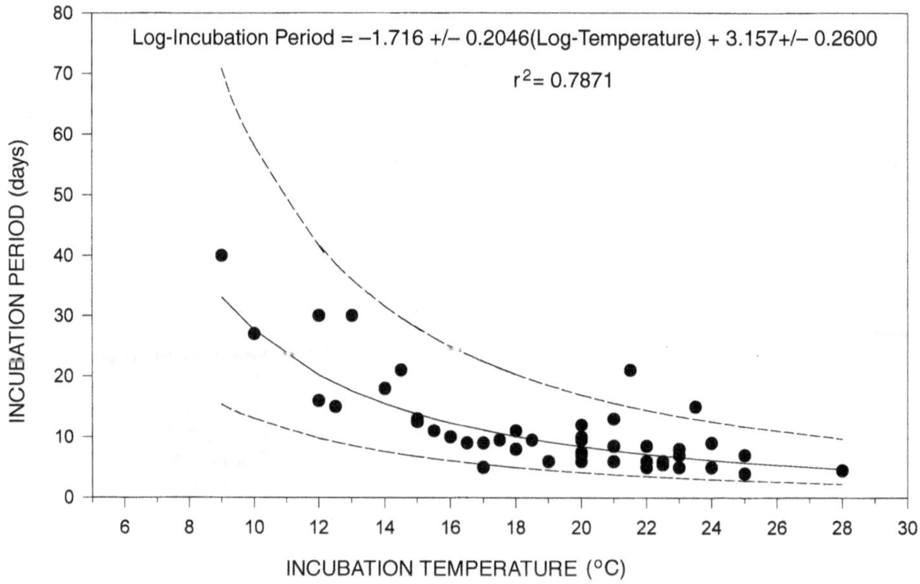

Fig. 12.2 Relationship between incubation period (fertilisation to hatching) and temperature. Data are from Table 12.2; median values were plotted when a range was given.

extreme and small, short-lived, early-maturing, semelparous species with low fecundity at the other extreme.

Based on studies of natural populations, maximum longevity varies from 1+ years in small species (e.g. *E. proeliare* and *E. microperca*) to 4+ years in large species (e.g. *P. sciera* and *E. blennioides*) (Page 1983). Individuals of *P. nasuta* kept in aquaria by Anderson *et al.* (1998) were described as '6 years or older'; however, it seems likely that darters rarely live as long as that under natural conditions.

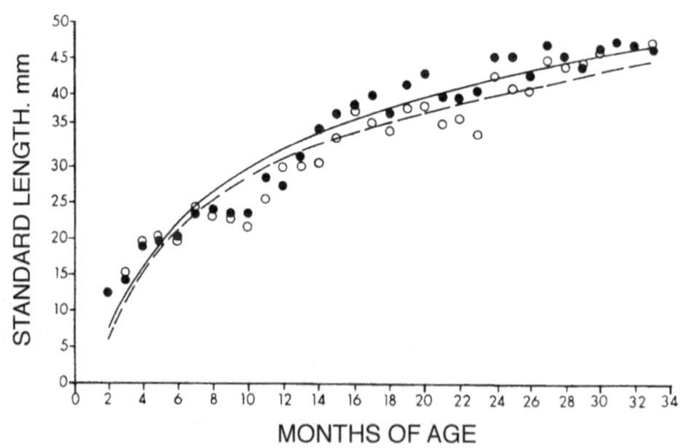

Fig. 12.3 Growth curves for *Etheostoma kennicotti* from Big Creek, Illinois (Page 1975). Standard length for males (•) = −3.11 + 33.02 log age (in months) and for females (○) = −3.37 + 31.55 log age; r = 0.939 and 0.911, respectively. Males are significantly larger than females after two years.

In most species of darters, males grow faster and are on average larger than females. This dimorphism in size is probably a consequence of the fact that females invest more energy in gamete production and put less energy into somatic growth. Also, males enjoy higher reproductive success when large because they are better able to attract females or to defend territories or space around females. The only darters in which the females are slightly larger than males are species of the subgenera *Boleichthys* and *Ozarka* which are thought to be non-territorial (Page 1983).

Most sex ratios among populations of darters studied have not deviated significantly from 1:1 (Page 1983). However, a few populations have been found to have ratios favouring males (Lachner *et al.* 1950; Schwartz 1965; Page & Smith 1971; Baker 1978; Widlak & Neves 1985) and in species of the subgenus *Catonotus* females often predominate (Lake 1936; Schwartz 1965; Page 1974b, 1975; Page & Burr 1976; Flynn & Hoyt 1979) in all but the oldest age class (in which males predominate). Females also predominate in at least one population of *E. histrio* Jordan & Gilbert (Kuhajda & Warren 1989).

Spacing

Many species of darters are extremely localised even though they range over large areas. Extreme examples are species of *Nothonotus*; e.g. *E. maculatum* ranges over six states but is known from only two rivers in Indiana, two in Ohio, three in Pennsylvania, one in New York, one in West Virginia and three in Kentucky. *Percina macrocephala* is found in only a few rivers even though it ranges over eight states. It is unreasonable to assume that the relatively few areas occupied by these species are the only areas within the ranges of the species that have suitable habitat. Instead, many populations of these species appear to have become extinct (for reasons that are not obvious) leaving extant populations isolated from one another.

The density of individuals varies interspecifically and, within species, varies geographically, seasonally and by habitat. With a few notable exceptions, e.g. *P. sciera* and *P. nigrofasciata* (Agassiz), species of *Percina* tend to have smaller populations than do species of *Etheostoma*. Some species of *Etheostoma* (e.g. *E. caeruleum*, *E. spectabile*, *E. olmstedi* and *E. nigrum*) tend to be common throughout their range and often are among the most common fishes in a stream. Species of *Ammocrypta* typically exist in small populations in the north and in relatively large populations in the south. The highest (postlarval) density recorded for a darter is for *E. microperca*, for which Burr & Page (1979) calculated a density of 33 individuals m^{-2} in a small stream in Illinois in October. *Etheostoma microperca* was most dense in late summer and autumn (12–33 individuals m^{-2}) when the water level was low and the number of juveniles was at a peak. Because of high mortality (*E. microperca* lives a maximum of only 1+ years) and a rising water level, the density declined throughout winter and early spring. In May the density began to increase.

Lachner *et al.* (1950), Schwartz (1965) and Reed (1968) found multispecific densities of 1.2, 1.4 and 10.8 darters m^{-2}, respectively, in various riffles in Pennsylvania. Wickliff (1941) calculated densities up to 16.8 darters m^{-2} in riffles in Ohio, or about one darter per 600 cm^{-2} of riffle.

Total population size has been determined for a few rare species of darters. Based on density estimates, Schenck & Whiteside (1976) calculated the total number of individuals of *E. fonticola*, a species restricted to the San Marcos Spring and River in Texas, to be 103 000. Pfleiger (1978) estimated the total population of *E. nianguae*, which occupies a small area of south-central Missouri, to be between 2300 and 27 000 individuals. Piller and Burr (1998) estimated the total number of *E. chienense* Page and Ceas, a species restricted to the Bayou du Chien, Kentucky, to be between 9500 and 31 000. *Etheostoma sellare* (Radcliffe & Welsh), a species with an extremely small range even when first discovered (Knapp 1976), has not been observed in recent years and may have been the first darter to become extinct as a result of anthropogenic changes in the environment.

Many species of darters defend territories during the breeding season and some are territorial at other times of the year (Winn 1958a,b; Page 1983). A few species (e.g. *E. fonticola*, *E. cragini* Gilbert and *E. proeliare*) do not appear to be territorial (Strawn 1956; Distler 1972; Burr & Page 1978) and at least one (*E. vitreum*) has been described as a communal spawner (Winn & Picciolo 1960). For males of some species, e.g. *E. blennioides*, the presence of other males at the time of egg-laying is necessary for the expression of territorial behaviour (Winn 1958b). Breeding males develop bright colours or other vivid patterns to attract females or to warn other males to maintain their distance. Females are much less likely than males to be territorial, but when they are, they may assume a colour pattern similar to that of a territorial male (Page 1975).

Darters, including those species with gas bladders, fusiform bodies and, presumably, ample swimming ability, are not known to undergo long-distance migrations. However, many species undergo an annual cycle of movements among habitats which may entail distances of several kilometres. A typical annual cycle consists of a pre-spawning move into the spawning habitat (often an area of stream with fast current and rocks or vegetation), a post-spawning residence in a relatively shallow habitat and a winter residence in a low-gradient, relatively deep habitat. The movement into the deep water for winter is usually downstream and that into the spawning habitat is usually upstream (Trautman 1957). Species occupying small streams in warm months are often absent in the same stream in winter, having moved downstream into deeper water. May (1969) observed that although most of the *E. variatum* remained on the same riffles where they had been found originally, some had moved about five kilometres downstream between autumn and winter. *Etheostoma histrio* is thought to move from smaller streams where it lives in riffles and debris in warm months to large rivers where it overwinters (Etnier & Starnes 1993). *Etheostoma spectabile* was noted to make upstream spawning migrations while syntopically occurring *E. flabellare* and *E. nigrum* did not (Ingersoll *et al.*

1984). Headwater species may survive stream desiccation by burrowing into the wet substrate and re-emerging when the stream bed refills with water (Tramer 1977).

Diet

The vast majority of darters feed on small crustaceans and aquatic insect larvae (Page 1983). They feed mainly on microcrustaceans (cladocerans, copepods and ostracods) as juveniles and on immature aquatic insects (mostly midge flies, black flies, mayflies and caddisflies) as adults. A second shift in food, from small insects (mainly midges and black flies) to large insects (usually caddisflies and mayflies) and large crustaceans (amphipods, isopods and crayfishes), is characteristic of large species (e.g. *E. squamiceps*, Page 1974b). Larger darters (species and individuals) eat a greater variety of food and larger food items than do small darters. Seasonal variations in diet (Hlohowskyj & White 1983) appear to accompany changes in prey availability. Darters sometimes constitute the largest portion of the stream insectivore community (Matthews 1998).

Unusual diets are those of spring-inhabiting species such as *E. tuscumbia* Gilbert & Swain and *E. ditrema* which feed heavily on amphipods, normally found abundantly in springs, and those of snail-eating species (including *P. tanasi* and *P. uranidea* (Jordan & Gilbert) and, to a lesser extent, *P. shumardi*, *P. ouachitae*, *E. sellare*, *E. blennioides*, *E. nuchale* and *E. cragini*). *Percina macrocephala*, a large species, feeds heavily on small crayfishes (Page 1983).

Darters appear to have keen vision and most are probably diurnal visual feeders (Page 1983; Greenberg 1991). Roberts and Winn (1962) demonstrated experimentally that *E. nigrum* feeds primarily by visual cues of movement and responds only slightly to olfactory stimuli. Studies on *P. maculata*, *P. phoxocephala*, *P. caprodes*, *E. flabellare*, *E. caeruleum*, *E. zonale* and *E. fonticola* found them to be diurnal feeders (Thomas 1970; Adamson & Wissing 1977; Schenck & Whiteside 1977a). *Percina nigrofasciata* is a diurnal feeder with peaks of activity in early morning and late afternoon (Mathur 1973a). Feeding activity in *E. microperca* and *E. zonale* increases markedly after sunrise, peaks at midday and decreases after sunset (Cordes & Page 1980); a similar pattern, but with a peak in late afternoon, was found for *P. roanoka* (Hobson 1979) and for *E. rufilineatum* (Cope) (Widlak and Neves 1985). *Etheostoma vitreum* forages at dawn and at dusk (Lee & Ashton 1979). *Etheostoma bellum* feeds mostly at dusk (Fisher 1990). *Crystallaria asprella*, a highly unusual species morphologically, may also be unusual in that it may be active nocturnally (Lindsay & Adam 1974); however, data of George *et al.* (1996) suggest that this species is diurnally active.

Most darters feed on similar food organisms, especially fly larvae, and it seems likely that dietary separations among syntopically occurring species are based on where darters feed rather than on what they feed (Page 1983; Schlosser & Toth 1984; Greenberg 1991). Some differences are suggested by morphological traits such as snout shape and mouth position and size. Pointed-snout species with terminal mouths (e.g. *E. caeruleum* and *P. phoxocephala*) extract food organisms from among stones. Blunt-snouted species with sub-terminal mouths feed on the sides and tops of boulders (e.g. *P. aurantiaca*) or pick food items from vegetation (e.g. *E. zonale*). *Etheostoma flabellare* and other species with upturned mouths remove organisms from the undersides of large stones. Species with large mouths are able to feed on larger food items than species with small mouths (Matthews *et al.* 1982; Paine *et al.* 1982).

Logperches (e.g. *P. caprodes*) use their bulbous snouts and flattened heads to flip stones and debris on the stream bed to expose food organisms (Keast & Webb 1966; Greenberg 1991). Although not its usual feeding behaviour, *P. maculata*, a mid-water species, was observed by Trautman (1957) feeding at the water surface and 'jumping into the air for flying insects'.

Predation

Most observations of predation on darters have involved other fishes (Page 1983). *Etheostoma olmstedi* was found to be the most commonly ingested fish in Ogden's (1970) study on the American eel, *Anguilla rostrata*, in New Jersey and in Raney & Lachner's (1942) study on the walleye, *Stizostedion vitreum*, in New York. *Percina caprodes* formed the bulk of fish remains found scattered among the nests of terns on an island in Lake Erie (McCormick 1892). White suckers, *Catostomus commersoni*, feed heavily on the eggs of *P. caprodes* in Douglas Lake, Michigan (Ellis & Roe 1917). Flynn & Hoyt (1979) recorded cannibalism in *E. barbouri*.

Acknowledgements

Photographs were provided by K. S. Cummings, P. A. Ceas and M. H. Sabaj. M. J. Dreslik carried out the statistical analysis for Figure 12.2. J. H. Knouft and T. J. Near made helpful comments on the manuscript.

Appendix

Table 12a List of all described species of darters, by genus and subgenus. Monophyletic species groups are recognised within large subgenera of *Etheostoma*.

Genus *Percina* – 40 species	
Subgenus *Odontopholis*	
P. cymatotaenia (Gilbert & Meek)	Bluestripe darter
P. stictogaster Burr & Page	Frecklebelly darter
Subgenus *Hypohomus*	
P. aurantiaca (Cope)	Tangerine darter
Subgenus *Alvordius*	
P. crassa Jordan & Brayton	Piedmont darter
P. gymnocephala Beckham	Appalachia darter
P. macrocephala (Cope)	Longhead darter
P. nevisense (Cope)	Neuse darter
P. maculata (Girard)	Blackside darter
P. notogramma (Raney & Hubbs)	Stripeback darter
P. pantherina (Moore & Reeves)	Leopard darter
P. peltata (Stauffer)	Shield darter
P. roanoka (Jordan & Jenkins)	Roanoke darter
P. species	Muscadine darter
Subgenus *Ericosma*	
P. evides (Jordan & Copeland)	Gilt darter
P. palmaris (Bailey)	Bronze darter
Subgenus *Cottogaster*	
P. aurora Suttkus & Thompson	Pearl darter
P. brevicauda Suttkus & Bart	Coal darter
P. copelandi (Jordan)	Channel darter
Subgenus *Imostoma*	
P. antesella Williams & Etnier	Amber darter
P. ouachitae (Jordan & Gilbert)	Saddleback darter
P. shumardi (Girard)	River darter
P. tanasi Etnier	Snail darter
P. uranidea (Jordan & Gilbert)	Stargazing darter
Subgenus *Hadropterus*	
P. aurolineata Suttkus & Ramsey	Goldline darter
P. lenticula Richards & Knapp	Freckled darter
P. nigrofasciata (Agassiz)	Blackbanded darter
P. sciera (Swain)	Dusky darter
Subgenus *Swainia*	
P. nasuta (Bailey)	Longnose darter
P. oxyrhynchus (Hubbs & Raney)	Sharpnose darter
P. phoxocephala (Nelson)	Slenderhead darter
P. squamata (Gilbert & Swain)	Olive darter
Subgenus *Percina*	
P. austroperca Thompson	Southern logperch
P. burtoni Fowler	Blotchside logperch

Cont.

Table 12a Contd.

P. caprodes (Rafinesque)	Logperch
P. carbonaria (Baird & Girard)	Texas logperch
P. jenkinsi Thompson	Conasauga logperch
P. kathae Thompson	Mobile logperch
P. macrolepida Stevenson	Bigscale logperch
P. rex (Jordan & Evermann)	Roanoke logperch
P. suttkusi Thompson	Gulf logperch
Genus *Crystallaria* – 1 species	
C. asprella (Jordan)	Crystal darter
Genus *Ammocrypta* – 6 species	
A. beanii Jordan	Naked sand darter
A. bifascia Williams	Florida sand darter
A. clara Jordan & Meek	Western sand darter
A. meridiana Williams	Southern sand darter
A. pellucida (Putnam)	Eastern sand darter
A. vivax Hay	Scaly sand darter
Genus *Etheostoma* – 126 species	
Subgenus *Allohistium*	
E. cinereum Storer	Ashy darter
Subgenus *Nothonotus*	
E. acuticeps Bailey	Sharphead darter
E. bellum Zorach	Orangefin darter
E. camurum Cope	Bluebreast darter
E. chlorobranchium Zorach	Greenfin darter
E. denoncourti Stauffer & van Snik	Golden darter
E. juliae Meek	Yoke darter
E. moorei Raney & Suttkus	Yellowcheek darter
E. rubrum Raney & Suttkus	Bayou darter
E. rufilineatum (Cope)	Redline darter
E. tippecanoe Jordan & Evermann	Tippecanoe darter
Greenbreast Darter Group	
E. chuckwachatte Mayden & Wood	Lipstick darter
E. douglasi Wood & Mayden	Tuskaloosa darter
E. etowahae Wood & Mayden	Etowah darter
E. jordani Gilbert	Greenbreast darter
Spotted Darter Group	
E. aquali Williams & Etnier	Coppercheek darter
E. microlepidum Raney & Zorach	Smallscale darter
E. maculatum Kirtland	Spotted darter
E. sanguifluum (Cope)	Blood darter
E. vulneratum (Cope)	Wounded darter
E. wapiti Etnier & Williams	Boulder darter
Subgenus *Litocara*	
E. nianguae Gilbert & Meek	Niangua darter
E. sagitta (Jordan & Swain)	Arrow darter

Contd.

Table 12a Contd.

New subgenus	
E. sellare (Radcliffe & Welsh)	Maryland darter
Subgenus *Poecilichthys*	
E. euzonum (Hubbs & Black)	Arkansas saddled darter
E. kanawhae (Raney)	Kanawha darter
E. osburni (Hubbs & Trautman)	Finescale saddled darter
E. tetrazonum (Hubbs & Black)	Missouri saddled darter
E. variatum Kirtland	Variegate darter
Subgenus *Etheostoma*	
Greenside Darter Group	
E. blennioides Rafinesque	Greenside darter
E. blennius Gilbert & Swain	Blenny darter
E. inscriptum (Jordan & Brayton)	Turquoise darter
E. swannanoa Jordan & Evermann	Swannanoa darter
E. thalassinum (Jordan & Brayton)	Seagreen darter
Snubnose Darter Group	
E. baileyi Page & Burr	Emerald darter
E. barrenense Burr & Page	Splendid darter
E. bellator Suttkus & Bailey	Warrior darter
E. brevirostrum Suttkus & Etnier	Holiday darter
E. chermocki Boschung, Mayden & Tomelleri	Vermilion darter
E. colorosum Suttkus & Bailey	Coastal darter
E. coosae (Fowler)	Coosa darter
E. duryi Henshall	Blackside snubnose darter
E. etnieri Bouchard	Cherry darter
E. flavum Etnier & Bailey	Saffron darter
E. histrio Jordan & Gilbert	Harlequin darter
E. lachneri Suttkus & Bailey	Tombigbee darter
E. lynceum Hay	Brighteye darter
E. rafinesquei Burr & Page	Kentucky snubnose darter
E. ramseyi Suttkus & Bailey	Alabama darter
E. raneyi Suttkus & Bart	Yazoo darter
E. rupestre Gilbert & Swain	Rock darter
E. scotti Bauer, Etnier & Burkhead	Cherokee darter
E. simoterum (Cope)	Tennessee snubnose darter
E. tallapoosae Suttkus & Etnier	Tallapoosa darter
E. zonale (Cope)	Banded darter
Subgenus *Doration*	
E. jessiae (Jordan & Brayton)	Blueside darter
E. stigmaeum (Jordan)	Speckled darter
Subgenus *Boleosoma*	
E. longimanum Jordan	Longfin darter
E. nigrum (Rafinesque)	Johnny darter
E. olmstedi Storer	Tessellated darter
E. podostemone Jordan & Jenkins	Riverweed darter
Subgenus *Ioa*	
E. vitreum (Cope)	Glassy darter

Contd.

Table 12a Contd.

Subgenus *Vaillantia*	
E. chlorosomum (Hay)	Bluntnose darter
E. davisoni Hay	Choctawhatchee darter
Subgenus *Fuscatelum*	
E. parvipinne Gilbert & Swain	Goldstripe darter
Subgenus *Belophlox*	
E. fricksium Hildebrand	Savannah darter
E. mariae (Fowler)	Pinewoods darter
E. okaloosae (Fowler)	Okaloosa darter
Subgenus *Villora*	
E. edwini (Hubbs & Cannon)	Brown darter
Subgenus *Ozarka*	
E. boschungi Wall & Williams	Slackwater darter
E. cragini Gilbert	Arkansas darter
E. pallididorsum Distler & Metcalf	Paleback darter
E. punctulatum (Agassiz)	Stippled darter
E. trisella Bailey & Richards	Trispot darter
E. tuscumbia Gilbert & Swain	Tuscumbia darter
E. species	Sunburst darter
Subgenus *Oligocephalus*	
E. asprigene (Forbes)	Mud darter
E. australe Jordan	Conchos darter
E. caeruleum Storer	Rainbow darter
E. collettei Birdsong & Knapp	Creole darter
E. ditrema Ramsey & Suttkus	Coldwater darter
E. grahami (Girard)	Rio Grande darter
E. hopkinsi (Fowler)	Christmas darter
E. lepidum (Baird & Girard)	Greenthroat darter
E. lugoi Norris & Minckley	Tufa darter
E. nuchale Howell & Caldwell	Watercress darter
E. pottsi (Girard)	Mexican darter
E. radiosum (Hubbs & Black)	Orangebelly darter
E. segrex Norris & Minckley	Rio Salado darter
E. swaini (Jordan)	Gulf darter
E. whipplei (Girard)	Redfin darter
Orangethroat Darter Group	
E. bison Ceas & Page	Buffalo darter
E. burri Ceas & Page	Brook darter
E. fragi Distler	Strawberry darter
E. spectabile (Agassiz)	Orangethroat darter
E. kantuckeense Ceas & Page	Highland Rim darter
E. luteovinctum Gilbert & Swain	Redband darter
E. tecumsehi Ceas & Page	Shawnee darter
E. uniporum Distler	Current darter
Subgenus *Catonotus*	
Spottail Darter Group	
E. chienense Page & Ceas	Relict darter
E. corona Page & Ceas	Crown darter

Contd.

Table 12a Contd.

E. crossopterum Braasch & Mayden	Fringed darter
E. forbesi Page & Ceas	Barrens darter
E. olivaceum Braasch & Page	Dirty darter
E. nigripinne Braasch & Mayden	Blackfin darter
E. neopterum Howell & Dingerkus	Lollypop darter
E. oophylax Ceas & Page	Guardian darter
E. pseudovulatum Page & Ceas	Egg-mimic darter
E. squamiceps (Jordan)	Spottail darter
Fantail Darter Group	
E. flabellare Rafinesque	Fantail darter
E. kennicotti (Putnam)	Stripetail darter
E. percnurum Jenkins	Duskytail darter
Barcheek Darter Group	
E. barbouri Kuehne & Small	Teardrop darter
E. obeyense Kirsch	Barcheek darter
E. smithi Page & Braasch	Slabrock darter
E. striatulum Page & Braasch	Striated darter
E. virgatum (Jordan)	Striped darter
Subgenus *Boleichthys*	
E. collis (Hubbs & Cannon)	Carolina darter
E. exile (Girard)	Iowa darter
E. gracile (Girard)	Slough darter
E. fusiforme (Girard)	Swamp darter
E. serrifer (Hubbs & Cannon)	Sawcheek darter
E. zonifer (Hubbs & Cannon)	Backwater darter
Least Darter Group	
E. fonticola (Jordan & Gilbert)	Fountain darter
E. microperca Jordan & Gilbert	Least darter
E. proeliare (Hay)	Cypress darter

Chapter 13
Other Percinae and Luciopercinae

13.1 Introduction

The Percinae, in addition to *Perca* species, contain two other genera naturally occurring in Eurasia, *Gymnocephalus* and *Percarina*. The Luciopercinae, as well as *Stizostedion* species, also include two further genera, *Romanichthys* and *Zingel*, which evolved in the River Danube (Chapter 1). The Danube basin may be the origin of percids in general. A common ancestor probably gave rise to *Romanichthys* and *Zingel*. Little has been published on their biology and ecology. They are endangered (Lelek 1987) and subject to loss through siltation and alterations to stream flow.

13.2 Percinae

Body form and habitat

The four species *of Gymnocephalus*: *G. cernuus* (ruffe), *G. acerinus* (Don ruffe), *G. schraetser* (striped ruffe) and *G. baloni* (Balon's ruffe) (Fig. 13.1), resemble the perch with the distinction that their dorsal fins are partially united. The ruffe is greenish brown in colour along the dorsal surface. The sides and belly of the fish are yellowish and the body is marked with numerous dark flecks. The Don ruffe is slightly lighter and the striped ruffe is lemon coloured. The striped ruffe is distinguished, as its name implies, by more or less distinct broken dark stripes lengthwise along its flanks. The head and body of Balon's ruffe is yellowish brown in colour, darker on the flanks and yellow on the belly. Along the back there are four to six spots which continue downwards and lose their intensity as in the perch (Holčik & Hensel 1974). Balon's ruffe is most closely related to the ruffe and both have short and deep bodies as compared to the Don and striped ruffes which have more elongate bodies. The species (apart from distinguishing between *G. cernuus* and *G. baloni*) can be identified by the number of lateral line scales and spines in the anterior dorsal fins (Table 13.1). Individuals can also be separated into the four species by biochemical and chromosomal markers (Rösch *et al.* 1996). Slechtová and Slechta (1996) were able to show marked interspecific differences between

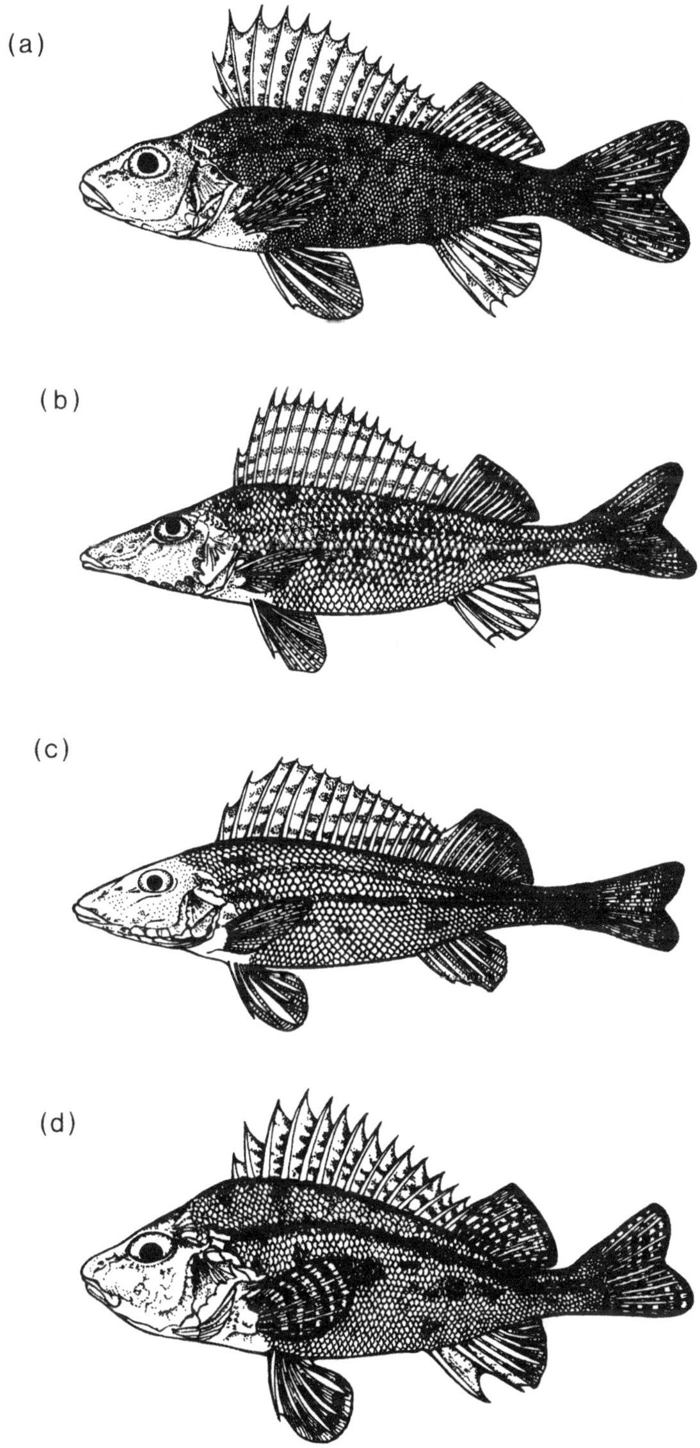

Fig. 13.1 *Gymnocephalus* species: (a) *G. cernuus*, (b) *G. acerinus*, (c) *G. schraetser* and (d) *G. baloni*.

Table 13.1 The number of lateral line scales and anterior dorsal fin spines in the four species of *Gymnocephalus*.

Species	Lateral line scales	Anterior dorsal fin spines
G. cernuus	35–40	11–16
G. baloni	35–39	14–16
G. acerinus	50–55	17–19
G. schraetser	55–62	17–19

ruffe, striped ruffe and Balon's ruffe and their hybrids using electrophoretic analysis.

The ruffe is found in lakes and slow flowing rivers throughout most of Europe and northern Asia (Fig. 1.4). It has been introduced into various new water bodies in this region as well as into the Laurentian Great Lakes of North America (see *Economic importance* below). It is a poor swimmer and lives close to the bottom. Its lateral line system is well developed for this mode of life, allowing detection of prey and predators (Disler & Smirnov 1977; Gray & Best 1989) and it can live in very turbid water. It is tolerant of brackish water such as the Baltic Sea (Vetemaa & Saat 1996). Balon's ruffe is found only in the Danube River basin and it lives in fast flowing water. The Don ruffe inhabits the drainage basins of rivers emptying into the northern Black Sea including the Don, Dniester and Dnieper. Like the ruffe, it lives in slow flowing water or in lakes. The striped ruffe is found in the Danube basin, including the estuary into the Black Sea, and normally lives in deep water over a substrate of sand or gravel.

Percarina is monotypic represented by the species *P. demidoffi* (percarina). The two dorsal fins are well separated in this species (Fig. 13.2) and the anterior dorsal fin has 9–11 spines. The body is yellow in colour and there is a row of black spots on the dorsal surface. The percarina is found in the northern area of the Black Sea, in the Sea of Azov and in the lower reaches of rivers entering these seas (Fig. 1.5). It is salinity tolerant.

Growth, food and feeding

The ruffe rarely grows to a length >250 mm (400 g) although 290 mm has been reported from the River Elbe (Hölker & Hammer 1994); the Don ruffe grows to about 180 mm and the striped ruffe to 200 mm. There are no recorded data for growth of Balon's ruffe There is some evidence that ruffe grow more slowly and live to a greater age the further north they are located. Female ruffe grow faster, grow to a greater ultimate length and live longer than males. Kålås (1995) who used otoliths to age the fish, found that males grew faster than females in their first year of life.

Gymnocephalus species feed actively through the day, usually with a peak towards the evening. At night they lie close to the bottom. Ogle *et al.* (1995) found that juvenile and adult ruffe would sometimes feed throughout 24 h and adults would

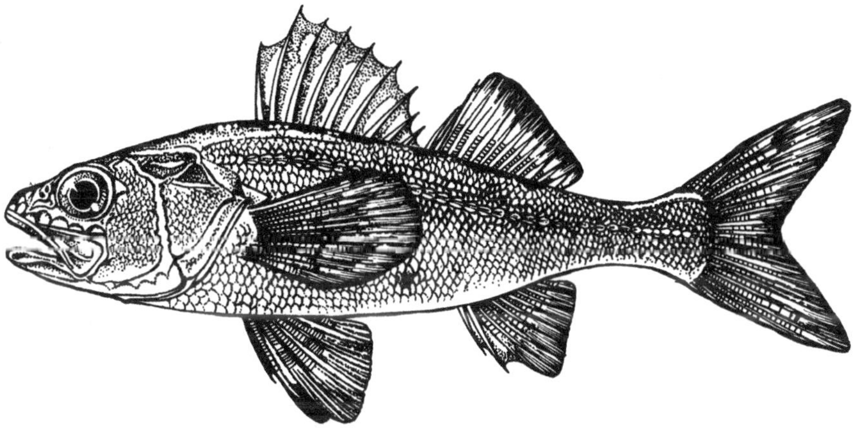

Fig. 13.2 *Percarina demidoffi.*

have fuller stomachs during the night than during the day. Members of the genus feed on insects and crustaceans and sometimes small fish. Chironomid larvae and pupae are probably the most important prey of ruffe (Kangur & Kangur 1996). The ruffe and striped ruffe also feed extensively on molluscs and eat fish eggs when they are available (see *Economic importance* below). The ruffe preys mainly on benthic animals throughout its life although exceptionally may feed predominantly on zooplankton (Kålås 1995). Interspecific competition may influence food selection.

Little is known of the ecology of the percarina. It reaches a length of 60–90 mm in three to four years and has a maximum length of *c.* 100 mm. It feeds on invertebrates and small fish, particularly copepods, cladocerans, mysids and gammarids. The larvae feed on zooplankton (Mikhman 1978).

Reproduction

Ruffe mature at two to three years of age at a length of 110–120 mm. In some warmer lakes they can mature earlier. For example in Lake Ijssel, the Netherlands, 50% mature at age one year at a size of 65–70 mm (Willemsen 1977b). Spawning normally occurs in all species between April and June although it can be extended over a longer period of time in colder areas. The ruffe undergoes spawning migrations from winter habitats which are deep areas in rivers or lakes to shallow areas which are stony and covered with aquatic macrophytes. The spawning migrations commence at a temperature of *c.* 4°C and spawning can occur at temperatures of 6–8°C although the normal range is 11.6–18.0°C (Hokanson 1977a). The number of eggs laid by the female depends on its size. In the Nadym River, in the former USSR, the average number of eggs laid by the ruffe is 17 800 per female (Kolomin 1977). However, the absolute fecundity can be very variable, *c.* 13 000–82 000 in the River Dabie, Poland (Neja 1988). The striped ruffe also spawns over stones and vegetation in shallow water but usually where the current is fairly rapid. About 10 000 eggs per female are laid by this species. There are no recorded observations of ruffe spawning behaviour but it is probably much like that of *Perca* species. However, unlike perch, the ruffe is an intermittent spawner and usually the eggs are laid in two or more batches (Leino & McCormick 1997). The eggs of the first batch are larger and have more yolk than those laid later. The eggs are deposited individually and are adhesive, sticking to stones and submerged vegetation. The average diameter of the egg is 1 mm. At a water temperature of 10–15°C the larval ruffe hatch in 8–10 days. Newly hatched larvae are 2.5–3.2 mm long (French & Edsall 1992). After the absorption of the yolk, the larval ruffe feed on rotifers and small crustaceans. The optimum temperature for growth of the larvae is in the range 25–30°C (Hokanson 1977a).

The percarina reaches sexual maturity normally after one year and spawns in June and July in shallow water which can be fresh, brackish, or seawater. The fecundity of the female is about 3000 eggs. Eggs are laid over sand and silt and hatch very quickly, normally within three days. The larval fish begin exogenous feeding on young stages of crustaceans.

Economic importance

Species of *Gymnocephalus* have usually been considered to have little economic importance. In eastern Europe the ruffe is fished locally for sport and sometimes commercially. It is used in the production of 'ruffe soup'! It may be important as a forage fish for more economically valuable fish such as the zander. However, in other situations it may conflict with other, more valuable food fish in sharing the same niche. The percarina has little economic importance except in some local net fisheries.

The ruffe may be an important link in the transmission of parasites to fish that prey on it, e.g. several helminth parasites are passed from ruffe, and other prey, to the predatory burbot, *Lota lota* (Linnaeus) (Valtonen & Julkunen 1995). The ruffe is a source of infection for *Anguillicola crassus*, a nematode that is found in the swimbladder of eels, *Anguilla anguilla* (Haenen & Van-Banning 1991; Hoeglund & Thomas 1992; Lehmann *et al.* 1996).

Little attention has been paid to the biology of this group of fishes although their effect on important species, e.g. predation on the eggs of coregonids by ruffe, has been known for some time (Pavlovskiy & Sterligova 1986). However, in recent years there have been several recorded introductions of ruffe, probably accidental by anglers using the fish as live bait, in Europe (Winfield *et al.* 1998a, b). The ruffe has also been introduced into the Laurentian Great Lakes of North America (Lake Superior in 1986 and Lake Huron in 1995), again accidentally, through ships' ballast water originating in Europe (Simon & Vondruska 1991; Pratt *et al.* 1992). These introductions have stimulated much interest in the ruffe including the production of a comprehensive bibliography (Winfield & McCulloch 1995) and an international symposium (Jensen 1997) resulting in several papers on its taxonomy, biology and ecology (Marsden *et al.* 1998). A major concern is the negative effects the ruffe may have on commercially important and/or rare species including other percids and coregonids (Winfield 1998a,b). However, these fish coexist naturally in many European lakes. The impact on recruitment of other fish caused by ruffe predation on their eggs may not be as significant as first feared (Winfield *et al.* 1998c). It is thought that the problem arises in water bodies where the ruffe has been introduced and where the population expands to high densities leading to unstable conditions amongst the community.

13.3 Luciopercinae

The asprete

The asprete, *R. valsanicola* (Dumitrescu *et al.* 1957) (Fig. 13.3) was first described in 1957. It is similar in build to the bullhead, *Cottus gobio* Linnaeus, and like the bullhead, the eyes are positioned dorsally on the head. It lives in the same habitat as the bullhead and often may have been confused with it. The asprete is commonly called the Romanian bullhead perch. The two dorsal fins are well separated and the posterior fin is larger than the anterior fin. The fish is brown in colour and the belly is lighter than the rest of the body. As an adaptation to living in fast flowing water, the asprete has no swim bladder.

The asprete was found in the Arges, Vilsan, Riul and Doamnei rivers in the Danube basin in Romania but has become extinct (Stanescu 1971) except in the Vilsan River (Fig. 1.8). It is probably nocturnal in habit and feeds on insect larvae, chiefly plecopterans, and small fishes. It normally grows in length to 80–120 mm with a maximum of 130 mm. Little is known about its reproduction. During the breeding season it develops breeding tubercles. Eggs are adhesive and hidden among stones.

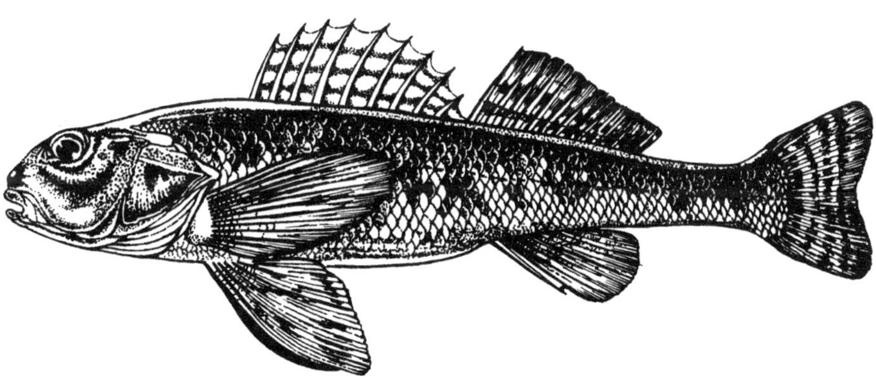

Fig. 13.3 *Romanichthys valsanicola.*

The zingels

Species of *Zingel*, *Z. Zingel* (zingel), *Z. asper* (apron) and *Z. streber* (streber), have slender, rounded bodies and the mouths are ventral in position (Fig. 13.4). The two dorsal fins are well separated in all three species. Scales cover the gill flaps. The fish have brown backs with dark mottling and blotches. The sides and belly are yellowish in colour. In zingel and apron the caudal peduncle is shorter than the base of the posterior dorsal fin, in streber it is longer. The zingel has 13–15 spines and 18–20 rays in the anterior and posterior dorsal fins respectively. Both the streber and the apron have 8–9 and 12–13 rays in the anterior and posterior dorsal fins respectively.

Zingel species, like *Romanichthys*, have no swim bladder and all three species are found in shallow, fast running water. They are bottom living rather inactive fish with flattened bellies and rather streamlined bodies which adapt them to their habitat. The streber generally occurs in deeper water than zingel. The zingel is found in the basin of the River Danube including the Rivers Morava and Turiec, Czech and Slovak Republics (Jurajda *et al.* 1994; Muzik 1996) and in the River Dniester. The streber also inhabits the Danube and is found in the Vardar River which flows into the Aegean Sea (Fig. 1.8). Šorić and Naumovski (1991) describe the systematic position of zingel and streber in Yugoslavia based on morphological characteristics. They suggest a division of streber into distinct subspecies: *Z. s. streber* Siebold in tributaries of the River Danube, *Z. s. balcanicus* (Karaman) from the River Vardar and *Z. s. nerensis* Bănărescu and Nalbant from the River Nera. Probably an ancestor of *Z. streber* entered the River Rhône and gave rise to *Z. asper* which is found there now. The current status of the apron in French rivers is given by Changeux and Pont (1995). In 1900 it had a continuous distribution throughout the River Rhône and its main tributaries. Currently it only occupies 17% of this area.

The zingel is the largest of the three species growing in length to 150–300 mm, streber grows to about 150–200 mm and the apron to about 120–150 mm. All three species are active at night but during the day they remain hidden and camouflaged between stones. They feed on benthic invertebrates and small fishes. The apron also feeds on fish eggs and fry.

All three species spawn from March to May and lay adhesive eggs amongst stones in fast flowing water (Patzner *et al.* 1994; Patzner & Glechner 1996). A female zingel may lay as many as 5000 eggs. These members of the Luciopercinae have no economic importance but are of biological interest in their adaptation to fast flowing water and their limited distribution.

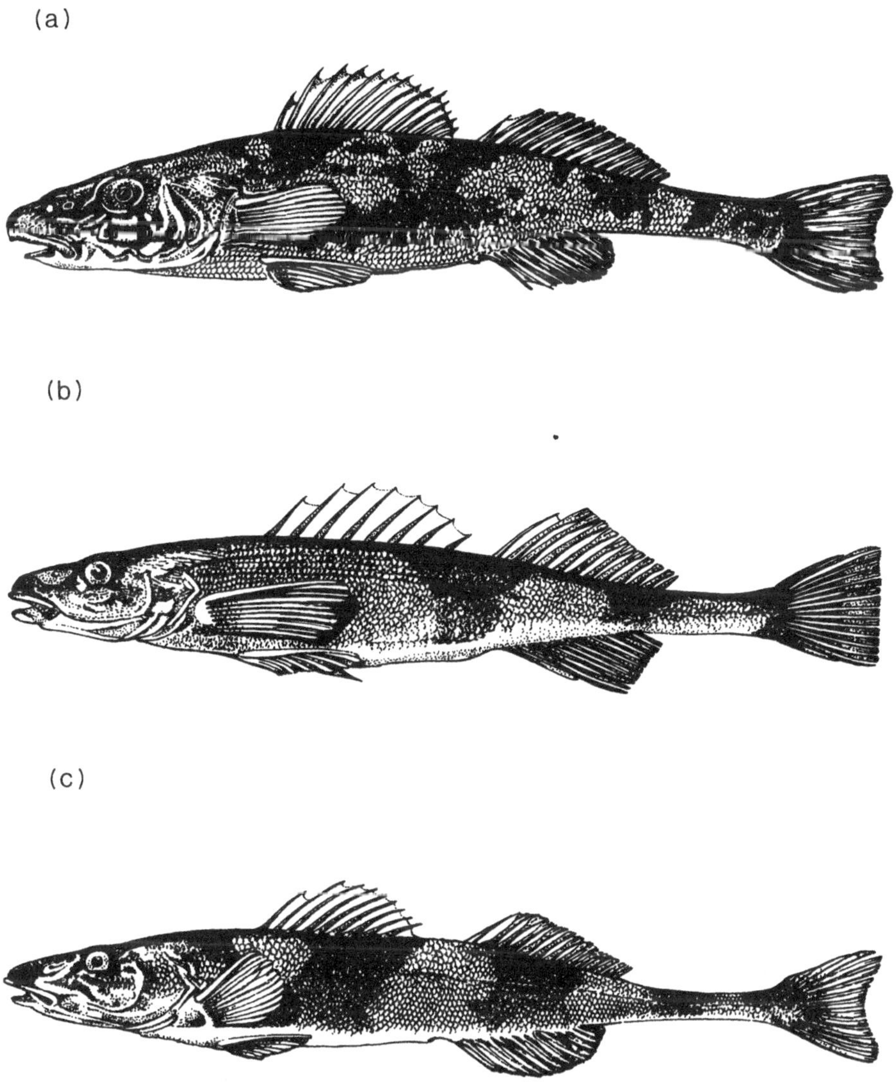

Fig. 13.4 *Zingel* species: (a) *Z. zingel*, (b) *Z. asper* and (c) *Z. streber*.

References

Aalto, S.K. & Newsome, G.E. (1990) Additional evidence supporting demic behaviour of a yellow perch (*Perca flavescens*) population. *Canadian Journal of Fisheries and Aquatic Sciences* **47**, 1959–62.

Abdu-Nabi, A.H. (1983) Factors related to selection of fish scales for growth studies of yellow perch, *Perca flavescens. Dissertation Abstracts International B Science and Engineering* **44** (6), 1703.

Abi-Ayad, S.M. (1998) *Etude expérimentale de la biologie de la reproduction de la perche fluviatile* (Perca fluviatilis). *Effet de la composition en acides gras de la série (n-3) de l'alimentation des géniteurs sur la qualité des oeufs et des larves.* Ph.D. Thesis, Université de Liège, Belgium.

Abi-Ayad, S.M., Mélard, C. & Kestemont, P. (1997) Effects of n-3 fatty acids in Eurasian perch broodstock diet on egg fatty acid composition and larvae stress resistance. *Aquaculture International* **5**, 161–8.

Adamson, S.W. & Wissing, T.E. (1977) Food habits and feeding periodicity of the rainbow, fantail and banded darters in Four-Mile Creek. *Ohio Journal of Science* **77**, 164–9.

Ahlbert, I.B. (1969) The organisation of the cone cells in the retinae of four teleosts with different feeding habits (*Perca fluviatilis* L., *Lucioperca lucioperca* L., *Acerina cernua* L., *Coregonus albula* L.). *Arkiv fur Zoologi* **22**, 445–81.

Akberali, H.B. & Earnshaw, M.J. (1984) Copperstimulated respiration in the unfertilized egg of the Eurasian perch, *Perca fluviatilis. Comparative Biochemistry and Physiology* **78C** (2), 349–52.

Aksiray, F. (1961) About sudak (*Lucioperca sandra* Cuv. and Val.) introduced into some of the lakes of Turkey. *Proceedings and technical papers of the General Fisheries Council of the Mediterranean* **6**, 335–43.

Akster, H.A. (1981) Ultrastructure of muscle fibres in head and axial muscles of the perch (*Perca fluviatilis* L.): A quantitative study. *Cell Tissue Research* **219** (1), 111–31.

Akster, H.A. & Osse, J.W.M. (1978) Muscle fibre types in head muscles of the perch, *Perca fluviatilis* L., Teleostei. A histochemical and electromyographical study. *Netherlands Journal of Zoology* **28** (1), 94–110.

Al-Absy, A.H. & Carlander, K.D. (1988) Criteria for selection of scale-sampling sites in growth studies of yellow perch. *Transactions of the American Fisheries Society* **117**, 209–12.

Albers, C. (1970) Acid–base balance. In: *Fish Physiology IV*, (eds W.S. Hoar & D.J. Randall), pp. 173–208. Academic Press, New York.

Aleev, Y.G. (1969) Function and gross morphology in fish. *Israel Program for Scientific Translations*, Jerusalem.

Alexander, R.M. (1967) *Functional Design in Fishes*. Hutchinson & Co., London.

Ali, M.A. & Anctil, M. (1968) Corrélation entre la structure rétinienne et l'habitat chez

Stizostedion vitreum vitreum et *S. canadense*. *Journal of the Fisheries Research Board of Canada* **25** (9), 2001–3.

Ali, M.A. & Anctil, M. (1977) Retinal structure and function in the walleye (*Stizostedion vitreum vitreum*) and sauger (*S. canadense*). *Journal of the Fisheries Research Board of Canada* **34** (10), 1467–74.

Ali, M.A., Ryder, R.A. & Anctil, M. (1977) Photoreceptors and visual pigments as related to behavioural responses and preferred habitats of perches (*Perca* spp.) and pikeperches (*Stizostedion* spp.). *Journal of the Fisheries Research Board of Canada* **34** (10), 1475–80.

Alm, G. (1946) Reasons for the occurrence of stunted fish populations with special regard to the perch. *Meddelanden från Statens undersöknings och försöksanstalt för sötvattensfisket (Reports from the Swedish State Institute of Freshwater Fishery Research, Drottningholm)* **25**, 1–146.

Amin, O.M. (1979) Lymphocystis disease in Wisconsin fishes. *Journal of Fish Diseases* **2**, 207–17.

Anders, K. & Darai, G. (1985) Genome analysis of fish lymphocystis disease virus. In: *Fish and Shellfish Pathology*, (ed. A.E. Ellis), pp. 301–6. Academic Press, London.

Anderson, D.P. (1974) *Fish Immunology*. T.F.H. Publications, Inc. Ltd, Neptune City, N.J.

Anderson, K.A., Rosenblum, P.M., Whiteside, B.G., Standage, R.W. & Brandt, T.M. (1998) Controlled spawning of longnose darters. *The Progressive Fish-Culturist* **60**, 137–45.

Andersson, P., Borg, H. & Kaerrhage, P. (1995) Mercury in fish muscle in acidified and limed lakes. *Water, Air and Soil Pollution* **80** (1–4), 889–92.

Andersson, P. & Haakanson, L. (1991) The resolution of mercury dose and lake sensitivity related to mercury content in fish. *Water, Air and Soil Pollution* **56**, 169–86.

Andersson, P. & Holm, K. (1995) Cadmium in water and perch (*Perca fluviatilis*) liver in limed L. Stensjoen in Tyresta National Park, Sweden. *Water, Air and Soil Pollution* **85** (2), 805–10.

Andrews, C.R. (1977) *The biology of the parasitic fauna of perch* (Perca fluviatilis *L.) from Llyn Tegid, North Wales*. Ph.D. Thesis, University of Liverpool.

Ansari, R.H. & Qadri, S.U. (1989) Individual variation in the foraging strategies of young yellow perch (*Perca flavescens*) from the Ottawa River. *Hydrobiologia* **174**, 207–12.

Antalfi, A. (1979) Propagation and rearing of pikeperch in pond culture. Workshop on mass rearing of fry and fingerling of freshwater fishes. The Hague, May 1979. (eds E.A. Huisman & H. Hogendoorn). *EIFAC Technical Paper* **35** *(Suppl. 1)*, 120–24. FAO, Rome.

Anthouard, M. & Fontaine, P. (in press). L'auto-alimentation chez la perche *Perca fluviatilis*. Adaptation à cette modalité de nourrissage et mise en évidence du rythme nycthéméral de la prise alimentaire. *Ichtyophysiologica Acta*, **21**, 1–13.

Armbruster, J.W. & Page, L.M. (1996) Convergence of a cryptic saddle pattern in benthic freshwater fishes. *Environmental Biology of Fishes* **45**, 249–57.

Armstrong, F.A.J. & Scott, D.P. (1979) Decrease in mercury content of fishes in Ball Lake, Ontario, since imposition of controls on mercury discharges. *Journal of the Fisheries Research Board of Canada* **36** (6), 670–72.

Atz, J.W. (1940) Reproductive behavior in the eastern johnny darter, *Boleosoma nigrum olmstedi* (Storer). *Copeia* **1940**, 100–106.

Auer, M.T. & Auer, N.A. (1990) Chemical suitability of substrates for walleye egg development in the Lower Fox River, Wisconsin. *Transactions of the American Fisheries Society* **119**, 871–6.

Avedikova, T.M. & D'Yakova, G.P. (1979) Produktivnost' populyatsij poluprokhodnykh rob pri menyayushchemsya rezhime Azovskogo Morya. Vliyanie sovremennykh vodokhozyajstvennrkh meropriyatij na biologicheskuyu produktivnost' Azovskogo i Kaspijskogo Morej, (ed. E.A. Yablonskaya). *Trudy Vniro Moskua* **133**, 70–83.

Awaïss, A., Kestemont, P. & Micha, J.C. (1992) Nutritional suitability of the rotifer *Brachionus calyciflorus* Pallas for rearing freshwater fish larvae. *Journal of Applied Ichthyology* **8**, 263–70.

Awaïss, A., Kestemont, P. & Micha, J.C. (1996) Fatty acid profiles of two freshwater fish larvae (gudgeon and perch) reared with *Brachionus calyciflorus* Pallas (Rotifera). *Aquaculture Research* **27**, 651–8.

Babaluk, J.A. & Campbell, J.S. (1987) Preliminary results of tetracycline labelling for validating annual growth increments in opercula of walleyes. *North American Journal of Fisheries Management* **7** (2), 138–41.

Babaluk, J.A., Craig, J.F. & Campbell, J.S. (1993) Age and growth estimation of walleye, *Stizostedion vitreum*, using opercula. *Canadian Manuscript Report of Fisheries and Aquatic Sciences* **2183**.

Baccante, D.A. & Reid, D.M. (1988) Fecundity changes in two exploited walleye populations. *North American Journal of Fisheries Management* **8** (2), 199–209.

Bagenal, T.B. (1972) The variability in numbers of perch, *Perca fluviatilis* L., caught in traps. *Freshwater Biology* **2**, 27–36.

Bagenal, T.B. & Tesch, F.W. (1978) Age and growth. In: *Methods for Assessment of Fish Production in Fresh Waters*. I.B.P. Handbook No. 3 (ed. T.B. Bagenal), 3rd edn. pp. 101–36. Blackwell Scientific Publications, Oxford.

Bailey, R.M. (1951) A check-list of the fishes of Iowa, with keys for identification. In: *Iowa Fish and Fishing*, (eds J.R. Harlan & E.B. Speaker), pp. 185–238. Iowa State Conservation Commission, Des Moines.

Bailey, R.M. & Gosline, W.A. (1955) Variation and systematic significance of vertebral counts in the American fishes of the family Percidae. *Miscellaneous Publications of the Museum of Zoology, University of Michigan* **93**, 1–44.

Bailey, R.M., Winn, H.E. & Smith, C.L. (1954) Fishes from the Escambia River, Alabama and Florida, with ecologic and taxonomic notes. *Proceedings of the Academy of Natural Sciences, Philadelphia* **106**, 109–64.

Bakanov, A.I., Kiyashko, V.I., Smetanin, M.M. & Strelnikov, A.S. (1987) Factors affecting fish growth. *Journal of Ichthyology* **27** (5), 124–32.

Baker, B.I. & Wigham, T. (1979) Endocrine aspects of metabolic coordination in teleosts. In: *Fish Phenology* (ed. P.J. Miller). *Symposia of the Zoological Society of London* **44**, 89–103. Academic Press, London.

Baker, J.R. (1978) The fantail darter *Etheostoma flabellare* in the Salt River drainage, Kentucky. *Transactions of the Kentucky Academy of Science* **39**, 150–59.

Balk, L., Foerlin, L., Soederstroem, M. & Larsson, A. (1993) Indications of regional and large-scale biological effects caused by bleached pulp mill effluents. *Chemosphere* **27** (4), 631–50.

Ball, I.R. (1967) The relative susceptibilities of some species of freshwater fish to poisons. II. Zinc. *Water Research* **1**, 777–83.

Ball, J.N. (1969) Prolactin (fish prolactin or paralactin) and growth hormone. In: *Fish Physiology II* (eds W.S. Hoar & D.J. Randall), pp. 207–40. Academic Press, London.

Balon, E.K. (1956) Mezdidruhová hybridižacia dunajských ryb. l. Oplodenie ikier ostrieza dunajského spermion zubačajného (*Perca fluviatilis* infraspecies *vulgaris* (Schäffer 1759) Pokrovskij 1951 × *Lucioperca lucioperca* (Linnaeus 1758)). *Polnohospodárstvo, Bratislava* **3** (5), 581–92.

Bangham, R.V. (1955) Studies on fish parasites of Lake Huron and Manitoulin Island. *American Midland Naturalist* **53** (1), 184–94.

Barlow, G.W. (1972) The attitude of fish eye-lines in relation to body shape and to stripes and bars. *Copeia* **1972**, 4–12.

Barrows, F.T. (1990) Walleye research summary; 1989 season. Annual Progress Report: Year 1. *Fisheries and Federal Aid Project* **81**. United States Fish & Wildlife Service, Denver, Colorado.

Barrows, F.T. (1996) Diet and nutrition. In: *Walleye Culture Manual* (ed. R.C. Summerfelt), pp. 315–21. *NCRAC Culture Series* **101**. North Central Regional Aquaculture Center Publications Office, Iowa State University, Ames.

Barrows, F.T., Lellis, W.A. & Nickum, J.G. (1988) Intensive culture of larval walleyes with dry or formulated feed: note on swim bladder inflation. *The Progressive Fish-Culturist* **50**, 160–66.

Barrows, F.T, Zitzow, R.E. & Kindschi, G.A. (1993) Effects of surface water spray, diet, and phase feeding on swim bladder inflation, survival, and cost of production of intensively reared larval walleyes. *The Progressive Fish-Culturist* **55**, 224–8.

Barry, T.P., Malison, J.A., Lapp, A.F. & Procarione, L.P. (1995) Effects of selected hormones and male cohorts on final oocyte maturation, ovulation and steroid production in walleye (*Stizostedion vitreum*). *Aquaculture* **138**, 331–47.

Bart, H.L., Jr. (1992) Spawning behavior of *Etheostoma davisoni* Hay. *Copeia* **1992**, 537–9.

Bart, H.L., Jr & Page, L.M. (1991) Morphology and adaptive significance of fin knobs in egg-clustering darters. *Copeia* **1991**, 80–86.

Bart, H.L., Jr & Page, L.M. (1992) The influences of size and phylogeny on life history variation in North American percids. In: *Systematics, Historical Ecology, and North American Freshwater Fishes*, (ed. R.L. Mayden), pp. 553–72. Stanford University Press.

Bartell, S.M., Breck, J.E. & Gardner, R.H. (1986) Individual parameter perturbation and error analysis of fish bioenergetics models. *Canadian Journal of Fisheries and Aquatic Sciences* **43**, 160–68.

Bauer, B.H., Etnier, D.A. & Burkhead, N.M. (1995) *Etheostoma* (*Ulocentra*) *scotti* (Osteichthyes: Percidae), a new darter from the Etowah River system in Georgia. *Bulletin of the Alabama Museum of Natural History* **17**, 1–16.

Bauer, O.N. (1962) Ecology of the parasites of freshwater fish. Interrelationships between the parasite and its habitat. *Israel Program for Scientific Translations* **622**.

Bauer, O.N., Musselius, V.A. & Strelkov, Yu. A. (1973) Diseases of pond fishes. *Israel Program for Scientific Translations* **60118**.

Beamish, F.W.H. (1990a) Swimming metabolism and temperature in juvenile walleye, *Stizostedion vitreum vitreum. Environmental Biology of Fishes* **27** (4), 309–14.

Beamish, F.W.H. (1990b) Metabolism and photoperiod in juvenile lake charr and walleye. *Environmental Biology of Fishes* **29** (3), 201–7.

Beamish, F.W.H. & MacMahon, P.D. (1988) Apparent heat increment and feeding strategy in walleye, *Stizostedion vitreum vitreum*. *Aquaculture* **68**, 73–82.

Becker, G.C. (1983) *Fishes of Wisconsin*. University of Wisconsin Press, Wisconsin.

Bein, R. & Ribi, G. (1994) Effects of larval density and salinity on the development of perch larvae (*Perca fluviatilis* L.). *Aquatic Sciences* **56** (2), 97–105.

Bekesi, L. & Kovacs-Gayer, E. (1986) Lymphosarcoma in pike-perch (*Stizostedion lucioperca* L.): a case report. *Acta Veterinaria Hungarica* **34** (1–2), 101–2.

Beklioglu, M. & Moss, B. (1995) The impact of pH on interactions among phytoplankton algae, zooplankton and perch (*Perca fluviatilis*) in a shallow, fertile lake. *Freshwater Biology* **33** (3), 497–509.

Belanger, S.E. & Hogler, S.R. (1982) Comparison of five ageing methodologies applied to walleye (*Stizostedion vitreum*) in Burt Lake, Michigan. *Journal of Great Lakes Research* **8** (4), 666–71.

Belogolova, L.A. (1987) Dynamics of abundance and distribution of young roach *Rutilus rutilus*, bream *Abramis brama* and pikeperch *Stizostedion lucioperca* in the northern Caspian Sea. *Journal of Ichthyology* **27** (6), 924–35.

Belyi, N.D. (1972) Downstream migration of pikeperch *Lucioperca lucioperca* and its feeding in the lower reaches of the Dnieper during early stages of development. *Journal of Ichthyology* **12** (3), 465–72.

Benndorf, J., Schultz, H., Benndorf, A., Ungèr, R., Penz, E., Kneschke, H., Kossatz, K., Dumke, R. *et al.* (1988) Food-web manipulation by enhancement of piscivorous fish stocks: long-term effects in the hypertrophic Bautzen reservoir. In: *Biomanipulation. Part 2: Studies of Integral Food Webs* (ed. J. Benndorf). *Limnologica* **19** (1), 97–110.

Bennett, D.H. & McArthur, T.J. (1990) Predicting success of walleye stocking programs in the United States and Canada. *Fisheries* **15** (4), 19–23.

Berg, L.S. (1905) *Ryby Turkestana.* St. Petersburg.

Berg, L.S. (1949) *Ryby presnykh vod SSSR. i Sopredelnykh stran.* **III**, 929–1382.

Berg, L.S. (1965) Freshwater fishes of the USSR and adjacent countries. *Israel Program for Scientific Translations* **743**.

Bergman, E. & Greenberg, L.A. (1994) Competition between a planktivore, a benthivore, and a species with ontogenetic diet shifts. *Ecology* **75** (5), 1233–45.

Bergquist, B.C. (1991) Extinction and natural recolonization of fish in acidified and limed lakes. *Nordic Journal of Freshwater Research* **66**, 50–62.

Berninger, K. & Pennanen, J. (1995) Heavy metals in perch (*Perca fluviatilis* L.) from two acidified lakes in the Salpausselkae esker area in Finland. *Water, Air and Soil Pollution* **81** (3–4), 283–94.

Bertalanffy, L. von. (1938) A quantitative theory of organic growth. *Human Biology* **10**, 181–213.

Berzins, B. (1949) On the biology of the Latvian perch (*Perca fluviatilis* L.). *Hydrobiologia* **2**, 64–71.

Best, C.D. (1981) *Initiation of artificial feeding and the control of sex differentiation in yellow perch, Perca flavescens.* M.Sc. Thesis, University of Wisconsin-Madison, Madison, Wisconsin, USA.

Beverton, R.J.H. (1987) Longevity in fish: some ecological and evolutionary considerations. In: *Evolution of Longevity in Animals. A Comparative Approach* (eds A.D. Woodhead & K.H. Thompson). Plenum, New York.

Beverton, R.J.H. & Holt, S.J. (1957) On the dynamics of exploited fish populations. *Fishery Investigations, Ministry of Agriculture, Fisheries and Food Series II* **19**. HMSO, London.

Beyerle, G.B. (1975) Summary of attempts to raise walleye fry and fingerlings on artificial diets, with suggestions on needed research and procedures to be used in future tests. *The Progressive Fish-Culturist* **37**, 103–5.

Billington, N. (1996) Geographical distribution of mitochondrial DNA (mtDNA) variation in walleye, sauger, and yellow perch. *Annales Zoologici Fennici* **33**, 699–706.

Billington, N., Barrette, R.J. & Hebert, P.D.N. (1992) Management implications of mitochondrial DNA variation in walleye stocks. *North American Journal of Fisheries Management* **12** (2), 276–84.

Billington, N., Danzmann, R.G., Hebert, P.D.N. & Ward, R.D. (1991) Phylogenetic relationships among four members of *Stizostedion* (Percidae) determined by mitochondrial DNA and allozyme analyses. *Journal of Fish Biology* **39** (Suppl. A), 251–8.

Billington, N., Hebert, P.D.N. & Ward, R.D. (1988) Evidence of introgressive hybridization in the genus *Stizostedion*: interspecific transfer of mitochondrial DNA between sauger and walleye. *Canadian Journal of Fisheries and Aquatic Sciences* **45** (11), 2035–41.

Billington, N., Hebert, P.D.N. & Ward, R.D. (1990) Allozyme and mitochondrial DNA variation among three species of *Stizostedion* (Percidae): phylogenetic and zoogeographical implications. *Canadian Journal of Fisheries and Aquatic Sciences* **47** (6), 1093–1102.

Biró, P. (1973) The food of pikeperch, *Lucioperca lucioperca* in Lake Balaton. *Annales Instituti Biologici, Tihany* **40**, 159–83.

Biró, P. (1985) Dynamics of the pikeperch, *Stizostedion lucioperca* (L.), in Lake Balaton. *Internationale Revue der Gesamten Hydrobiologie* **70** (4), 471–90.

Biró, P. (1997) Temporal variation in Lake Balaton and its fish populations. *Ecology of Freshwater Fish* **6** (4), 196–216.

Blaxter, J.H.S. (1968) Light intensity, vision and feeding in young plaice. *Journal of Experimental Marine Biology and Ecology* **2**, 293–307.

Blaxter, J.H.S. (1980) Vision and feeding of fishes. In: *Fish Behavior and its use in the Capture and Culture of Fishes* (eds J.E. Bardach, J.J. Magnuson, R.C. May & J.M. Reinhart). *ICLARM Conference Proceedings* **5**, 32–56.

Bodaly, R.A. (1980) Pre- and post-spawning movements of walleye, *Stizostedion vitreum*, in Southern Indian Lake, Manitoba. *Canadian Technical Report of Fisheries and Aquatic Sciences* **931**.

Bodaly, R.A., Hecky, R.E. & Fudge, R.J.P. (1984) Increases in fish mercury levels in lakes flooded by the Churchill River diversion northern Manitoba, Canada. *Canadian Journal of Fisheries and Aquatic Sciences* **41** (4), 682–91.

Boehling, P. & Lehtonen, H. (1985) Effect of environmental factors on migrations of perch (*Perca fluviatilis* L.) tagged in the coastal waters of Finland. *Finnish Fisheries Research* **1985** (5), 31–40.

Böhling, P., Hudd, R., Lehtonen, H., Karås, P., Neuman, E. & Thoresson, G. (1991) Variations in year-class strengths of different perch (*Perca fluviatilis*) populations in the Baltic Sea with special reference to temperature and pollution. *Canadian Journal of Fisheries and Aquatic Sciences* **48**, 1181–7.

Boisclair, D. & Leggett, W.C. (1989a) Among-population variability of fish growth: I. Influence of the quantity of food consumed. *Canadian Journal of Fisheries and Aquatic Sciences* **46**, 457–67.

Boisclair, D. & Leggett, W.C. (1989b) Among-population variability of fish growth: II. Influence of prey type. *Canadian Journal of Fisheries and Aquatic Sciences* **46**, 468–82.

Boisclair, D., & Leggett, W.C. (1989c). Among-population variability of fish growth: III. Influence of fish community. *Canadian Journal of Fisheries and Aquatic Sciences* **46**, 1539–50.

Boisclair, D. & Leggett, W.C. (1989d) The importance of activity in bioenergetics models applied to actively foraging fishes. *Canadian Journal of Fisheries and Aquatic Sciences* **46**, 1859–67.

Boisclair, D. & Rasmussen, J.B. (1996) Empirical analysis of the influence of environmental variables associated with lake eutrophication on perch growth, consumption, and activity rates. *Annales Zoologici Fennici* **33**, 507–15.

Bonar, A. (1977) Relations between exploitation, yield, and community structure in Polish pikeperch (*Stizostedion lucioperca*) lakes, 1966–71. *Journal of the Fisheries Research Board of Canada* **34** (10), 1576–80.

Boschung, H. (1979) *Report on the breeding habits of the slackwater darter (Percidae:* Etheostoma boschungi) *in the Cypress Creek watershed.* Report to US Department of Agriculture Soil Conservation Service, Auburn, Alabama.

Botsjarnikowa, A.W. (1952) Dannye no biologiirasmnozheniya i razvitiya kubanskogo sudako. *Zoologichesky Zhurnal* **31** (1), 122.

Bowser, P.R., Casey, J.W., Vogt, V.M., Poulet, F.M., Quackenbush, S. & Martineau, D. (1994) Walleye dermal sarcoma virus: a novel retrovirus that causes tumors in fish. In: *International Symposium on Aquatic Animal Health: Program and Abstracts.* University of California, Davis, California, USA.

Bowser, P.R., Martineau, D. & Wooster, G.A. (1990) Effects of water temperature on experimental transmission of dermal sarcoma in fingerling walleyes. *Journal of Aquatic Animal Health* **2** (3), 157–61.

Bowser, P.R., Wolfe, M.J., Forney, J.L. & Wooster, G.A. (1988) Seasonal prevalence of skin tumors from walleye (*Stizostedion vitreum*) from Oneida Lake, New York. *Journal of Wildlife Diseases* **24** (2), 292–8.

Bowser, P.R. & Wooster, G.A. (1991) Regression of dermal sarcoma in adult walleyes. *Journal of Aquatic Animal Health* **3** (2), 147–50.

Bowser, P.A. & Wooster, G.A. (1994) Ether sensitivity of the walleye dermal sarcoma virus. *Journal of Aquatic Animal Health* **6** (2), 178–9.

Braasch, M.E. & Mayden, R.L. (1985) Review of the subgenus *Catonotus* (Percidae) with descriptions of two new darters of the *Etheostoma squamiceps* species group. *Occasional Paper of the Museum of Natural History, University of Kansas* **119**, 1–83.

Braasch, M.E. & Smith, P.W. (1967) The life history of the slough darter, *Etheostoma gracile* (Pisces, Percidae). *Illinois Natural History Survey, Biological Notes* **58**, 1–12.

Brabrand, Å. (1995) Intra-cohort cannibalism among larval stages of perch (*Perca fluviatilis*). *Ecology of Freshwater Fish* **4**, 70–76.

Brabrand, Å. & Faafeng, B. (1993) Habitat shift in roach (*Rutilus rutilus*) induced by pike-perch (*Stizostedion lucioperca*) introduction: predation risk versus pelagic behaviour. *Oecologia* **95** (1), 38–46.

Brandt. S.B., Mason, D.M., MacNeill, D.B., Coates, T. & Gannon, J.E. (1987) Predation by alewives on larvae of yellow perch in Lake Ontario. *Transactions of the American Fisheries Society* **116** (4), 641–5.

Breder, C.M., Jr. (1926) The locomotion of fishes. *Zoologica (N.Y.)* **4**, 159–297.

Breder, C.M., Jr. (1959) Studies on social groupings in fishes. *Bulletin of the American Museum of Natural History* **117**, 393–483.

Breder, C.M. & Rosen, D.E. (1966) Subfamily percinae. In: *Modes of Reproduction in Fishes*, pp. 440–45. T.F.H. Publications, Jersey City, N.J.

Brett, J.R. (1971) Satiation time, appetite and maximum food intake of sockeye salmon (*Oncorhynchus nerka*). *Journal of the Fisheries Research Board of Canada* **28**, 409–15.

Brett, J.R., Shelbourn, J.E. & Shoop, C.T. (1969) Growth rate and body composition of fingerling sockeye salmon, *Oncorhynchus nerka*, in relation to temperature and ration size. *Journal of the Fisheries Research Board of Canada* **26**, 2363–94.

Brezonik, P.L., Eaton, J.G., Frost, T.M., Garrison, P.J., Kratz, T.K., Mach, C.E., McCormick, J.H., Perry, J.A., Rose, W.A. *et al.* (1993) Experimental acidification of Little Rock Lake, Wisconsin: chemical and biological changes over the pH range 6.1 to 4.7. *Canadian Journal of Fisheries and Aquatic Sciences* **50** (5), 1101–21.

Bristow, B.T. & Summerfelt, R.C. (1994) Performance of larval walleye cultured intensively in clear and turbid water. *Journal of the World Aquaculture Society* **25** (3), 454–64.

Bristow, B.T., Summerfelt, R.C. & Clayton, R.D. (1996) Comparative performance of intensively cultured larval walleye in clear, turbid, and colored water. *The Progressive Fish-Culturist* **58**, 1–10.

Broenmark, C. (1994) Effects of tench and perch on interactions in a freshwater, benthic food chain. *Ecology* **75** (6), 1818–28.

Bronte, C.R., Selgeby, J.H. & Swedberg, D.V. (1993) Dynamics of a yellow perch population in western Lake Erie. *North American Journal of Fisheries Management* **13** (3), 511–23.

Brooks, R.C., Heidinger, R.C. & Kohler, C.C. (1994) Mass-marking otoliths of larval and juvenile walleyes by immersion in oxytetracycline, calcein, or calcein blue. *North American Journal of Fisheries Management* **14** (1), 143–50.

Brousseau, C.S. & Armstrong, E.R. (1987) The role of size limits in walleye management. *Fisheries* **12** (1), 2–5.

Brown, G.G. (1996) Comparative storage methods and fertility studies of walleye semen. In: *Walleye Culture Manual* (ed. R.C. Summerfelt), pp. 45–50. *NCRAC Culture Series* **101**. North Central Regional Aquaculture Center Publications Office, Iowa State University, Ames.

Brown, P.B., Dabrowski, K. & Garling, D. (1996) Nutrition and feeding of yellow perch (*Perca flavescens*). *Journal of Applied Ichthyology* **12** (3–4), 171–4.

Brown, P.B., Wilson, K.A. & Twibell, R.G. (1998) Dietary sulfur amino acid requirement and cysteine replacement value for juvenile yellow perch. In: World Aquaculture Society Symposium: *Aquaculture '98*. Las Vagas, Nevada, 15–19 February 1998.

Brunelli, G. & Rizzo, L. (1928) Chizndala esocrina, ovario impari ed ermafroditismo nella, '*Perca fluviatilis*'. *Atti della Academia Nazionale dei Lincei, Classe di Scienze Fisiche Matematiche e Naturali, Rendiconti* **7**, 865–7.

Brunet, M.R. (1957) Le sandre et le brochet. *Rivières et Forêts* **6**, 33–8.

Bruylants, B., Vandelannoote, A. & Verheyen, R. (1986) The movement pattern and density distribution of perch, *Perca fluviatilis* L., in a channelized lowland river. *Aquaculture and Fisheries Management* **17** (1), 49–57.

Bryan, S.D., Soupir, C.A., Duffy, W.G. & Freiburger, C.E. (1996) Caloric densities of three predatory fishes and their prey in Lake Oahe, South Dakota. *Journal of Freshwater Ecology* **11** (2), 153–61.

Bryant, R.T. (1979) The life history and comparative ecology of the sharphead darter, *Etheostoma acuticeps*. *Tennessee Wildlife Research Agency Technical Report* **79–50**.

Bucke, D., Cawley, G.D., Craig, J.F., Pickering, A.D. & Willoughby, L.G. (1979) Further studies of an epizootic of perch, *Perca fluviatilis* L., of uncertain aetiology. *Journal of Fish Diseases* **2** (4), 297–311.

Buijse, A.D. (1992) *Dynamics and exploitation of unstable percid populations*. Ph.D. Thesis, Agricultural University of Wageningen, The Netherlands.

Buijse, A.D. & Houthuijzen, R.P. (1992) Piscivory, growth, and size-selective mortality of age 0 pikeperch (*Stizostedion lucioperca*). *Canadian Journal of Fisheries and Aquatic Sciences* **49** (5), 894–902.

Buijse, A.D., Schaap, L.A. & Bult, T.P. (1992) Influence of water clarity on the catchability of six freshwater fish species in bottom trawls. *Canadian Journal of Fisheries and Aquatic Sciences* **49**, 885–93.

Bulkowski, L. & Meade, J.W. (1983) Changes in phototaxis during early development of walleye. *Transactions of the American Fisheries Society* **112** (3), 445–7.

Burkhardt, D.A. & Hassin, G. (1983) Quantitative relations between colour-opponent response of horizontal cells and action spectra of cones. *Journal of Neurophysiology* **49** (4), 961–75.

Burkhead, N.M. & Williams, J.D. (1992) The boulder darter: a conservation challenge. *Endangered Species Technical Bulletin* **17**, 4–6.

Burr, B.M. (1978) Systematics of the percid fishes of the subgenus *Microperca*, genus *Etheostoma*. *Bulletin of the Alabama Museum of Natural History* **4**, 1–53.

Burr, B.M. (1979) Systematics and life history aspects of the percid fish *Etheostoma blennius* with description of a new subspecies from Sequatchie River, Tennessee. *Copeia* **1979**, 191–203.

Burr, B.M. & Page, L.M. (1978) The life history of the cypress darter, *Etheostoma proeliare*, in Max Creek, Illinois. *Illinois Natural History Survey Biological Notes* **106**, 1–15.

Burr, B.M. & Page, L.M. (1979) The life history of the least darter, *Etheostoma microperca*, in the Iroquois River, Illinois. *Illinois Natural History Survey Biological Notes* **112**, 1–15.

Bykhovskaya-Pavlovskaya, I.E., Gusev, A.V., Dubinina, M.N., Izyumova, N.A., Smirnova, T.S., Sokolovskaya, I.L., Shtein, G.A., Shul'man, S.S. & Epshtein, V.M. (1964) Key to parasites of freshwater fish of the USSR. *Israel Program for Scientific Translations* **1136**.

Cadwallader, P.L. & Backhouse, G.N. (1983) *A Guide to the Freshwater Fish of Victoria.* Victorian Government Printing Office, Melbourne.

Cai, Y. & Summerfelt, R.C. (1992) Effects of temperature and size on oxygen consumption and ammonia excretion by walleye. *Aquaculture* **104** (1–2), 127-38.

Calbert, H.A. (1975) Aquaculture: raising perch for the midwest market. *University of Wisconsin Sea Grant College Program Advisory Report* **13**, 7–8. Madison, WI.

Calbert, H.A. (1977) The production of yellow perch fingerlings. In: *Perch Fingerling Production for Aquaculture* (ed. R.W. Soderberg). Proceedings of a conference held at the University of Wisconsin, December 12, 1977. *University of Wisconsin Sea Grant Program Advisory Report* **421**, 1–5.

Campbell, A.D. (1974) The parasites of fish in Loch Leven. *Proceedings of the Royal Society of Edinburgh (B)* **74**, 347–64.

Campbell, J.S. & Babaluk, J.A. (1979) Age determination of walleye, *Stizostedion vitreum vitreum* (Mitchill), based on the examination of eight different structures. *Technical Report Fisheries and Marine Service Canada* **849**.

Campbell, R.N.B. (1992) Food of an introduced population of pikeperch, *Stizostedion lucioperca* L., in Lake Egirdir, Turkey. *Aquaculture and Fisheries Management* **23** (1), 71–85.

Cangelosi, A.A. (1997) Preventing transfer of viable organisms in the ballast water of commercial vessels: developments in policy and technology. In: *Symposium Abstracts* (ed. D.A. Jensen), pp. 21–23. *International Symposium on Biology and Management of Ruffe*, Ann Arbor, Michigan, March 1997.

Cannon, L.R.G. (1973) Diet and intestinal helminths in a population of perch, *Perca flavescens*. *Journal of Fish Biology* **5**, 447–58.

Carlander, K.D. (1961) Variations on rereading walleye scales. *Transactions of the American Fisheries Society* **90** (2), 230–31.

Carlander, K.D. & Cleary, R.E. (1949) The daily activity patterns of some freshwater fishes. *American Midland Naturalist* **41** (2), 447–52.

Carney, D.A. & Burr, B.M. (1989) Life histories of the bandfin darter, *Etheostoma zonistium*, and the firebelly darter, *Etheostoma pyrrhogaster*, in western Kentucky. *Illinois Natural History Survey Biological Notes* **134**, 1–16.

Cartwright, D.D., Watkins, B., Burgess, J., Warner, K., White, R., Jonker, Y.M. & Brown, P.B. (1998) Dietary lipid requirements of juvenile yellow perch. In: *Aquaculture'98. Book of Abstracts*. WAS symposium, Las Vegas, 15–19 Feb. 1998. Nevada, USA.

Chabot, F. & Maly, E.J. (1986) Variation in the diet of yellow perch (*Perca flavescens*) in a Quebec reservoir. *Hydrobiologia* **137** (2), 117–24.

Changeux, T. & Pont, D. (1995) Current status of the riverine fishes of the French Mediterranean Basin. *Biological Conservation* **72** (2), 137–58.

Chauderon, L. (1970) *Pisciculture en étangs*. Club Halieutique Interdépartemental, France.

Cheshire, W.F. & Steele, K.L. (1972) Hatchery rearing of walleyes using artificial food. *The Progressive Fish-Culturist* **34**, 96–9.

Chevey, P. (1922) Observation sur une perche hermaphrodite (*Perca fluviatilis* Linn). *Bulletin Société Zoologique de France* **47**, 3–4.

Chiasson, R.B. (1966) *Laboratory Anatomy of the Perch*. Wm. C. Brown Company Publishers, Dubuque. (2nd edn 1980.)

Chin, B.H., Sullivan, L.J. & Eldridge, J.E. (1979) In vitro metabolism of carbaryl by liver explants of bluegill, catfish, perch, goldfish and kissing gourami. *Journal of Agricultural and Food Chemistry* **27** (6), 1395–8.

Chipps, S.R., Perry, W.B. & Perry, S.A. (1994) Patterns of microhabitat use among four species of darters in three Appalachian streams. *American Midland Naturalist* **131**, 175–80.

Chirkova, Z.N. (1955) O raspredelenii i roste segoletkov okunya v Rybinskom vodokhranilishche. *Trudy Biologicheskoi Stantsii Borok imeni N.A. Morozova* **2**, 191–9.

Christensen, B. & Persson, L. (1993) Species-specific antipredatory behaviours: effects on prey choice in different habitats. *Behavioral Ecology and Sociobiology* **32** (1), 1–9.

Christie, G.C. & Regier, H.A. (1988) Measures of optimal thermal habitat and their relationship to yields for four commercial fish species. *Canadian Journal of Fisheries and Aquatic Sciences* **45** (2), 301–14.

Chubb, J.C. (1964) Observations on the occurrence of the plerocercoids of *Triaenophorus nodulosus* (Pallas, 1781) (Cestoda: Pseudophyllidea) in the perch, *Perca fluviatilis* L., of Llyn Tegid (Bala Lake), Merionethshire. *Parasitology* **54**, 481–91.

Chubb, J.C. (1977) Seasonal occurrence of helminths in freshwater fishes. Part I. Monogenea. *Advances in Parasitology* **15**, 133–99.

Chubb, J.C. (1979) Seasonal occurrence of helminths in freshwater fishes. Part II. Trematoda. *Advances in Parasitology* **17**, 141–313.

Chubb, J.C. (1980) Seasonal occurrence of helminths in freshwater fishes. Part III. Larval Cestoda and Nematoda. *Advances in Parasitology* **18**, 1–120.

Chubb, J.C. (1982) Seasonal occurrence of helminths in freshwater fishes. Part IV. Adult Cestoda, Nematoda and Acanthocephala. *Advances in Parasitology* **20**, 1–292.

Churchill, W.S. (1957) Conclusions from a ten year creel census on a lake with no angling restrictions. *Journal of Wildlife Management* **21** (2), 182–8.

Ciereszko, A., Ramseyer, L. & Dabrowski, K. (1993) Cryopreservation of yellow perch semen. *The Progressive Fish-Culturist* **55**, 261–4.

Ciereszko, K. & Dabrowski, K. (1993) Estimation of sperm concentration of rainbow trout, whitefish and yellow perch using a spectrophotometric technique. *Aquaculture* **109**, 367–73.

Ciereszko, R.E., Dabrowski, K., Ciereszko, A. & Ottobre, J.S. (in press) Plasma concentrations of steroid hormones in male yellow perch, *Perca flavescens*: the effects of age and photothermal manipulation. *Environmental Biology of Fishes*.

Ciereszko, R.E., Dabrowski, K., Ciereszko, A., Ebeling, J. & Ottobre, J.S. (1997) Effects of temperature and photoperiod on reproduction of female yellow perch *Perca flavescens*: plasma concentrations of steroid hormones, spontaneous and induced ovulation, and quality of eggs. *Journal of the World Aquaculture Society* **28** (4), 344–56.

Cǐhăr, J. (1975). Geographical and ecologic variability of perch (*Perca fluviatilis* L.) and history of its distribution from Eurasia to North America. *Acta Musei Nationalis Pragae* **XXXI B** (1–2), 5–89.

Clady, M.D. (1978) Structure of fish communities in lakes that contain yellow perch, sauger and walleye populations. In: *Selected Coolwater Fishes of North America* (ed. R. L. Kendall). *American Fisheries Society Special Publication* **11**, 100–108.

Clady, M. & Hutchinson, B. (1975) Effect of high winds on eggs of yellow perch, *Perca flavescens*, in Oneida Lake, New York. *Transactions of the American Fisheries Society* **104** (3), 524–5.

Clayton, J.W., Harris, R.E.K. & Tretiak, D.N. (1974) Geographic distribution of alleles for supernatant malate dehydrogenase in walleye (*Stizostedion vitreum vitreum*) populations from Western Canada. *Journal of the Fisheries Research Board of Canada* **31** (3), 342–5.

Clugston, J.P., Oliver, J.L. & Ruelle, R. (1978) Reproduction, growth and standing crops of yellow perch in southern reservoirs. In: *Selected Coolwater Fishes of North America* (ed. R.L. Kendall). *American Fisheries Society Special Publication* **11**, 89–99.

Coburn, M.M., & Gaglione, J.I. (1992) A comparative study of percid scales (Teleostei, Perciformes). *Copeia* **1992** (4), 986–1001.

Cohen, Y., Radomski, P. & Moen., R. (1993) Assessing the interdependence of assemblages from Rainy Lake fisheries data. *Canadian Journal of Fisheries and Aquatic Sciences* **50**, 402-9.

Colby, P.J. & Lehtonen, H. (1994) Suggested causes for the collapse of zander *Stizostedion lucioperca* (L.), populations in northern and central Finland through comparisons with North American walleye, *Stizostedion vitreum* (Mitchill). *Aqua Fennica* **24** (1), 9–20.

Colby, P.J., McNicol, R.E. & Ryder, R.A. (1979) Synopsis of biological data on the walleye, *Stizostedion v. vitreum* (Mitchill 1818). *FAO Fisheries Synopsis* **119**.

Colby, P.J. & Nepszy, S.J. (1981) Variation among stocks of walleye (*Stizostedion vitreum*

vitreum): management implications. *Canadian Journal of Fisheries and Aquatic Sciences* **38** (12), 1814–31.

Cole, C.F. (1967) A study of the eastern johnny darter, *Etheostoma olmstedi* Storer (Teleostei, Percidae). *Chesapeake Science* **8**, 28–51.

Coles, T.F. (1981) The distribution of perch, *Perca fluviatilis* L., throughout their first year of life in Llyn Tegid, North Wales. *Journal of Fish Biology* **18** (1), 15–30.

Colesante, R.T. (1996) Intensive culture of walleye using brine shrimp and formulated diets. In: *Walleye Culture Manual* (ed. R.C. Summerfelt), pp. 191–4. *NCRAC Culture Series* **101**. North Central Regional Aquaculture Center Publications Office, Iowa State University, Ames.

Colesante, R.T. & Youmans, N.B. (1983) Water hardening walleye eggs with tannic acid in a production hatchery. *The Progressive Fish-Culturist* **45**, 126–7.

Colesante, R.T., Youmans, N.B. & Ziolkoski, B. (1986) Intensive culture of walleye fry with live food and formulated diets. *The Progressive Fish-Culturist* **48**, 33–7.

Collette, B.B. (1963) The subfamilies, tribes, and genera of the Percidae (Teleostei). *Copeia* **1963** (4), 615–23.

Collette, B.B. (1965) Systematic significance of breeding tubercles in fishes of the family Percidae. *Proceedings of the United States National Museum* **117**, 567–614.

Collette, B.B. & Bănărescu, P. (1977) Systematics and zoogeography of the fish family Percidae. *Journal of the Fisheries Research Board of Canada* **34** (10), 1450–63.

Collette, B.B. & Yerger, R.W. (1962) The American percid fishes of the subgenus *Villora*. *Tulane Studies in Zoology* **9**, 213–30.

Collvin, L. (1985) The effect of copper on growth, food consumption and food conversion of perch, *Perca fluviatilis*, offered maximal food rations. *Aquatic Toxicology* **2**, 105–14.

Confer, J.L., Mills, E.L. & O'Bryan, L. (1990) Influence of prey abundance on species and size selection by young yellow perch (*Perca flavescens*). *Canadian Journal of Fisheries and Aquatic Sciences* **47**, 882–7.

Conover, M.C. (1986) Stocking cool-water species to meet management needs. In: *Fish Culture in Fisheries Management* (ed. R.H. Stroud), pp. 31–9. Fish Culture Section and Fisheries Management Section of the American Fisheries Society, Bethesda, Maryland.

Constantz, G.D. (1979) Social dynamics and parental care in the tessellated darter (Pisces: Percidae). *Proceedings of the Academy of Natural Sciences of Philadelphia* **131**, 131–8.

Cook, F.A. (1959) *Freshwater Fishes in Mississippi*. Mississippi Game and Fish Commission. Jackson, Mississippi.

Cooper, C.L., Ashmead, R.R. & Crites, J.L. (1977) Prevalence of certain endoparasitic helminths of the yellow perch from Western Lake Erie. *Proceedings of the Helminthological Society of Washington* **44** (1), 96.

Cooper, J.E. (1978) Eggs and larvae of the logperch, *Percina caprodes* (Rafinesque). *American Midland Naturalist* **99** (2), 257–69.

Cooper, J.E. (1979) Description of eggs and larvae of fantail (*Etheostoma flabellare*) and rainbow (*E. caeruleum*) darters from Lake Erie tributaries. *Transactions of the American Fisheries Society* **108** (1), 46–56.

Cope, W.G., Wiener, J.G. & Rada, R.G. (1990) Mercury accumulation in yellow perch in Wisconsin seepage lakes: relation to lake characteristics. *Environmental Toxicology and Chemistry* **9** (7), 931–40.

Corazza, L. & Nickum, J.G. (1981) Positive phototaxis during initial feeding stages of walleye larvae. In: *The Early Life History of Fish: Recent Studies* (eds R. Lasker & K. Sherman), pp. 492–4. *Rapports et Procès-verbaux Conseil international pour l'Exploration de la Mer* **178**.

Corbett, B.W. & Powles, P.M. (1986) Spawning and larva drift of sympatric walleyes and white suckers in an Ontario stream. *Transactions of the American Fisheries Society* **115** (1), 41–6.

Cordes, L.E. & Page, L.M. (1980) Feeding chronology and diet composition of two darters (Percidae) in the Iroquois River system, Illinois. *American Midland Naturalist* **104**, 202–6.

Craciun, V., Craciun, M., Neacsu, I & Trandafirescu, I. (1982) Na super (+) – K super (+) pump activity in the zander during acclimatisation to salinity conditions in the Black Sea. *Cercetari Marine (Recherches Marines)* 15, 227–33.

Craig, J.F. (1973) *The population dynamics of perch (Perca fluviatilis* L.) in Slapton Ley. M.Phil. Thesis, University of London.

Craig, J.F. (1974a) Population dynamics of perch, *Perca fluviatilis* L., in Slapton Ley, Devon. 1. *Freshwater Biology* **4**, 417–31.

Craig, J.F. (1974b) Population dynamics of perch, *Perca fluviatilis* L., in Slapton Ley, Devon. 2. *Freshwater Biology* **4**, 433–44.

Craig, J.F. (1977a) The body composition of adult perch, *Perca fluviatilis* L., in Windermere, with reference to seasonal changes and reproduction. *Journal of Animal Ecology* **46**, 617–32.

Craig, J.F. (1977b) Seasonal changes in the day and night activity of adult perch, *Perca fluviatilis* L. *Journal of Fish Biology* **11** (2), 161–6.

Craig, J.F. (1978) A study of the food and feeding of perch, *Perca fluviatilis* L., in Windermere. *Freshwater Biology* **8** (1), 59–68.

Craig, J.F. (1979) *Some aspects of the feeding, growth and population dynamics of perch,* Perca fluviatilis *L., in Windermere*. Ph.D. Thesis, University of Lancaster.

Craig, J.F. (1980a) Growth and production of the 1955 to 1972 cohorts of perch, *Perca fluviatilis* L., in Windermere. *Journal of Animal Ecology* **49** (1), 291–315.

Craig, J.F. (1980b) Sampling with traps. In: *Guidelines for Sampling Fish in Inland Waters* (eds T. Backiel & R.L. Welcomme). *EIFAC Technical Paper* **33**, 55–70.

Craig, J.F. (1982) Population dynamics of Windermere perch. *Annual Report of the Freshwater Biological Association* **50**, 49–59.

Craig, J.F. (1985) Aging in fish. *Canadian Journal of Zoology* **63** (1), 1–8.

Craig, J.F., Babaluk, J.A., Stevenson, S.G. & Williams, P.C. (1995) Variation in growth and reproduction of walleye (*Stizostedion vitreum*) in three Manitoba lakes. *Canadian Journal of Zoology* **73**, 367–72.

Craig, J.F. & Fletcher, J.M. (1982) The variability in the catches of charr, *Salvelinus alpinus* L., and perch, *Perca fluviatilis* L., from multi-mesh gillnets. *Journal of Fish Biology* **20** (5), 517–26.

Craig, J.F. & Kipling, C. (1983) Reproduction effort versus the environment: case histories of Windermere perch, *Perca fluviatilis* L., and pike, *Esox lucius* L. *Journal of Fish Biology* **22** (6), 713–27.

Craig, J.F., Kipling, C., Le Cren, E.D. & McCormack, J.C. (1979) Estimates of the numbers, biomass and year-class strengths of perch (*Perca fluviatilis* L.) in Windermere from 1967 to 1977 and some comparisons with earlier years. *Journal of Animal Ecology* **48** (1), 315–25.

Craig, J.F. & Smiley, K. (1986) Walleye (*Stizostedion vitreum*) and northern pike (*Esox lucius*) populations in three Alberta lakes. *Journal of Fish Biology* **29**, 67–85.

Cross, S.X. (1938) A study of the fish parasite relationships in the Trout Lake region of Wisconsin. *Transactions of the Wisconsin Academy of Science* **32**, 439–56.

Crowder, L.B., Magnuson, J.J. & Brandt, S.B. (1981) Complementarity in the use of food and thermal habitat by Lake Michigan fishes. *Canadian Journal of Fisheries and Aquatic Sciences* **38** (6), 662–8.

Culver, D.A. (1996) Fertilization procedures for pond culture of walleye and saugeye. In: *Walleye Culture Manual* (ed. R.C. Summerfelt), pp. 115–22. *NCRAC Culture Series* **101**. North Central Regional Aquaculture Center Publications Office, Iowa State University, Ames.

Culver, D.A., Madon, S.P. & Qin, J. (1993) Percid pond production techniques: timing, enrichment, and stocking density manipulation. *Journal of Applied Aquaculture* **2**, 9–31.

Czesny, S., Kolkovski, S., Yackey, C. & Dabrowski, K. (1998) The effects of (n-3) HUFA enriched *Artemia* nauplii on growth, survival, and quality of walleye *Stizostedion vitreum* fry. In: *Aquaculture'98, Book of Abstracts.* WAS symposium, Las Vegas, 15–19 February, 1998. Nevada, USA.

Dabrowski, K., Ciereszko, R.E., Ciereszko, A., Czesny, S. & Ottobre, J.S. (1994b) Reproductive function of female yellow perch can be stimulated by manipulation of temperature and photoperiod. *Biology of Reproduction* **50**, 111.

Dabrowski, K., Ciereszko, R.E., Ciereszko, A., Toth, G.P., Christ, S.A., El-Saidy, D. & Ottobre, J. (1996) Reproductive physiology of yellow perch (*Perca flavescens*): environmental and endocrinological cues. *Journal of Applied Ichthyology* **12**, 139–48.

Dabrowski, K., Ciereszko, A., Ramseyer, L., Culver, D. & Kestemont, P. (1994a) Effects of hormonal treatment on induced spermiation and ovulation in the yellow perch (*Perca flavescens*). *Aquaculture* **120**, 171–80.

Dabrowski, K., Culver, D.A., Brooks, C.L., Voss, A.C., Binkowski, F.P., Yeo, S.E. & Balogun, A.M. (1993) Biochemical aspects of the early life history of yellow perch (*Perca flavescens*). In: *Fish Nutrition in Practice* (eds S.J. Kaushik & P. Luquet), pp. 531–9. Biarritz, France, June 24–27, 1991. Les Colloques, 61, INRA, Paris.

Daniels, R.A. (1989) Significance in burying in *Ammocrypta pellucida. Copeia* **1989**, 29–34.

Danilenko, T.P. (1992) Sexual cycle of the Volga zander, *Stizostedion volgensis*, in Zaporozhye Reservoir. *Hydrobiological Journal* **28** (4), 1–10.

Danzmann, R.G. (1979) The karyology of eight species of fish belonging to the family Percidae. *Canadian Journal of Zoology* **57**, 2055–60.

Daoust, P.Y., Larson, B.E. & Johnson, G.R. (1989) Mycobacteriosis in yellow perch (*Perca flavescens*) from two lakes in Alberta. *Journal of Wildlife Diseases* **25** (1), 31–7.

Day, F. (1880) *The Fishes of Great Britain and Ireland, Vol. 1.* William and Norgate, London.

Day, F. (1886) On the hybridisation of Salmonidae at Howietown. *Report of the British Association for the Advancement of Science* **55**, 1059–63.

Dechtiar, A.O. (1972a) Parasites of fish from Lake of the Woods, Ontario. *Journal of the Fisheries Research Board of Canada* **29** (3), 275–83.

Dechtiar, A.O. (1972b) New parasite records for Lake Erie fish. *Great Lakes Fishery Commission Technical Report* **17**.

Dechtiar, A.O. & Loftus, K.H. (1965) Two new hosts for *Cyathocephalus truncatus* (Pallas 1781) (Cestoda, Cyathocephalidae) in Lake Huron. *Canadian Journal of Zoology* **43**, 407–8.

Deelder, C.L. (1951) A contribution to the knowledge of the stunted growth of perch (*Perca fluviatilis* L.) in Holland. *Hydrobiologia* **3**, 357–78.

Deelder, C.L. & Willemsen, J. (1964) Synopsis of biological data on pikeperch, *Lucioperca lucioperca* (Linnaeus) 1758. *FAO Fisheries Synopsis 28.*

Dence, W.A. (1938) Hermaphroditism in a walleyed pike (*Stizostedion vitreum*). *Copeia* **1938** (2), 95.

Denoncourt, R.F. (1976) Sexual dimorphism and geographic variation in the bronze darter, *Percina palmaris* (Pisces: Percidae). *Copeia* **1976**, 54–9.

Derback, B. (1947) The adverse effect of cold weather upon the successful reproduction of pickerel, *Stizostedion vitreum*, at Heming Lake, Manitoba in 1947. *Canadian Fish-Culturist* **2** (1), 22–3.

Deufel, J. (1961) Barschsterben im Bodensee. *Allgemeine Fischwirtschaftszeitung* **86** (16), 1.

De Vault, D.S., Hesselberg, R., Rodgers, P.W. & Feist, T.J. (1996) Contaminent trends in lake trout and walleye from the Laurentian Great Lakes. *Journal of Great Lakes Research* **22** (4), 884–95.

Diamond, M., & Wakefield, P.M. (1986) The consumption of eggs of the perch, *Perca fluviatilis* L., by macroinvertebrates in the field. *Freshwater Biology* **16** (3), 373–6.

Diehl, S. (1995) Direct and indirect effects of omnivory in a littoral lake community. *Ecology* **76** (6), 1727–40.

Diehl, S. & Ekloev, P. (1995) Effects of piscivore-mediated habitat use on resources, diet, and growth of perch. *Ecology* **76** (6), 1712–26.

Dimmick, W.W. & Page, L.M. (1992) Systematic significance of lactate dehydrogenase variation at the generic level in percids. *Copeia* **1992**, 535–7

Disler, N.N. (1950) Development of the lateral line sense organs in perch and ruff. *Trudy Instituta Morfologii Zhivotnykh, Akademiya Nauk SSSR* **2**, 85–139. (In Russian.)

Disler, N.N. & Smirnov, S.A. (1977) Sensory organs of the lateral-line canal system in two percids and their importance in behaviour. *Journal of the Fisheries Research Board of Canada* **34** (10), 1492–1503.

Distler, D.A. (1972) Observations on the reproductive habits of captive *Etheostoma cragini* Gilbert. *Southwestern Naturalist* **16**, 439–41.

Dlugosz, M. (1983) Histological observations of the development of ovaries in perch (*Perca fluviatilis*) originating from heated Goslavice Lake. Ocz. *Rocznik Nauk Rolniczych, Seria H (Rybackai)* **100**, 57–72 (in Polish).

Dobie, J. (1956) Walleye pond management in Minnesota. *The Progressive Fish-Culturist* **17**, 51–7.

Dodd, J.M. & Kerr, T. (1963) Comparative morphology and histology of the hypothalamo-neurohypophysial system. *Symposia of the Zoological Society of London* **9**, 5–27.

Dolinin, V.A. (1974) Effect of environmental conditions on the main parameters of the respiratory function in fish having different mobility and oxyphily. *Journal of Ichthyology* **14** (1), 122–32.

Donaldson, E.M., Fagerlund, U.H.M., Higgs, D.A. & McBride, J.R. (1979) Hormonal enhancement of growth. In: *Fish Physiology VIII* (eds W.S. Hoar, D.J. Randall & J.R. Brett), pp. 456–598. Academic Press, New York.

Dorr, J.A. (1982) Substrate and other environmental factors in reproduction of the yellow perch (*Perca flavescens*). *Dissertation Abstracts International B Science and Engineering* **43** (6), 1703.

Douglas, N.H. (1968) A new record size for darters. *Proceedings of the Louisiana Academy of Sciences* **31**, 41–2.

Dreyer, S. (1987) *Der europaïsche Flußbarsch* (Perca fluviatilis) *als objekt der Aquakultur.* Ph.D. Thesis, University of Hohenheim, Germany.

Driedzic, W.R., Bailey, J.R. & Sephton, D.H. (1996) Cardiac adaptations to low temperature in non-polar teleost fish. *Journal of Experimental Zoology* **275** (2–3), 186–95.

Driscoll, C.T., Blette, V., Yan, C., Schofield, C.L., Munson, R. & Holsapple, J. (1995) The role of dissolved organic carbon in the chemistry and bioavailability of mercury in remote Adirondack lakes. *Water, Air and Soil Pollution* **80** (1–4), 499–508.

Driver, E.A. & Garside, E.T. (1966) Meristic numbers of yellow perch in saline lakes in Manitoba. *Journal of the Fisheries Research Board of Canada* **23**, 1815–7.

Dukravets, G.M. & Biryukov, Yu.A. (1976) Ichthyo-fauna of the Nura River basin in the Central Kazakh-SSR, USSR. *Journal of Ichthyology* **16** (2), 271–6.

Dukravets, G.M. & Mamilov, R.Sh. (1993) Morphometric and biological data on percids of the Chu River basin. *Journal of Ichthyology* **33** (3), 76–85.

Dumitrescu, M., Bănărescu, P. & Stoica, N. (1957) *Romanichthys valsanicola* nov. gen. nov. sp. (Pisces, Percidae). *Travaux du Museum d'Histoire Naturelle 'Grigore Antipa'* **1**, 225–44.

Dunel-Erb, S., Bailly, Y. & Laurent, P. (1993) Pattern of gill innervation in two teleosts, the perch and the trout. *Canadian Journal of Zoology* **71**, 18–25.

Eaton, B.T., Hyatt, A.D. & Hengstberger, S. (1991) Epizootic haematopoietic necrosis virus: Purification and classification. *Journal of Fish Diseases* **14** (2), 157–69.

Economon, P.P. (1978) Myofibrogranuloma, a muscular dystrophy-like anomaly of walleye (*Stizostedion vitreum vitreum*). In: *Selected Coolwater Fishes of North America* (ed. R.L. Kendall). *American Fisheries Society Special Publication* **11**, 226–34.

Edgren, M. & Notter, M. (1980) Cadmium uptake by fingerlings of perch (*Perca fluviatilis*) studied by Cd–115M at two different temperatures. *Bulletin of Environmental Contamination and Toxicology* **24** (5), 647–51.

Edgren, M., Olsson, M. Reutergardh, L. (1981) A one year study of the seasonal variations of SDDT and PCB levels in fish from heated and unheated areas near a nuclear power plant. *Chemosphere* **10** (5), 447–52.

Egloff, M. (1996) Failure of swim bladder inflation of perch, *Perca fluviatilis* L. found in natural populations. *Aquatic Sciences* **58** (1), 15–23.

Einhouse, D. & Winter, J. (1981) Movement patterns and habitat utilisation of radio-tagged walleye in Chautauqua Lake, New York. *Underwater Telemetry Newsletter* **11** (2), 1–3.

Eklöv, A.G., Greenberg, L.A. & Kristiansen, H. (1994) The effect of depth on the interaction between perch (*Perca fluviatilis*) and minnow (*Phoxinus phoxinus*). *Ecology of Freshwater Fish* **3**, 1–8.

Ekloev, P. & Hamrin, S.F. (1989) Predatory efficiency and prey selection: interactions between pike *Esox lucius*, perch *Perca fluviatilis* and rudd *Scardinus erythrophthalmus*. *Oikos* **56** (2), 149–56.

Ekloev, P. & Persson, L. (1996) The response of prey to the risk of predation: proximate cues for refuging juvenile fish. *Animal Behaviour* **51** (1), 105–15.

Ellenberger, S.A., Baumann, P.C. & May, T.W. (1994) Evaluation of effects caused by high copper concentrations in Torch Lake, Michigan, on reproduction of yellow perch. *Journal of Great Lakes Research* **20** (3), 531–6.

Elliott, J.M. (1976a) Body composition of brown trout (*Salmo trutta* L.) in relation to temperature and ration size. *Journal of Animal Ecology* **45**, 273–89.

Elliott, J.M. (1976b) Energy losses in the waste products of brown trout (*Salmo trutta* L.). *Journal of Animal Ecology* **45**, 561–80.
Elliott, J.M. (1979) Energetics of freshwater teleosts. In: *Fish Phenology* (ed. P.J. Miller). *Symposia of the Zoological Society of London* **44**, 29–61. Academic Press, London.
Elliott, J.M. & Mann, K.H. (1979) A key to the British freshwater leeches. *Freshwater Biological Association Special Publication* **40**.
Elliott, J.M. & Persson, L. (1978) The estimation of daily rates of food consumption for fish. *Journal of Animal Ecology* **47** (3), 977–91.
Ellis, D.V. & Giles, M.A. (1965) The spawning behaviour of the walleye, *Stizostedion vitreum* (Mitchill). *Transactions of the American Fisheries Society* **94** (4), 358–62.
Ellis, M.M. & Roe, G.C. (1917) Destruction of log perch eggs by suckers. *Copeia* **47**, 69–71
Ellison, D.G. & Franzin, W.G. (1992) Overview of the symposium on walleye stocks and stocking. *North American Journal of Fisheries Management* **12** (2), 271–5.
Elsey, C.A. & Thomson, R.T. (1977) Exploitation of walleye (*Stizostedion vitreum vitreum*) in Lac des Mille Lacs, northern Ontario, by commercial and sport fisheries, 1958–75. *Journal of the Fisheries Research Board of Canada* **34** (10), 1769–73.
Elshoud-Oldenhave, M.J.W. (1979) Prey capture in the pikeperch, *Stizostedion lucioperca* (Teleostei, Percidae): A structural and functional analysis. *Zoomorphologie* **93** (1), 1–32.
Enger, P.S., Karlsen, H.E., Knudsen, F.R. & Sand, O. (1993) Detection and reaction of fish to infrasound. In: *Fish Behaviour in Relation to Fishing Operations* (eds C.S. Wardle & C.E. Hollingworth). *International Council for the Exploration of the Sea, Marine Science Symposium* **196**, 108–12.
Epple, A. (1969) The endocrine pancreas. In: *Fish Physiology II* (eds W.S. Hoar & D.J. Randall), pp. 275–319. Academic Press, New York.
Erickson, C.M. (1983) Age determination of Manitoban walleyes using otoliths, dorsal spines, and scales. *North American Journal of Fisheries Management* **3** (2), 176–81.
Eriksson, L-O. (1978) A laboratory study of diel and annual activity rhythms and vertical distribution in the perch, *Perca fluviatilis,* at the Arctic Circle. *Environmental Biology of Fishes* **3** (3), 301–7.
Eschmeyer, P.H. (1950) The life history of the walleye, *Stizostedion vitreum vitreum* (Mitchill), in Michigan. *Bulletin of Michigan Department of Conservation, Institute of Fisheries Research* **3**.
Eslami, A. & Mokhayer, B. (1977) Nematode larvae of medical importance found in market fish in Iran. *Pahlavi Medical Journal* **8** (3), 345–6.
Etnier, D.A. & Starnes, W.C. (1993) *The Fishes of Tennessee.* The University of Tennessee Press, Knoxville.
Evans, D.H. (ed.). (1993) *The Physiology of Fishes.* CRC Press, Boca Raton.
Ewing, R.R. (1989) *Production and flow cytometry identification of triploid walleye.* M.Sc. Thesis, South Dakota State University, Brookings, South Dakota, USA.
Fabricius, E. (1956) Hur abborren leker. *Zoologist Revy.* **18**, 48–55.
Fahy, W.E. (1954) The life history of the northern greenside darter, *Etheostoma blennioides blennioides* Rafinesque. *Journal of the Elisha Mitchell Scientific Society* **70**, 139–205.
Fänge, R. & Grove, D. (1979) Digestion. In: *Fish Physiology VIII* (eds W.S. Hoar, D.J. Randall & J.R. Brett), pp. 162–260. Academic Press, New York.
Fänge, R. & Holmgren, S. (1982) Choline acetyltransferase activity in the fish swimbladder. *Journal of Comparative Physiology* **B 146** (1), 57–61.

FAO (1999) FAO Yearbook. Fishery Statistics. Capture production. *FAO Fisheries Series* **52**. *FAO Statistics Series* **147**. Vol. **84**, 1997.

Farmer, G.J. & Beamish, F.W.H. (1969) Oxygen consumption of *Tilapia nilotica* in relation to swimming speed and salinity. *Journal of the Fisheries Research Board of Canada* **26**, 2807–21.

Faurot, M.W. & White, R.G. (1994) Feeding ecology of larval fisheries in Lake Roosevelt, Washington. *Northwest Science* **68** (3), 189–96.

Fedorova, G.V. & Drozzhina, K.S. (1982) Daily feeding rhythm of pikeperch, *Stizostedion lucioperca*, and perch, *Perca fluviatilis*, from Lake Ladoga. *Journal of Ichthyology* **22** (2), 52–60.

Ferroni, J.M., Bau, F., Parent, J.P. & Vellas, F. (1992) Variations of somatic components in perch (*Perca fluviatilis*) from Pareloup Lake (Aveyron, France). *Ichtyophysiologica Acta* **15**, 69–78.

Filatov, G.P. & Duplakov, S.N. (1926) Materialen zur kenntuiss der fische des Aral Sees. *Bulletin de l'Université de l'Asie Centrale, Tashkent* **24**, 203–30.

Findlay, D.L., Kasian, S.E.M., Hendzel, L.L., Regehr, G.W., Schindler, E.U. & Shearer, J.A. (1994) Biomanipulation of Lake 221 in the Experimental Lakes Area (ELA): effects on phytoplankton and nutrients. *Canadian Journal of Fisheries and Aquatic Sciences* **51**, 2794–807.

Fiogbé, E.D. (1996) *Contribution à l'étude des besoins nutritionnels de la perche* Perca fluviatilis *L.* Ph.D. Thesis, Presses Universitaires de Namur, Namur, Belgium.

Fiogbé, E.D., Kestemont, P., Mélard, C. & Micha, J.-C. (1996) Effects of dietary crude protein on growth and body composition of Eurasian perch *Perca fluviatilis*. *Aquaculture* **144**, 239–49.

Fiogbé, E.D., Kestemont, P., Micha, J.-C. & Mélard, C. (1994) Some physiological and biochemical characteristics of perch (*Perca fluviatilis*) development in a fertilized pond. Preliminary results. In: *Measures for success – Metrology and instrumentation in aquaculture management* (eds J. Muir & F. Sevilla). *EAS Special Publication* **21**, 150–1.

Fiogbé, E.D., Kestemont, P., Micha, J.-C. & Mélard, C. (1995) Comparative growth of *Perca fluviatilis* larvae fed with enriched or standard *Artemia metanauplii*, frozen *Artemia nauplii* or dry food. In: *Larvi' 95. Fish and Shellfish Larviculture Symposium* (eds P. Lavens, E. Jaspers & I. Roelands). *EAS Special Publication* **24**, 166–9.

Fischthal, J.H. (1952) Parasites of northwest Wisconsin fishes III. The 1946 survey. *Transactions of the Wisconsin Academy of Sciences* **41**, 17–58.

Fish, G.R. (1960) The comparative activity of some digestive enzymes in the alimentary canal of tilapia and perch. *Hydrobiologia* **15**, 161–79.

Fisher, W.L. (1990) Life history and ecology of the orangefin darter *Etheostoma bellum* (Pisces: Percidae). *American Midland Naturalist* **123**, 268–81.

Fisher, W.L. & Pearson, W.D. (1987) Patterns of resource utilization among four species of darters in three central Kentucky streams. In: *Community and Evolutionary Ecology of North American Stream Fishes* (ed. W.J. Matthews & D.C. Heins), pp. 69–76. University of Oklahoma Press, Norman.

Fitter, R.S.R. (1959) *The Ark in Our Midst*. Collins, London.

Fitzsimmons, J.D., Leach, J.H., Nepszy, S.J. & Cairns, V.W. (1995) Impacts of zebra mussel on walleye (*Stizostedion vitreum*) reproduction in western Lake Erie. *Canadian Journal of Fisheries and Aquatic Sciences* **52** (3), 578–86.

Fletcher, A.M. (1957) A rare darter-spawning. *Aquarium* **26**, 202–3.

Flock, A. (1971) The lateral line organ mechanoreceptors. In: *Fish Physiology V* (eds W.S. Hoar & D.J. Randall), pp. 241–63. Academic Press, New York.

Flowers, D.D. (1996) Fingerling production in drainable ponds at White Lake fish culture station. In: *Walleye Culture Manual* (ed. R.C. Summerfelt), pp. 123–8. *NCRAC Culture Series* **101**. North Central Regional Aquaculture Center Publications Office, Iowa State University, Ames.

Flynn, R.B. & Hoyt, R.D. (1979) The life history of the teardrop darter, *Etheostoma barbouri* Kuehne and Small. *American Midland Naturalist* **101**, 127–41.

Foerlin, L., Andersson, T., Balk, L. & Larsson, A. (1995) Biochemical and physiological effects in fish exposed to bleached kraft mill effluents. *Ecotoxicology and Environmental Safety* **30** (2), 164–70.

Fontaine, P., Anthouard, M., Gardeur, J.N. & Terver, D. (1997b) Effect of feeding method on growth and feeding behaviour of Eurasian perch *Perca fluviatilis* L. reared at low densities in a recirculating system. In: *Island Aquaculture and Tropical Aquaculture* (eds L. Creswell & Y. Harache), pp. 331–2. Second International Workshop on Aquaculture of Percid Fish. *EAS Special Publication.* Oostende, Belgium.

Fontaine, P., Gardeur, J.N., Kestemont, P. & Georges, A. (1997a) Influence of feeding level on growth, intraspecific weight variability and sexual growth dimorphism of Eurasian perch *Perca fluviatilis* L. reared in a recirculation system. *Aquaculture* **157**, 1–9.

Fontaine, P., Tamazouzt, L. & Capdeville, B. (1996) Growth of the Eurasian perch (*Perca fluviatilis* L.) reared in floating cages and in water recirculated system: first results. *Journal of Applied Ichthyology* **12** (3–4), 181–4.

Fontaine, P., Vlavonou, R., Tamazouzt, L., Masson, G. & Capdeville, B. (1995) Weaning and post-weaning growth performances of pond harvested common perch *Perca fluviatilis* larvae, fry and juveniles. In: *Abstracts Percis II. Second International Percid Fish Symposium.* Finnish Government Fisheries Research Institute, Helsinki, Finland.

Forney, J.L. (1963) Distribution and movement of marked walleyes in Oneida Lake, New York. *Transactions of the American Fisheries Society* **92** (1), 47–52.

Forney, J.L. (1966) Factors affecting first-year growth of walleyes in Oneida Lake, New York. *New York Fish and Game Journal* **13** (2), 146–67.

Forney, J.L. (1967) Estimates of biomass and mortality rates in a walleye population. *New York Fish and Game Journal* **14** (2), 176–92.

Forney, J.L. (1976) Year-class formation in the walleye (*Stizostedion vitreum vitreum*) population of Oneida Lake, New York, 1966–73. *Journal of the Fisheries Research Board of Canada* **33** (4), 783–92.

Forney, J.L. (1977a) Evidence of inter and intra-specific competition as factors regulating walleye (*Stizostedion vitreum vitreum*) biomass in Oneida Lake, New York. *Journal of the Fisheries Research Board of Canada* **34** (10), 1812–20.

Forney, J.L. (1977b) Reconstruction of yellow perch (*Perca flavescens*) cohorts from examination of walleye (*Stizostedion vitreum vitreum*) stomachs. *Journal of the Fisheries Research Board of Canada* **34**, 925–32.

Forney, J.L. (1977c) Population dynamics of walleye and yellow perch in Oneida Lake. *US Federal Aid in Fish Restoration. Report* **F-17-R**. New York State.

Forney, J.L. (1980) Evolution of a management strategy for the walleye in Oneida Lake, New York. *New York Fish and Game Journal* **27**, 105–41.

Fox, M.G. (1989) Effect of prey density and prey size on growth and survival of juvenile walleye (*Stizostedion vitreum*). *Canadian Journal of Fisheries and Aquatic Sciences* **46** (8), 1323–8.

Fox, M.G. (1991) Food consumption and bioenergetics of young-of-the-year walleye (*Stizostedion vitreum vitreum*): Model predictions and population density effects. *Canadian Journal of Fisheries and Aquatic Sciences* **48** (3), 434–41.

Fox, M.G., Flowers, D.D. & Watters, C. (1992) The effects of supplementary inorganic fertilization on juvenile walleye (*Stizostedion vitreum*) reared in organically fertilized ponds. *Aquaculture* **106**, 27-40.

Francis, J.T., Robillard, S.R. & Marsden, J.E. (1996) Yellow perch management in Lake Michigan: a multi-jurisdictional challenge. *Fisheries* **21** (2), 18–20.

Frank, R., Van Hove Holdrinet, M., Desjardine, R.L. & Dodge, D.P. (1978) Organochlorine and mercury residues in fish from Lake Simcoe, Ontario 1970–76. *Environmental Biology of Fishes* **3** (3), 275–85.

Frankiewicz, P., Dabrowski, K. & Zalewski, M. (1996) Mechanism of establishing bimodality in a size distribution of age-0 pikeperch, *Stizostedion lucioperca* (L.) in the Sulejów Reservoir, Central Poland. *Annales Zoologici Fennici* **33**, 321–7.

Frankiewicz, P., Zalewski, M., Schiemer, F. & Dabrowski, K. (1997) Vertical distribution of planktivorous 0+ pikeperch, *Stizostedion lucioperca* (L.), in relation to particulate or filter feeding. *Fisheries Management and Ecology* **4** (2), 93–101.

French, J.R.P. & Edsall, T.A. (1992) Morphology of ruffe (*Gymnocephalus cernuus*) protolarvae from the St. Louis River, Lake Superior. *Journal of Freshwater Ecology* **7**, 59–68.

Friedmann, A.S., Watzin, M.C., Brinck-Johnsen, T. & Leiter, J.C. (1996) Low levels of dietary methylmercury inhibit growth and gonadal development in juvenile walleye (*Stizostedion vitreum*). *Aquatic Toxicology* **35** (3–4), 265–78.

Fry, F.E.J. (1947) Effects of environmental factors on animal activity. *University of Toronto Studies in Biology Series* **55**. *Publication of Ontario Fisheries Research Laboratory* **68**.

Fry, F.E.J. (1957) Aquatic respiration of fish. In: *The Physiology of Fishes I* (ed. M.E. Brown), pp. 1–63. Academic Press, London.

Fryer, G. (1982) The parasitic Copepoda and Branchiura of British freshwater fishes. *Freshwater Biological Association Special Publication* **46**.

Fuller, R.C. (1998a) Sperm competition affects male behaviour and sperm output in the rainbow darter. *Proceedings of the Royal Society of London B* **265**, 2365–71.

Fuller, R.C. (1998b) Fecundity estimates for rainbow darters, *Etheostoma caeruleum*, in southwestern Michigan. *Ohio Journal of Science* **98**, 2–5.

Furnass, T.I. (1979) Laboratory experiments on prey selection by perch fry, *Perca fluviatilis*. *Freshwater Biology* **19** (1), 33–43.

Gagne, J.L. (1977) *Le sandre et sa pisciculture*. Toulouse University, France.

Gale, W.F. & Deutsch, W.G. (1985) Fecundity and spawning frequency of captive tessellated darters – fractional spawners. *Transactions of the American Fisheries Society* **114**, 220–9.

Gantverg, A.N. & Rozengart, V.I. (1984) Carbophos decomposition rate in fish liver in-vitro. *Journal of Ichthyology* **24** (5), 161–2.

Garling, D.L. (1991) NCRAC research programs to enhance the potential of yellow perch culture in the north central region. In: *NCRAC-91, Proceedings of the North Central Aquaculture Conference*, pp. 253–5. Kalamazoo, Michigan, March 1991.

Gee, J.H., Machniak, K. & Chalanchuk, S.M. (1974) Adjustment of buoyancy and excess internal pressure of swimbladder gases in some North American freshwater fishes. *Journal of the Fisheries Research Board of Canada* **31**, 1139–41.

Gee, J.H., Tallmann, R.F. & Smart, H.J. (1978) Reactions of some Great Plains fishes to progressive hypoxia. *Canadian Journal of Zoology* **56** (9), 1962–6.

George, S.G., Slack, W.T. & Douglas, N.H. (1996) Demography, habitat, reproduction, and sexual dimorphism of the crystal darter, *Crystallaria asprella* (Jordan), from south-central Arkansas. *Copeia* **1996**, 68–78.

Geredeaux, D. (1990) Fisheries management in an international lake: Lake Geneva. In: *Management of Freshwater Fisheries* (eds W.L.T. Van Densen, B. Steinmetz & R.H. Hughes), pp. 168–81. EIFAC Symposium, Gothenburg, Sweden, 1988.

Gill, H.S., Weatherley, A.H., Lee, R. & Legere, D. (1989) Histochemical characterization of myotomal muscle of five teleost species. *Journal of Fish Biology* **34** (3), 375–86.

Gillet, C. & Dubois, J.P. (1995) A survey of the spawning of perch (*Perca fluviatilis*), pike (*Esox lucius*), and roach (*Rutilus rutilus*), using artificial spawning substrates in lakes. In: *Space Partition within Aquatic Ecosystems* (ed. G. Balvay). *Hydrobiologia* **300–301**, 409–15.

Gillet, C., Dubois, J.P. & Bonnet, S. (1995) Influence of temperature and size of females on the timimg of spawning of perch, (*Perca fluviatilis*), in Lake Geneva from 1984 to 1993. *Environmental Biology of Fishes* **42**, 355–63.

Gislason, G.S., MacMillan, J.A. & Craven, J.W. (1982) *The Manitoba Commercial Freshwater Fishery: an Economic Analysis*. University of Manitoba Press, Manitoba.

Glenn, C.L. & Mathias, J.A. (1985) Circuli development on body scales of young pond-reared walleye (*Stizostedion vitreum*). *Canadian Journal of Zoology* **63** (4), 912–15.

Godemar, B. (1994) *Experimental study of nitrogenous excretion of perch* (Perca fluviatilis *L. 1758*). MSc Thesis, Centre de recherche ecology de l'université de Metz, France.

Goetz, F.W. (1983) Hormonal control of oocyte maturation and ovulation in fishes. In: *Fish Physiology IX B. Behavior and Fertility Control* (eds W.S. Hoar, D.J. Randall & E.M. Donaldson), pp. 51–403. Academic Press, San Diego.

Goetz, F.W. & Bergman, H.L. (1978) The *in vitro* effects of mammalian and piscine gonadotropin and pituitary preparations on final maturation in yellow perch (*Perca flavescens*) and walleye (*Stizostedion vitreum*). *Canadian Journal of Zoology* **56** (2), 348–50.

Goetz, F.W. & Theofan, G. (1979) *In vitro* stimulation of germinal vesicle breakdown and ovulation of yellow perch (*Perca flavescens*) oocytes. Effects of 17α- hydroxy – 20β-dihydroprogesterone and prostaglandins. *General and Comparative Endocrinology* **37**, 273–85.

Goldspink, C.R. (1990) The distribution and abundance of young (I+ – II+) perch, *Perca fluviatilis* L. in a deep eutrophic lake, England. *Journal of Fish Biology* **36** (3), 439–47.

Golovanenko, L.F., Shuvatova, T.F., Putina, Ye.P., Fedorova, L.S. & Arakelova, A.L. (1970) A physiological and biochemical description of Don pikeperch females at different stages of the sexual cycle. *Journal of Ichthyology* **10**, 260–7.

Gould, A.R., Hyatt, A.D., Hengstberger, S.H., Whittington, R.J. & Coupar, B.E.H. (1995) A polymerase chain reaction (PCR) to detect epizootic haematopoietic necrosis virus and Bohle iridovirus. *Diseases of Aquatic Organisms* **22** (3), 211–15.

Grabda-Kazubska, B. (1974) *Clinostomum complanatum* (Rudolphi, 1819) and *Euclinostomum heterostomum* (Rudolphi, 1809) (Trematoda, Clinostomatidae), their occurrence and possibility of acclimatisation in artificially heated lakes in Poland. *Acta Parasitologica Polonica* **22**, 285–93.

Graham, D.M. & Sprules, W.G. (1992) Size and species selection of zooplankton by larval and juvenile walleye (*Stizostedion vitreum vitreum*) in Oneida Lake, New York. *Canadian Journal of Zoology* **70** (10), 2059–67.

Gray, J.A.B. & Best, A.C.G. (1989) Patterns of excitation of the lateral line of the ruffe. *Journal of the Marine Biological Association of the United Kingdom* **69**, 289–306.

Greenberg, L.A. (1991) Habitat use and feeding behavior of thirteen species of benthic stream fishes. *Environmental Biology of Fishes* **31**, 389–401.

Greenberg, L.A. & Stiles, R.A. (1993) A descriptive and experimental study of microhabitat use by young-of-the-year benthic stream fishes. *Ecology of Freshwater Fish* **2**, 40–9

Greenwood, P.H. (1975) *A History of Fishes.* 3rd edn. Ernest Benn, London.

Griffiths, D. & Kirkwood, R.C. (1995) Seasonal variation in growth, mortality and fat stores of roach and perch in Lough Neagh, Northern Ireland. *Journal of Fish Biology* **47** (3), 537–54.

Griffiths, W.E. (1976) Feeding and gastric evacuation in perch (*Perca fluviatilis* L.). *Mauri Ora* **4**, 19–34.

Grignard, J.C., Mélard, C., Baras, E., Poirier, A., Philippart, J.C. & Bussers, J.C. (1996b) Occurrence and impact of *Heteropolaria* sp. (Protozoa, Ciliophora) on intensively cultured perch (*Perca fluviatilis*). *Annales Zoologici Fennici* **33**, 653–7.

Grignard, J.C., Mélard, C. & Kestemont, P. (1996a) A preliminary study of parasites and diseases of perch in an intensive culture system. *Journal of Applied Ichthyology* **12** (3–4), 195–200.

Grinstead, B.G. (1971) Reproduction and some aspects of the early life history of walleye, *Stizostedion vitreum* (Mitchill), in Canton Reservoir, Oklahoma. *Special Publication of the American Fisheries Society* **8**, 41–51.

Gripp, K. & Beyle, M. (1937) Das Interglazial von Billstedt (Öjendorf). *Mitteilungen Geologisch Staatsinstitut Hamburg* **16**, 19–36.

Grizzle, J.M. & Curd, M.R. (1978) Posthatching histological development of the digestive system and swim bladder of logperch, *Percina caprodes*. *Copeia* **1978**, 448–55.

Guiney, P.D. & Peterson, R.E. (1980) Distribution and elimination of a polychlorinated biphenyl after acute dietary exposure in yellow perch and rainbow trout. *Archives of Environmental Contamination* and *Toxicology* **9** (6), 667–74.

Guma'a, S.A. (1978a) On the early growth of 0+ perch, *Perca fluviatilis*, in Windermere. *Freshwater Biology* **8** (3), 213–20.

Guma'a, S.A. (1978b) The effects of temperature on the development and mortality of eggs of perch, *Perca fluviatilis*. *Freshwater Biology* **8** (3), 221–7.

Guma'a, S.A. (1978c) The food and feeding habits of young perch, *Perca fluviatilis*, in Windermere. *Freshwater Biology* **8** (2), 177–87.

Guma'a, S.A. (1982) Retinal development and retinomotor responses in perch, *Perca fluviatilis* L. *Journal of Fish Biology* **20** (5), 611–18.

Gunn, J.M. (1982) Acidification of lake trout (*Salvelinus namaycush*) lakes near Sudbury, Ontario. In: *Acid Rain/Fisheries* (ed. R.E. Johnson), p. 351. American Fisheries Society, Bethesda, MD (USA), Northeastern Division.

Gunn, J.M., Hamilton, J.G., Booth, G.M., Wren, C.D., Beggs, G.L., Rietveld, H.J. & Munro, J.R. (1990) Survival, growth, and reproduction of lake trout (*Salvelinus namaycush*) and yellow perch (*Perca flavescens*) after neutralization of an acidic lake near Sudbury, Ontario. *Canadian Journal of Fisheries and Aquatic Sciences* **47** (2), 446–53.

Gunther, A. (1859) *Catalogue of the Acanthopterygian Fishes in the Collection of the British Museum I.* British Museum, London.

Guseva, T.V. (1974) The ecology of spawning and embryonic development of the sea zander, *Lucioperca marina. Izvestia Akademiia Nauk Turkmenskoi SSR, Seriia Biologicheskikh Nauk* **2,** 87–91. (In Russian.)

Guti, G. (1992) The population density of perch, *Perca fluviatilis* L. in the Cikola backwater system of the river Danube, Hungary. *Hydrobiologia* **242**, 195–8.

Guti, G. (1993) Mortality, growth and diet of perch *Perca fluviatilis* L. in the Cikola branch system of the Szigetköz area, river Danube. *Archiv für Hydrobiologie* **128**, 317–27.

Haaparanta, A., Tellervo-Valtonen, E. & Hoffmann, R.W. (1994) Pathogenicity and seasonal occurrence of *Henneguya creplini* (Protozoa, Myxosporea) on the gills of perch (*Perca fluviatilis*) in central Finland. *Diseases of Aquatic Organisms* **20** (1), 15–20.

Haenen, O.L.M. & Van Banning, P. (1990) Detection of larvae of *Anguillicola crassus* (an eel swimbladder nematode) in freshwater fish species. *Aquaculture* **87**, 103–9.

Haines, T.A., Komov, V. & Jagoe, C.H. (1992) Lake acidity and mercury content of fish in Darwin National Reserve, Russia. *Environmental Pollution* **78** (1–3), 107–12.

Hale, J.G. & Carlson, A. R. (1972) Culture of the yellow perch in the laboratory. *The Progressive Fish-Culturist* **34** (4), 195–8.

Hamley, J.M. (1980) Sampling with gillnets. In: *Guidelines for Sampling Fish in Inland Waters* (eds T. Backiel & R. Welcomme). *EIFAC Technical Report* **33**, 37–53. FAO, Rome.

Hamley, J.M. & Howley, T.P. (1985) Factors affecting variability of trapnet catches. *Canadian Journal of Fisheries and Aquatic Sciences* **42** (6), 1079–87.

Hamley, J.M. & Regier, H.A. (1973) Direct estimates of gillnet selectivity to walleye (*Stizostedion vitreum vitreum*). *Journal of the Fisheries Research Board of Canada* **30** (6), 817–30.

Hanson, J.M. & Leggett, W.C. (1985) Experimental and field evidence for interspecific and intraspecific competition in two fresh water fishes. *Canadian Journal of Fisheries and Aquatic Sciences* **42** (2), 280–6.

Hansson, S. (1985) Local growth differences in perch (*Perca fluviatilis* L.) in a Baltic archipelago. *Hydrobiologia* **121** (1), 3–10.

Hansson, S., Arrhenius, F. & Nellbring, S. (1997) Diet and growth of pikeperch (*Stizostedion lucioperca* L.) in a Baltic Sea area. *Fisheries Research* **31**, 163–7.

Hanzelova, V., Snabel, V. & Spakulova, M. (1996) On the host specificity of fish tapeworm *Proteocephalus exiguus* la Rue, 1911 (Cestoda). *Parasite* **3** (3), 253–7.

Hara, T.J. (1971) Chemoreception. In: *Fish Physiology V* (eds W.S. Hoar & D.J. Randall), pp. 79–120. Academic Press, New York.

Harding, J.M., Burky, A.J. & Way, C.M. (1998) Habitat preferences of the rainbow darter, *Etheostoma caeruleum*, with regard to microhabitat velocity shelters. *Copeia* **1998**, 988–97.

Harris, R.C. & Snodgrass, W.J. (1993) Bioenergetic simulations of mercury uptake and retention in walleye (*Stizostedion vitreum*) and yellow perch (*Perca flavescens*). *Water Pollution Research Journal of Canada* **28** (1), 217–36.

Hartig, J.H. & Jude, D.J. (1984) Opportunistic cyclopoid predation on fish larvae. *Canadian Journal of Fisheries and Aquatic Sciences* **41** (3), 526–32.

Hartman, K.J. & Margraf, F.J. (1992) Effects of prey and predator abundances on prey consumption and growth of walleyes in western Lake Erie. *Transactions of the American Fisheries Society* **121**, 245–60.

Hartman, K.J. & Margraf, F.J. (1993) Evidence of predatory control of yellow perch (*Perca flavescens*) recruitment in Lake Erie, U.S.A. *Journal of Fish Biology* **43** (1), 109–19.

Hartmann, J. (1993) Cyprinid dynamics of Lake Constance. *Österreichs Fischerei* **46** (8–9), 201–7.

Hartmann, J. (1994) Fish biomass/fish yield of Lake Constance. *Österreichs Fischerei* **47** (1), 13–19.

Hartmann, J. (1995) Reduced effort for determination of year-class strength. *Österreichs Fischerei* **48** (7), 161–5.

Hartmann, J. & Knoepfler, G. (1986) Growth of perch in Lake Constance. *Österreichs Fischerei* **39** (2–3), 45–51.

Hasler, A.D. & Bardach, J.E. (1949) Daily migrations of perch in Lake Mendota, *Wisconsin Journal of Wildlife Management* **13** (1), 40–51.

Hasler, A.D. & Villemonte, J.R. (1953) Observations on the daily movements of fishes. *Science (Washington D.C.)* **118**, 321.

Hatch, R.W., Nepszy, S.J. & Rawson, M.R. (1990) Management of percids in Lake Erie, north America. In: *Management of Freshwater Fisheries* (eds W.L.T. Van Densen, B. Steinmetz & R.H. Hughes), pp. 624–36. EIFAC symposium, Gothenburg, Sweden, 1988.

Hayes, D.B. & Taylor, W.W. (1994) Changes in the composition of somatic and gonadal tissues of yellow perch following white sucker removal. *Transactions of the American Fisheries Society* **123** (2), 204–16.

Hayes, D.B., Taylor, W.W. & Schneider, J.C. (1992) Response of yellow perch and the benthic invertebrate community to a reduction in the abundance of white suckers. *Transactions of the American Fisheries Society* **121** (1), 36–53.

Hayward, R.S. & Margraf, F.J. (1987) Eutrophication effects on prey size and food available to yellow perch in Lake Erie. *Transactions of the American Fisheries Society* **116** (2), 210–23.

Healey, M.C. (1984) Multiattribute analysis and the concept of optimum yield. *Canadian Journal of Fisheries and Aquatic Sciences* **41**, 1393–1406.

Heath, D. & Roff, D.A. (1987) Test of genetic differentiation in growth of stunted and nonstunted populations of yellow perch and pumpkinseed. *Transactions of the American Fisheries Society* **116** (1), 98–102.

Hedrick, R.P., McDowell T.S., Ahne, W., Torhy, C. & de Kinkelin, P. (1992) Properties of three iridovirus-like agents associated with systemic infections of fish. *Diseases of Aquatic Organisms* **13** (3), 203–9.

von Hegi, H.R. & Geiger, W. (1979) Schwermetalle (Hg, Cd, Cu, Pb, Zn) in lebern und muskulatur des Flussbarsches (*Perca fluviatilis*) aus Bielersee und Walensee. *Schweizerische Zeitschrift für Hydrologie* **41** (1), 94–107.

Heidinger, R.C. & Clodfelter, K. (1987) Validity of the otolith for determining age and growth of walleye, striped bass, and smallmouth bass in power plant cooling ponds. In: *Age and Growth in Fish* (eds R.C. Summerfelt & G.E. Hall), pp. 241–51. Iowa State University Press, Ames.

Heidinger, R.C. & Kayes, T.B. (1986) Yellow perch. In: *Culture of Non-Salmonid Freshwater Fishes* (ed. R.R. Stickney), pp. 103–13. CRC Press Inc., Boca Raton, Florida.

Heidinger, R.C., Tetzlaff, B.L. & Brooks, R.C. (1996) Hormone induced spawning of walleye. In: *Walleye Culture Manual* (ed. R.C. Summerfelt), pp. 59–62. *NCRAC Culture Series* **101**. North Central Regional Aquaculture Center Publications Office, Iowa State University, Ames.

Heidinger, R.C., Tetzlaff, B.L., Brooks, R.C. & Bickers, C.A. (1990) *Walleye culture study* **F-60-R**. Illinois Department of Conservation.

Heins, D.C. (1985) Life history traits of the Florida sand darter *Ammocrypta bifascia*, and comparisons with the naked sand darter *Ammocrypta beani*. *American Midland Naturalist* **113**, 209–16.

Heins, D.C. & Baker, J.A. (1989) Growth, population structure, and reproduction of the percid fish *Percina vigil*. *Copeia* **1989**, 727–36.

Heins, D.C. & Baker, J.A. (1993) Clutch production in the darter *Etheostoma lynceum* and its implications for life-history study. *Journal of Fish Biology* **42**, 819–29.

Heins, D.C., Baker, J.A. & Dunlap, W.P. (1992) Yolk loading in oocytes of darters and its consequences for life-history study. *Copeia* **1992**, 404–12.

Heins, D.C. & Rooks, J.R. (1984) Life history of the naked sand darter, *Ammocrypta beani* in southeastern Mississippi. In: *Environmental Biology of Darters* (ed. D.G. Lindquist & L.M. Page), pp. 61–9. *Developments in Environmental Biology of Fishes* **4**. Dr W. Junk Publishers, The Hague.

Heisler, N. (1984) Acid–base regulation in fishes. In: *Fish Physiology XA* (eds W.S. Hoar & D.J. Randall), pp. 315–401. Academic Press, Orlando, Florida.

Helal, H. & Culver, D.A. (1991) N:P ratio and plankton production in fish ponds. *Verhandlungen der Internationalen Vereinigung für Limnologie* **24**, 1508–11.

Helfman, G.S. (1979) Twilight activities of yellow perch, *Perca flavescens*. *Journal of the Fisheries Research Board of Canada* **36** (2), 173–9.

Helfman, G.S. (1984) School fidelity in fishes: the yellow perch pattern. *Animal Behaviour* **32** (3), 673–89.

Helfman, G.S. (1986) Fish behaviour by day, night and twilight. In: *Behaviour of Teleost Fishes* (ed. T.J. Pitcher), pp. 366–87. Croom Helm, London.

Henderson, B.A. (1985) Factors affecting growth and recruitment of yellow perch, *Perca flavescens* Mitchill, in South Bay, Lake Huron. *Journal of Fish Biology* **26** (4), 449–58.

Henderson, B.A. & Nepszy, S.J. (1994) Reproductive tactics of walleye (*Stizostedion vitreum*) in Lake Erie. *Canadian Journal of Fisheries and Aquatic Sciences* **51** (5), 986–97.

Hengstberger, S.G., Hyatt, A.D., Speare, R. & Coupar, B.E.H. (1993) Comparison of epizootic haematopoietic necrosis and Bohle iridoviruses, recently isolated Australian iridiviruses. *Diseases of Aquatic Organisms* **15** (2), 93-107.

Hergenrader, G.L. (1969) Spawning behaviour of *Perca flavescens* in aquaria. *Copeia* **1969**, 839–41.

Hergenrader, G.L. & Hasler, A.D. (1966) Diel activity and vertical distribution of yellow perch (*Perca flavescens*) under the ice. *Journal of the Fisheries Research Board of Canada* **23**, 499–509.

Hergenrader, G.L. & Hasler, A.D. (1967) Seasonal changes in swimming rates of yellow perch in Lake Mendota as measured by sonar. *Transactions of the American Fisheries Society* **96**, 373–82.

Hergenrader, G.L. & Hasler, A.D. (1968) Influence of changing seasons on schooling behaviour of yellow perch. *Journal of the Fisheries Research Board of Canada* **25**, 711–16.

Hewett, S.W. & Johnson, B.L. (1992) A generalised bioenergetics model of fish for microcomputers, 2nd edn. *University of Wisconsin Sea Grant Institute Technical Report* **WIS-SG-87-245**.

Hickman, C.P. & Trump, B.F. (1969) The kidney. In: *Fish Physiology I* (eds W.S. Hoar & D.J. Randall), pp. 91–239. Academic Press, New York.

Hildén, M. & Hirvi, J.P. (1987) The survival of larval perch, *Perca fluviatilis* L., under different combinations of acidity and duration of acid conditions, analysed with a generalized linear model. *Journal of Fish Biology* **30** (6), 667–77.

Hilge, V. (1990) Beobachtungen zür Aufzucht von Zandern (*Stizostedion lucioperca* L.) im Labor. *Archiv für Fischerei Wissenschaft* **40**, 167–73.

Hilge, V. & Steffens, W. (1996) Aquaculture of fry and fingerling of pike-perch (*Stizostedion lucioperca* L.) – a short review. *Journal of Applied Ichthyology* **12** (3–4), 167–70.

Hinegardner, R.T. (1968) Evolution of cellular DNA content in teleost fishes. *American Naturalist* **102**, 517–23.

Hinegardner, R.T. & Rosen, D.E. (1972) Cellular DNA content and the evolution of teleostean fishes. *American Naturalist* **106**, 621–44.

Hinshaw, M. (1985) Effects of illumination and prey contrast on survival and growth of larval yellow perch (*Perca flavescens*). *Transactions of the American Fisheries Society* **114**, 540–5.

Hinshaw, M. (1986) *Factors affecting survival and growth of larval and early juvenile perch* (Perca flavescens, *Mitchill*). Ph.D. Thesis, North Carolina State University, USA.

Hirji, K.N. (1983) Observations on the histology and histochemistry of the oesophagus of the perch, *Perca fluviatilis* L. *Journal of Fish Biology* **22** (2), 145–52.

Hirji, K.N. & Courtney, W.A.M. (1979) Pear-shaped cells in the digestive tract of the perch, *Perca fluviatilis* L. *Journal of Fish Biology* **15** (4), 469–72.

Hlohowskyj, I. & White, A.M. (1983) Food resource partitioning and selectivity by the greenside, rainbow, and fantail darters (Pisces: Percidae). *Ohio Journal of Science* **83**, 201–8.

Hlohowskyj, I. & Wissing, T.E. (1986) Substrate selection by fantail (*Etheostoma flabellare*), greenside (*E. blennioides*), and rainbow (*E. caeruleum*) darters. *Ohio Journal of Science* **86**, 124–9.

Hoar, W.S. & Randall, D.J. (eds) (1969) *Fish Physiology II*. Academic Press, New York.

Hobson, J.F. (1979) Life history aspects of a percid fish, *Percina crassa roanoka,* the Piedmont darter. *Bios* **50**, 148–57.

Hoeglund, J. & Thomas, K. (1992) The black goby *Gobius niger* as a potential paratenic host for the parasitic nematode *Anguillicola crassus* in a thermal effluent of the Baltic. *Diseases of Aquatic Organisms* **13**, 175–80.

Hoeglund, J. & Thulin, J. (1992) Identification of *Diplostomum* spp. in the retina of perch *Perca fluviatilis* and the lens of roach *Rutilus rutilus* from the Baltic Sea, an experimental study. *Systematic Parasitology* **21** (1), 1–19.

Hoestlandt, H. (1979) Namisme de la perche en France est-il un mythe? *Bulletin du Centre d'Études et du Recherche Scientifiques, Biarritz* **12** (3), 453–70.

Hoffman, G.L. (1967) *Parasites of North American Freshwater Fishes.* University of California Press, Berkeley.

Hofmann, J., Geldhauser, F. & Gerstner, P. (1987) *Der Teichwirt*, **6** Auflage. P. Parey, Hamburg und Berlin.

Hokanson, K.E.F. (1977a) Temperature requirements of some percids and adaptations to the seasonal temperature cycle. *Journal of the Fisheries Research Board of Canada* **34** (10), 1524–50.

Hokanson, K.E.F. (1977b) Optimum culture requirements of the early life phases of the yellow perch. In: *Perch Fingerling Production for Aquaculture* (ed. R.W. Soderberg), pp. 24–40. *University of Wisconsin Sea Grant Program Advisory Report* **421**.

Hokanson, K.E.F. & Kleiner, C.F. (1974) Effects of constant and rising temperatures on survival and developmental rates of embryonic and larval yellow perch, *Perca flavescens* (Mitchill). In: *The Early Life History of Fish* (ed. J.H.S. Blaxter), pp. 437–48. Springer-Verlag, New York.

Hokanson, K.E.F. & Lien, G.J. (1986) Effects of diet on growth and survival of larval walleyes. *The Progressive Fish-Culturist* **48**, 250–8.

Holčik, J. & Hensel, K. (1974) A new species of *Gymnocephalus* (Pisces: Percidae) from the Danube, with remarks on the genus. *Copeia* **1974**, (2), 471–86.

Holčik, J. & Mihalik, J. (1970) *Freshwater Fishes*, 3rd edn. Hamlyn Publishing Group Ltd, London.

Hölker, F. & Hammer, C. (1994) Growth and food of ruffe (*Gymnocephalus cernuus* L.) in the Elbe estuary. *Archive of Fishery and Marine Research* **42**, 47–62.

Holloway, H.L. Jr. & Smith, C.E. (1982) A myopathy in North Dakota walleye, *Stizostedion vitreum* (Mitchell). *Journal of Fish Diseases* **5** (6), 527–30.

Holt, C.S., Grant, G.D.S., Oberstar, G.P., Oakes, C.C. & Bradt, D.W. (1977) Movement of walleye *Stizostedion vitreum*, in Lake Bemidji, Minnesota, as determined by radio-biotelemetry. *Transactions of the American Fisheries Society* **106** (2), 163–200.

Holtze, K.E. & Hutchinson, N.J. (1989) Lethality of low pH and A1 to early life stages of six fish species inhabiting Precambrian Shield waters in Ontario. *Canadian Journal of Fisheries and Aquatic Sciences* **46** (7), 1188–1202.

Hontela, A., Daniel, C. & Rasmussen, J.B. (1997) Structural and functional impairment of the hypothalamo-pituitary-interrenal axis in fish exposed to bleached kraft mill effluent in the St Maurice River, Quebec. *Ecotoxicology* **6** (1), 1–12.

Hontela, A., Dumont, P., Duclos, D. & Fortin, R. (1995) Endocrine and metabolic dysfunction in yellow perch, *Perca flavescens*, exposed to organic contaminants and heavy metals in the St Lawrence River. *Environmental Toxicolology and Chemistry* **14** (4), 725–31.

Hontela, A., Rasmussen, J.B., Audet, C. & Chevalier, G. (1992) Impaired cortisol stress response in fish from environments polluted by PAHs, PCBs and mercury. *Archives of Environmental Contamination and Toxicology* **22** (3), 278–83.

Horvath, L., Tamas, G. & Tölg, I. (1984) *Special methods in pond fish husbandry*. J.E. Halver Corporation, Seattle.

Houde, E.D. (1969a) Distribution of larval walleyes and yellow perch in a bay of Oneida Lake and its relation to water currents and zooplankton. *New York Fish and Game Journal* **16**, 184–205.

Houde, E.D. (1969b) Sustained swimming ability of larvae of walleye (*Stizostedion vitreum vitreum)* and yellow perch (*Perca flavescens*). *Journal of the Fisheries Research Board of Canada* **26** (6), 1647–59.

Howell, J.F. (1971) *The life history, behavior, and ecology of* Percina aurantiaca (*Cope*). Ph.D. Dissertation, University of Tennessee, USA.

Howell, W.M. (1968) Taxonomy and distribution of the percid fish, *Etheostoma stigmaeum* (Jordan), with the validation and redescription of *Etheostoma davisoni* Hay. *Dissertation Abstracts* **29B**, 812–13.

Hubbs, C. (1958) Geographic variations in egg complement of *Percina caprodes* and *Etheostoma spectabile. Copeia* **1958**, 102–5.

Hubbs, C. (1961a) Developmental temperature tolerances of four etheostomatine fishes occurring in Texas. *Copeia* **1961**, 195–8.
Hubbs, C. (1961b) Differences in the incubation period of two populations of *Etheostoma lepidum*. *Copeia* **1961**, 198–200.
Hubbs, C. (1967) Geographic variations in survival of hybrids between etheostomatine fishes. *Texas Memorial Museum Bulletin* **13**, 1–72.
Hubbs, C. (1971) Survival of intergroup percid hybrids. *Journal of Ichthyology* **18** (2), 65–75.
Hubbs, C. & Armstrong, N.E. (1962) Developmental temperature tolerance of Texas and Arkansas-Missouri *Etheostoma spectabile* (Percidae: Osteichthyes). *Ecology* **43**, 742–4.
Hubbs, C. & Bryan, C. (1975) Ontogenetic rates and tolerances of the channel darter, *Hadropterus copelandi*. *Texas Journal of Science* **26**, 623–4.
Hubbs, C. & Johnson, M.V. (1961) Differences in the egg complement of *Hadropterus scierus* from Austin and San Marcos. *Southwestern Naturalist* **6**, 9–12.
Hubbs, C. & Strawn, K. (1957) The effects of light and temperature on the fecundity of the greenthroat darter, *Etheostoma lepidum*. *Ecology* **38**, 596–602.
Hubbs, C., Stevenson, M.M. & Peden, A.E. (1968) Fecundity and egg size in two central Texas darter populations. *Southwestern Naturalist* **13**, 301–23.
Hubbs, C.L. (1926) A check-list of the fishes of the Great Lakes and tributary waters, with nomenclatorial notes and analytical keys. *University of Michigan Museum of Zoology Miscellaneous Publications* **15**, 1–77.
Hubbs, C.L. (1941) The relation of hydrological conditions to speciation in fishes. In: *A Symposium on Hydrobiology*, pp. 182–95. University of Wisconsin Press, Madison.
Hubert, W.A. & Sandheinrich, M.B. (1983) Patterns of variation in gill-net catch and diet of yellow perch in a stratified Iowa lake. *North American Journal of Fisheries Management* **3** (2), 156–62
Huet, M. (1970) *Traité de Pisciculture*. C. de Wingaert, Brussels.
Hughes, G.M. (1984) General anatomy of gills. In: *Fish Physiology XA* (eds W.S. Hoar & D.J. Randall), pp. 1–72. Academic Press, Orlando.
Hulsman, P.F., Powles, P.N. & Gunn, J.M. (1983) Mortality of walleye eggs and rainbow trout yolk-sac larvae in low-pH waters of the La Cloche Mountain area, Ontario. *Transactions of the American Fisheries Society* **112** (5), 680–8.
Hurley, D.A. (1972) Observations on incubating walleye eggs. *The Progressive Fish-Culturist* **34** (1), 49–54.
Huuskonen, S. & Lindstroem-Seppae, P. (1995) Hepatic cytochrome P4501A and other biotransformation activities in perch (*Perca fluviatilis*): the effects of pulp mill effluents. *Aquatic Toxicology* **31** (1), 27–41.
Hynes, H.B.N. (1970) *The Ecology of Running Waters*. Liverpool University Press, Liverpool.
Ihssen, P.E., McKay, L.R., McMillan, I. & Phillips, R.B. (1990) Ploidy manipulation. and gynogenesis in fishes: cytogenetic and fisheries applications. *Transactions of the American Fisheries Society* **119**, 698–717.
Imbrock, F., Appenzeller, A. & Eckmann, R. (1996) Diel and seasonal distribution of perch in Lake Constance: a hydroacoustic study and *in situ* observations. *Journal of Fish Biology* **49** (1), 1–13.
Ingersoll, C.G., Hlohowskyj, I. & Mundahl, N.D. (1984) Movements and densities of the darters *Etheostoma flabellare*, *E. spectabile*, and *E. nigrum* during spring spawning. *Journal of Freshwater Ecology* **2**, 345–51.

Isaia, J. (1984) Water and nonelectrolyte permeation. In: *Fish Physiology XA* (eds W.S. Hoar & D.J. Randall), pp. 1–38. Academic Press, Orlando.

Ivanc, A., Maletin, S., Djukic, N., Pujin, V., Miljanovic, B. & Zhenjun, S. (1996) Haematology of Percidae from the Yugoslav section of the River Danube. In: *The Ecology of Large Rivers* (ed. D. Sutcliffe). *Archiv für Hydrobiologie Supplementband Large Rivers* **113** (1–4), 529–34.

Jachner, A. (1991) Food and habitat partitioning among juveniles of three fish species in the pelagial of a mesotrophic lake. *Hydrobiologia* **226** (2), 81–9.

Jackson, J.J., Willis, D.W. & Fielder, D. (1992) Food habits of young-of-the-year walleyes in Okobojo Bay of Lake Oahe, South Dakota. *Journal of Freshwater Ecology* **7** (3), 329–41.

Jäger, T., Nellen, W. & Sell, H. (1984) Beleuchtete Netzgehegeanlagen zür Aufzucht von Fischbrut bis zur Stezlingsgröße. Eine Bauanleitung und Aufzuchtbeschreibung. *Berichte Institut für Meereskunde, Christian-Albrechts Universität, Kiel* **126**.

Jagoe, C.H., Matey, V.E., Haines, T.A. & Komov, V.T. (1993) Effect of beryllium in acid water is analogous to aluminum toxicity. *Aquatic Toxicology* **24** (3–4), 241–56.

James, P.W. & Maughan, E.O. (1989) Spawning behavior and habitat of the threatened leopard darter, *Percina pantherina. Southwestern Naturalist* **34**, 298–301.

James, P.W. & Taber, C.A. (1986) Reproductive biology and age and growth of the yoke darter, *Etheostoma juliae. Copeia* **1986**, 536–40.

Jansen, W.A. & Mackay, W.C. (1992) Foraging in yellow perch, *Perca flavescens*; biological and physical factors affecting diel periodicity in feeding, consumption, and movement. *Environmental Biology of Fishes* **34** (3), 287–303.

Janssen, J. (1997) Comparison of response distance to prey via the lateral line in the ruffe and yellow perch. *Journal of Fish Biology* **51** (5), 921–30.

Jansson, B-O. & Olsson, R. (1960) The cytology of the caecal epithelial cells of *Perca. Acta Zoologica (Stockholm)* **41**, 267–76.

Jastrzebski, M. (1984) Polymorphism of flagellates, *Trypanosoma*, occurring in blood of crucian carp, *Carassius auratus gibelio*, perch, *Perca fluviatilis*, and stone perch, *Acerina cernua. Wiadomosci Parazytologiczne* **30** (2), 172–82.

Jellyman, D.J. (1976) Hermaphrodite European perch, *Perca fluviatilis* L. *New Zealand Journal of Marine and Freshwater Research* **10** (4), 721–3.

Jenkins, R.E. (1971) Nuptial tuberculation and its systematic significance in the percid fish *Etheostoma (Ioa) vitreum. Copeia* **1971**, 735–8.

Jenkins, R.E. (1980) *Etheostoma podostemone* Jordan and Jenkins, riverweed darter. In: *Atlas of North American Freshwater Fishes* (eds D.S. Lee, C.R. Gilbert, C.H. Hocutt, R.E. Jenkins, D.E. McAllister, & J.R. Stauffer, Jr), p. 682. North Carolina State Museum of Natural History, Raleigh.

Jenkins, R.E. & Burkhead, N.M. (1994) *Freshwater fishes of Virginia.* American Fisheries Society, Bethesda, Maryland.

Jennings, M.J., Claussen, J.E. & Philipp, D.P. (1996) Evidence for heritable preferences for spawning habitat between two walleye populations. *Transactions of the American Fisheries Society* **125** (6), 978–82.

Jensen, A.L. (1993) Dynamics of fish populations with different compensatory processes when subjected to random survival of eggs and larvae. *Ecological Modelling* **68** (3–4), 249–56.

Jensen, D.A. (ed.) (1997) *International Symposium on Biology and Management of Ruffe: Symposium Abstracts.* Minnesota Sea Grant Publication **X48**.

Jezierska, B. (1974) The effect of various types of food on the growth and chemical composition of the body of perch (*Perca fluviatilis* L.) in laboratory conditions. *Polskie Archiwum Hydrobiologii* **21** (3–4), 467–79.

Johnson, B.M., Luecke, C., Stewart, R.S., Staggs, M.D., Gilbert, S.J. & Kitchell, J.F. (1992b) Forecasting effects of harvest regulations and stocking of walleyes on prey fish communities in Lake Mendota, Wisconsin. *North American Journal of Fisheries Management* **12** (4), 797–807.

Johnson, B.M., Milliman, S.R., Bishop, R.C. & Kitchell, J.F. (1992a) Evaluating fishery rehabilitation under uncertainty: a bioeconomic analysis of quota management for the Green Bay yellow perch fishery. *North American Journal of Fisheries Management* **12** (4), 703–20.

Johnson, G.D. (1993) Percomorph phylogeny – progress and problems. *Bulletin of Marine Science* **52** (1), 3–28.

Johnson, G.D. & Patterson, C. (1993) Percomorph phylogeny – a survey of acanthomorphs and a new proposal. *Bulletin of Marine Science* **52** (1), 554–626.

Johnson, J.E. (1971) Maturity and fecundity of threadfin shad, *Dorosoma petenense* (Gunther), in central Arizona reservoirs. *Transactions of the American Fisheries Society* **100**, 74–85.

Johnson, M.G., Leach, J.H., Minns, C.K. & Olver, C.H. (1977) Limnological characteristics of Ontario lakes in relation to associations of walleye (*Stizostedion vitreum vitreum*), northern pike (*Esox lucius*), lake trout (*Salvelinus namaycush*), and smallmouth bass (*Micropterus dolomieui*). *Journal of the Fisheries Research Board of Canada* **34** (10), 1592–1601.

Johnson, T. & Müller, K. (1978) Different phase position of activity in juvenile and adult perch. *Naturwissenschaften* **65**, 392.

Johnston, C.E. (1990) Spawning in the eastern sand darter, *Ammocrypta pellucida* (Pisces: Percidae), with comments on the phylogeny of *Ammocrypta* and related taxa. *Transactions of the Illinois State Academy of Science* **82**, 163–7.

Johnston, C.E. (1994) Spawning behavior of the goldstripe darter (*Etheostoma parvipinne* Gilbert and Swain) (Percidae). *Copeia* **1994**, 823–5.

Johnston, C.E. (1995) Spawning behavior of the paleback darter, *Etheostoma pallididorsum*, (Percidae). *Southwestern Naturalist* **40**, 422–5.

Johnston, C.E. & Haag, W.R. (1996) Life history of the Yazoo darter (Percidae: *Etheostoma raneyi*), a species endemic to north-central Mississippi. *Tulane Studies in Zoology and Botany* **30**, 47–60.

Johnston, C.E. & Shute, J.R. (1997) Observational notes on the spawning behavior of the blue shiner (*Cyprinella caeurlea*) and the holiday darter (*Etheostoma brevirostrum*), two rare fishes of the Conasauga River, Georgia and Tennessee. *Southeastern Fishes Council Proceedings* **35**, 1–2.

Johnston, T.A. (1997) Within population variability in egg characteristics of walleye (*Stizostedion vitreum*) and white sucker (*Catostomus commersoni*). *Canadian Journal of Fisheries and Aquatic Sciences* **54** (5), 1006–14.

Johnston, T.A., Bodaly, R.A. & Mathias, J.A. (1991) Predicting fish mercury levels from physical characteristics of boreal reservoirs. *Canadian Journal of Fisheries and Aquatic Sciences* **48** (6), 1468–75.

Johnston, T.A. & Mathias, J.A. (1994) Feeding ecology of walleye, *Stizostedion vitreum*, larvae: effects of body size, zooplankton abundance, and zooplankton community composition. *Canadian Journal of Fisheries and Aquatic Sciences* **51**, 2077–89.

Johnston, T.A. & Mathias, J.A. (1996) Maintenance food requirements and response to short-term food deprivation of walleye larvae. *Transactions of the American Fisheries Society* **125** (2), 211–23.

Johnston, T.B. & Kitchell, J.F. (1996) Long-term changes in zooplanktivorous fish community composition: implications for food webs. *Canadian Journal of Fisheries and Aquatic Sciences* **53**, 2792–803.

Jokela, J., Valtonen, E.T. & Lappalainen, M. (1991) Development of glochidia of *Anodonta piscinalis* and their infection of fish in a small lake in northern Finland. *Archiv für Hydrobiologie* **120** (3), 345–55.

Jones, D.R., Kiceniuk, J.W. & Bamford, O.S. (1974) Evaluation of the swimming performance of several fish species from the MacKenzie River. *Journal of the Fisheries Research Board of Canada* **31** (10), 1641–7.

Jones, D.R. & Randall, D.J. (1978). The respiratory and circulatory systems during exercise. In: *Fish Physiology VII* (eds W.S. Hoar & D.J. Randall), pp. 425–501. Academic Press, New York.

Jones, J.R.E. (1964) *Fish and River Pollution*. Butterworths, London.

Jordon, D.S. & Gilbert, C.H. (1877) On the genera of North American freshwater fishes. *Proceedings of the Academy of Natural Sciences Philadelphia* **29**, 83–104.

Jovanovic, M. (1970) Comparative life histories of the North American and European walleyes. *Michigan Department of Natural Resources Research and Development Report* **201**, 1–71.

Juell, J.E. (1991) Hydroacoustic detection of food waste. A method to estimate maximum food intake of fish populations in seacages. *Aquacultural Engineering* **10**, 207–17.

Jurajda, P., Gelnar, M. & Koubkova, B. (1994) Occurrence of zingel (*Zingel zingel*) in the River Morava with notes on its parasites. *Folia Zoologica* **43** (1), 93–6.

Kålås, S. (1995) The ecology of ruffe *Gymnocephalus cernuus* (Pisces: Percidae) introduced to Mildevatn, western Norway. *Environmental Biology of Fishes* **42**, 219–32.

Kamler, E., Lewkowicz, M., Lewkowicz, S., Uchmanski, J. & Urban-Juvierka, E. (1986) Gravimetric techniques for measuring consumption of live feeds and artificial diets by fish larvae. *Aquaculture* **54**, 109–22.

Kammerer, P. (1907) Bastardierung von Flussbarsch (*Perca fluviatilis* L.) und Kaulbarsch (*Acerina cernua* L.). *Archiv für Entwicklungsmechanik der Organismen* **23**, 511–51.

Kangur, K. & Kangur, A. (1996) Feeding of ruffe (*Gymnocephalus cernuus*) in relation to the abundance of benthic organisms in Lake Võrtsjärv (Estonia). *Annales Zoologici Fennici* **33**, 473–80.

Kapoor, B.G., Smit, H. & Verighina, I.A. (1975) The alimentary canal and digestion in teleosts. *Advances in Marine Biology* **13**, 109–239.

Karås, P. (1990) Seasonal changes in growth and standard metabolic rate of juvenile perch, *Perca fluviatilis* L. *Journal of Fish Biology* **37** (6), 913–20.

Karås, P. & Hudd, R. (1993) Reproduction areas of fresh-water fish in the northern quark (Gulf of Bothnia). *Aqua Fennica* **23** (1), 39–49.

Karås, P., Neuman, E. & Sandstrom, O. (1991) Effects of a pulp mill effluent on the population dynamics of perch, *Perca fluviatilis*. *Canadian Journal of Fisheries and Aquatic Sciences* **48** (1), 28–34.

Karås, P. & Thoresson, G. (1992) An application of a bioenergetics model to Eurasian perch (*Perca fluviatilis* L.). *Journal of Fish Biology* **41** (2), 217–30.

Katula, R.S. (1991) Observations of spawning methods of the creole darter & the finescale saddled darter (*Etheostoma collettei* & *Etheostoma osburni*). *American Currents* (Summer 1991), 15–16.

Kayes, T.B. (1977) Reproductive biology and artificial propagation methods for adult perch. In: *Perch Fingerling Production for Aquaculture* (ed. R.W. Soderberg), pp. 6–23. *University of Wisconsin Sea Grant Program Advisory Report* **421**.

Kayes, T.B. & Calbert, H.E. (1979) Effects of photoperiod and temperature on the spawning of yellow perch (*Perca flavescens*). *Proceedings of the World Mariculture Society* **10**, 306–16.

Kazanskiy, A.B. (1985) A simulation model for optimum reproductive strategy of fish populations. *Journal of Ichthyology* **25** (4), 1–11.

Kearns, P.K. & Atchison, G.J. (1979) Effects of trace metals on growth of yellow perch (*Perca flavescens*) as measured by RNA-DNA ratios. *Environmental Biology of Fishes* **4** (4), 383–7.

Keast, A. & Fox, M.G. (1990) Fish community structure, spatial distribution and feeding ecology in a beaver pond. *Environmental Biology of Fishes* **27** (3), 201–14.

Keast, A. & Webb, D. (1966) Mouth and body form relative to feeding ecology in the fish fauna of a small lake, Lake Opinicon, Ontario. *Journal of the Fisheries Research Board of Canada* **23**, 1845–74.

Keevin, T.L., Page, L.M. & Johnston, C.E. (1989) The spawning behavior of the saffron darter (*Etheostoma flavum*). *Transactions of the Kentucky Academy of Science* **50**, 55–8.

Kelly, R.K., Nielsen, O., Mitchell, S.C. & Yamamoto, Y. (1983) Characterisation of *Herpesvirus vitreum* isolated from hyperplastic epidermal tissue of walleye, *Stizostedion vitreum vitreum* (Mitchill). *Journal of Fish Diseases* **6**, 249–60.

Kelso, J.R.M. (1972) Conversion, maintenance and assimilation for walleye, *Stizostedion vitreum vitreum*, as affected by size, diet and temperature. *Journal of the Fisheries Research Board of Canada* **29** (8), 1181–92.

Kelso, J.R.M. (1976) Diel movement of walleye, *Stizostedion vitreum vitreum*, in West Blue Lake, Manitoba, as determined by ultrasonic tracking. *Journal of the Fisheries Research Board of Canada* **33** (9), 2070–72.

Kelso, J.R.M. & Johnson, M.G. (1991) Factors related to the biomass and production of fish communities in small, oligotrophic lakes vulnerable to acidification. *Canadian Journal of Fisheries and Aquatic Sciences* **48** (12), 2523–32.

Kelso, J.R.M. & Ward, F.J. (1977) Unexploited percid populations of West Blue Lake, Manitoba, and their interactions. *Journal of the Fisheries Research Board of Canada* **34** (10), 1655–69.

Kennedy, C.R. (1974) A checklist of British and Irish freshwater fish parasites with notes on their distribution. *Journal of Fish Biology* **6**, 613–44.

Kennedy, C.R. (1981) Long term studies on the population biology of two species of eyefluke *Diplostomum gasterostei* and *Tylodelphys clavata* (Digenea: Diplostomatidae), concurrently infecting the eyes of perch, *Perca fluviatilis*. *Journal of Fish Biology* **19** (2), 221–36.

Kerr, S.R. (1971) Prediction of fish growth efficiency in nature. *Journal of the Fisheries Research Board of Canada* **28**, 809–14.

Kerr, S.R. (1982) Estimating the energy budgets of actively predatory fishes. *Canadian Journal of Fisheries and Aquatic Sciences* **39**, 371–9.

Kessler, R.K., Casper, A.F. & Weddle, G.K. (1995) Temporal variation in microhabitat use and spatial relations in the benthic fish community of a stream. *American Midland Naturalist* **134**, 361–70.

Kessler, R.K. & Thorp, J.H. (1993) Microhabitat segregation of the threatened spotted darter (*Etheostoma maculatum*) and closely related orangefin darter (*E. bellum*). *Canadian Journal of Fisheries and Aquatic Sciences* **50**, 1084–91.

Kestemont, P. (1995) L'élevage des espèces d'accompagnement de la carpe en étang. In: *Les Carpes: Biologie et Elevage* (ed. R. Billard), pp. 348–70. INRA, Paris.

Kestemont, P., Cooremans, J., Abi-Ayad, A. & Mélard, C. (1999a) Cathepsin L in eggs and larvae of perch *Perca fluviatilis*: variations with developmental stage and spawning period. *Fish Physiology and Biochemistry* **21**, 59–64.

Kestemont, P. & Mélard, C. (1996) Mise au point et développement d'une technologie d'élevage intensif des percidés en région wallonne. *Rapport annuel 1. Ministère de la Région wallonne*. DGRNE, Belgium.

Kestemont, P. & Mélard, C. (1997) Mise au point et développement d'une technologie d'élevage intensif des percidés en région wallonne. *Rapport annuel 2. Ministère de la Région wallonne*. DGRNE, Belgium.

Kestemont, P., Mélard, C. Etienne, F. & Baras, E. (1998b) Bases biologiques de l'élevage intensif de percidés en cage. *Rapport annuel 3. Ministère Région. wallonne*. DGTRE, Belgium.

Kestemont, P., Mélard, C., Fiogbé, E., Vlavonou, R. & Masson, G. (1996) Nutritional and animal husbandry aspects of rearing early life stages of Eurasian perch *Perca fluviatilis*. *Journal of Applied Ichthyology* **12** (3–4), 157–66.

Kestemont, P., Mélard, C., Fontaine, P., Anthouard, M., Boujard, T. & Kentouri, M. (1999b) An integrated study of interindividual competition and its relationships with feeding physiology and behaviour in cultured predatory fish. *CE-FAIR 96-1572, 2nd annual report*. European Commission, Brussels.

Kestemont, P., Mélard, C., Houbart, M. & Fontaine, P. (1998a) An overview of major factors affecting growth heterogeneity and sibling cannibalism during early life stages of Eurasian perch *Perca fluviatilis*. World Aquaculture Society: *Aquaculture'98*. Las Vegas, Nevada, 15–19 February, 1998.

Kestemont, P., Vandeloise, E., Brown, P.B., Mélard, C. & Fontaine, P. (1997) Effects of casein-based diets containing different levels of lipids on survival, growth and nutritional status of Eurasian perch *Perca fluviatilis*. In: *Island Aquaculture and Tropical Aquaculture* (eds L. Creswell & Y. Harache), pp. 333–4. *EAS Publication*. European Aquaculture Society, Oostende, Belgium.

Ketola, H.G. (1978) Nutritional requirements and feeding of selected coolwater fishes: a review. *The Progressive Fish-Culturist* **40** (4), 127–32.

Keup, L. & Bayless, J. (1964) Fish distribution at varying sediments in Neuse River Basin, North Carolina. *Chesapeake Science* **5**, 119–23.

Kipling, C. (1976) Year-class strengths of perch and pike in Windermere. *Report of the Freshwater Biological Association* **44**, 68–75.

Kipling, C. (1984) A study of perch (*Perca fluviatilis* L.) and pike (*Esox lucius* L.) in Windermere from 1941 to 1982. *Journal du Conseil Internationale pour l'Exploration de la Mer* **41**, 259–67.

Kipling, C. & Le Cren, E.D. (1984) Mark–recapture experiments on fish in Windermere, 1943–1982. *Journal of Fish Biology* **24**, 395–414.

Kirsipuu, A. (1967) Differences in the protein composition of the blood serum in two forms of perch in the Lake Vortsjarv. *Eesti NSV Teaduste Akedeemia Toimetised Biologia* **16** (1), 37–40. (In Russian with English summary.)

Kirsipuu, A. (1971) Some physiological processes and environmental conditions affecting the protein composition of the blood serum in fish. *Rapports Proces-verbaux Conseil international pour l'Exploration de la Mer* **161**, 154–7.

Kitchell, J.F., Stewart, D.J. & Weininger, D. (1977) Applications of a bioenergetics model to yellow perch (*Perca flavescens*) and walleye (*Stizostedion vitreum vitreum*). *Journal of the Fisheries Research Board of Canada* **34** (10), 1922–35.

Kjellman, J., Hudd, R. & Urho, L. (1996) Monitoring 0+ perch (*Perca fluviatilis*) abundance in respect to time and habitat. *Annales Zoologici Fennici* **33**, 363–70.

Klein Breteler, J.G.P. (1989) Intensive culture of pikeperch fry with live food. In: *Aquaculture – a Biotechnology in Progress Vol. 1* (eds N. De Pauw, G. Persoone & E. Jaspers), pp. 203–7. *European Aquaculture Society Special Publication* **13**. Bredene, Belgium.

Knapp, L.W. (1976) Redescription, relationships and status of the Maryland darter, *Etheostoma sellare* (Radcliffe and Welsh), an endangered species. *Proceedings of the Biological Society of Washington* **89**, 99–117.

Knapp, R.A. & Sargent, R.C. (1989) Egg-mimicry as a mating strategy in the fantail darter, *Etheostoma flabellare*: Females prefer males with eggs. *Behavioral Ecology and Sociobiology* **25**, 321–6.

Knight, J.G. & Ross, S.T. (1992) Reproduction, age and growth of the bayou darter *Etheostoma rubrum* (Pisces, Percidae): An endemic of Bayou Pierre. *American Midland Naturalist* **127**, 91–105.

Knight, R.L., Margraf, F.J. & Carline, R.F. (1984) Piscivory by walleyes, *Stizostedion vitreum vitreum*, and yellow perch, *Perca flavescens*, in western Lake Erie, USA. *Transactions of the American Fisheries Society* **113** (6), 677–93.

Knight, R.L. & Vondracek, B. (1993) Changes in prey fish populations in western Lake Erie, 1969–88, as related to walleye, *Stizostedion vitreum vitreum*, predation. *Canadian Journal of Fisheries and Aquatic Sciences* **50** (6), 1289–98.

Koch, L.M. (1978) *Aspects of life history of* Etheostoma (Psychromaster) tuscumbia *in Buffler Spring, Lauderdale County, Alabama*. M.Sc. Thesis, University of Northern Alabama, Florence, USA.

Koenig, S.D., Kayes, T.B. & Calbert, H.E. (1978) Preliminary observations on the sperm of yellow perch. In: *Selected Coolwater Fishes of North America* (ed. R.L. Kendall). *American Fisheries Society Special Publication* **11**, 177–86.

Kokurewicz, B. (1969) The influence of temperature on the embryonic development of the perches: *Perca fluviatilis* L. and *Lucioperca lucioperca* (L.). *Zoologica Poloniae* **19** (1), 47–66.

Kolkovski, S., Czesny, S., Yackey, C., Moreau, R. & Dabrowski, K. (1998) The effects of vitamins C and E in (n-3) HUFA enriched *Artemia* nauplii on growth, survival, and stress resistance of walleye *Stizostedion vitreum* fry. In: World Aquaculture Society Symposium *Aquaculture'98*, Las Vegas, Nevada, 15–19 February, 1998.

Kolkovski, S., Dabrowski, K. & Yackey, C. (1997) Larval rearing of yellow perch *Perca flavescens* spawned out-of-the-season. In: *Island Aquaculture and Tropical Aquaculture* (eds L. Creswell & Y. Harache), pp. 335–6. Second International Workshop on Aquaculture of Percid Fish. European Aquaculture Society, Oostende, Belgium.

Kolomin, Yu. M. (1977) The Nadym River ruffe, *Acerina cernua*. *Journal of Ichthyology* **17** (3), 345–9.

Konstantinov, K.G. (1957) Sravitelnyi analiz morfologii i biologii okunya, sudaka i bersha na raznykh etapakh razritiya. *Trudy Institut Morfologii Zhivotnyth Akademiya Nauk, SSSR*. **16**, 181–236.

Koonce, J.F., Busch, W.D.N. & Czapla, T. (1996) Restoration of Lake Erie: contribution of water quality and natural resource management. *Canadian Journal of Fisheries and Aquatic Sciences* **53** (Suppl. 1), 105–12.

Kopp, R.L., Wissing, T.E. & Guttman, S.I. (1994) Genetic indicators of environmental tolerance among fish populations exposed to acid deposition. *Biochemical Systematics and Ecology* **22** (5), 459–75.

Korhonen, P., Virtanen, M. & Schultz, T. (1995) Bioenergetic calculation of mercury accumulation in fish. *Water, Air and Soil Pollution* **80** (1–4), 901–4.

Kornman, L.E. (1980) *Life history aspects of the striped darter*, Etheostoma virgatum (*Jordan*) *1880, from Clear Creek, Rockcastle County, Kentucky*. M.S. Thesis, Eastern Kentucky University, USA.

Koulikov, A.O. (1996) Physiological and ecological factors influencing the radiocaesium contamination of fish species from Kiev reservoir. *Science of the Total Environment* **177**, 125–35.

Koupal, K.D. (1996) Field utilization of extended semen. In: *Walleye Culture Manual* (ed. R.C. Summerfelt), pp. 55–8. *NCRAC Culture Series* **101**. North Central Regional Aquaculture Center Publications Office, Iowa State University, Ames.

Koupal, K.D., Satterfield J.R., Jr. & Flickinger, S.A. (1995) Evaluation of walleye semen: extender ratios in relation to hatching percentage. *The Progressive Fish-Culturist* **57**, 188–91.

Kouril, J., Linhart, O. & Relot, P. (1997) Induced spawning of perch by means of a GnRH analogue. *Aquaculture International* **5**, 375–7.

Krise, W.F. (1988) Optimum protease exposure time for removing adhesiveness of walleye eggs. *The Progressive Fish-Culturist* **50**, 126–7.

Krise, W.F., Bulkowski-Cummings, L., Shellman, A.D., Kraus, K.A. & Gould, R.W. (1986) Increased walleye egg hatch and larval survival after protease treatment of eggs. *The Progressive Fish-Culturist* **48**, 95–100.

Krise, W.F. & Meade, J.W. (1986) Review of the intensive culture of walleye fry. *The Progressive Fish-Culturist* **48**, 81–9.

Kubečka, J. (1994) Models for comparing average first-year growth in length of freshwater fish. *Fisheries Management and Ecology* **1** (1), 45–55.

Kucharczyk, D., Kujawa, R. & Mamcarz, A. (1996b) New experimental incubation unit for eggs of the perch *Perca fluviatilis*. *The Progressive Fish-Culturist* **58**, 281–3.

Kucharczyk, D., Kujawa, R., Mamcarz, A., Skrzypczak, A. & Wyszomirska, E. (1996a) Induced spawning in perch *Perca fluviatilis* L. using carp pituitary extract and HCG. *Aquaculture Research* **27**, 847–52.

Kudrinskaya, O.M. (1970) Food and temperature as factors affecting the growth, development and survival of pikeperch and perch larvae. *Journal of Ichthyology* **10**, 779–88.

Kuhajda, B.R. & Warren, M.L., Jr (1989) Life history aspects of the harlequin darter, *Etheostoma histrio*, in western Kentucky. *ASB (Association of Southeastern Biologists) Bulletin* **36**, 66–7.

Kukuradze, A.M. (1974) A description of the spawning stock and reproduction of the pikeperch *Lucioperca lucioperca,* of the Danube Delta and waters in the Danubian Region. *Journal of Ichthyology* **14** (3), 385–92.

Kuzmina, V.V. (1984a) Relative enzyme activity of the fish intestinal lumen and mucosa. *Journal of Ichthyology* **24** (3), 140–44.

Kuzmina, V.V. (1984b) Effect of temperature on pH – function of phosphatases active in fish intestine. *Journal of Ichthyology* **24** (1), 136–42.

Kuzmina, V.V. (1989). Seasonal dynamics in the activity of some carbohydrase and alkaline phosphotase enzymes in the intestinal mucosa of fish. *Journal of Ichthyology* **28** (6) 156–61.

Kuznetsov, V.A. (1982) Vliyanie uslovij nagula na plodovitost' i kachestvo irky bersha, *Stizostedion volgensis* (Gmelin) (Percidae), Kujbyshevskogo Vodokhtanilishcha. *Voprosy Ikhtiologii* **22** (4), 599–607.

Kuznetsov, V.A. (1986) Dynamics of fecundity, egg quality and morphophysiological characteristics of female pike perch as dependent on feeding conditions in the Kujbyshev Reservoir. *Biologicheskie Nauki*, **8**, 32–7.

Laarman, P.W. (1978) Case histories of stocking walleyes, *Stizostedion vitreum vitreum*, in inland lakes, impoundments and the Great Lakes – 100 years with walleyes. In: *Selected Coolwater Fishes of North America* (ed. R.L. Kendall). *American Fisheries Society Special Publication* **11**, 254–60.

Laarman, P.W. (1981) Vital statistics of a Michigan fish population, with special emphasis on the effectiveness of stocking 15 cm walleye fingerlings. *North American Journal of Fisheries Management* **1** (2), 177–85.

Lachner, E.A., Westlake, E.F., & Handwerk, P.S. (1950) Studies on the biology of some percid fishes from western Pennsylvania. *American Midland Naturalist* **43**, 92–111.

Lahnsteiner, F., Berger, B., Weismann, T. & Patzner, R. (1995) Fine structure and motility of spermatozoa and composition of the seminal plasma in the perch. *Journal of Fish Biology* **47** (3), 492–508.

Lake, C.T. (1936) The life history of the fan-tailed darter *Catonotus flabellaris flabellaris* (Rafinesque). *American Midland Naturalist* **17**, 816–30.

Lang, C. (1987) Mortality of perch, *Perca fluviatilis* L., selected from the size and abundance of egg strands. *Journal of Fish Biology* **31** (5), 715–20.

Lang, I. (1981) Electron microscopic and histochemical study of the postovulatory follicles of *Perca fluviatilis* L. (Teleostei). *General and Comparative Endocrinology* **45**, 219–33.

Langdon, J.S., Humphrey, J.D., Williams, L.M., Hyatt, A.D. & Westbury, H.A. (1986) First virus isolation from Australian fish: an iridovirus-like pathogen from redfin perch, *Perca fluviatilis* L. *Journal of Fish Diseases* **9**, 263–8.

Lapkin, V.V., Svirskij, A.M. & Golovanov, V.K. (1981) Vozrastnaya dinamika izbiraemykh i letal 'nykh temperatur ryb. *Zoologichesky Zhurnal* **60** (12), 1792–1801. (In Russian.)

Lappalainen, A., Rask, M. & Vuorinen, P.J. (1988) Acidification affects the perch, *Perca fluviatilis*, populations in small lakes of southern Finland. *Environmental Biology of Fishes* **21** (3), 231–9.

Lappalainen, J., Erm, V. & Lehtonen, H. (1995) Pikeperch, *Stizostedion lucioperca* (L.) catch relation to juvenile density and water temperature in Parnu Bay, Estonia. *Fisheries Management and Ecology* **2** (2), 113–20.

Larsson, A., Haux, C., Sjoebeck, M.-L. & Lithner, G. (1984) Physiological effects of an additional stressor on fish exposed to a simulated heavy-metal-containing effluent from a sulphide ore smeltery. *Ecotoxicology and Environmental Safety* **8** (2), 118–28.

Lauder, G.V. (1983) Functional design and evolution of the pharyngeal jaw apparatus in euteleostean fishes. *Zoological Journal of the Linnean Society* **77** (1), 1–38.

Lawler, G.H. (1969) Aspects of the biology of *Triaenophorus nodulosus* in yellow perch, *Perca flavescens*, in Heming Lake, Manitoba. *Journal of the Fisheries Research Board of Canada* **26**, 821–31.

Layman, S.R. (1984) The duskytail darter, *Etheostoma* (*Catonotus*) sp., confirmed as an egg-clusterer. *Copeia* **1984**, 992–4.

Layman, S.R. (1993) Life history of the Savannah darter, *Etheostoma fricksium*, in the Savannah River drainage, South Carolina. *Copeia* **1993**, 959–68.

Leary, R. & Booke, H.E. (1982) Genetic stock analysis of yellow perch from Green Bay and Lake Michigan. *Transactions of the American Fisheries Society* **111** (1), 52–7.

Lebedev, V.D. (1952) Fishes from alatepaleolithic settlement at Murzak-Koba in Crimea. *Bulletin of the Society of Natural History Moscow* **51** (6), 46–51. (In Russian.)

Le Cren, E.D. (1947) The determination of the age and growth of the perch (*Perca fluviatilis*) from the opercular bone. *Journal of Animal Ecology* **16**, 188–204.

Le Cren, E.D. (1958) Observations on the growth of perch (*Perca fluviatilis* L.) over twenty-two years with special reference to the effects of temperature and changes in population density. *Journal of Animal Ecology* **27**, 287–334.

Le Cren, E.D. (1965) Some factors regulating the size of populations of freshwater fishes. *Mitteilungen der Internationale Vereinigung für Limnologie* **13**, 88–105.

Le Cren, E.D. (1987) Perch (*Perca fluviatilis*) and pike (*Esox lucius*) in Windermere from 1940 to 1985: studies in population dynamics. *Canadian Journal of Fisheries and Aquatic Sciences* **44** (Suppl. 2), 216–28.

Le Cren, E.D. (1992) Exceptionally big individual perch (*Perca fluviatilis* L.) and their growth. *Journal of Fish Biology* **40** (4), 599–625.

Lee, D.S. & Ashton, R.E., Jr (1979) Seasonal and daily activity patterns of the glassy darter, *Etheostoma vitreum* (Percidae). *ASB (Association of Southeastern Biologists) Bulletin* **26**, 36.

Lehmann, J., Mock, D., Klinger, H. & Kriwet, T. (1996) The ruffe (*Gymnocephalus cernuus*) as probably the most important infestation and reinfestation source with swimbladder parasite *Anguillicola crassus* for older eels in Nordrhein-Westfalen. *Fischer und Teichwirt* **47** (11), 442–5.

Lehtinen, K.-J. & Oikari, A. (1980) Sublethal effects of kraft pulp mill waste water on the perch, *Perca fluviatilis*, studied by rotaryflow and histological techniques. *Annales Zoologici Fennici* **17** (4), 255–9.

Lehtonen, H. & Lappalainen, J. (1995) The effects of climate on the year-class variations of certain freshwater fish species. *Canadian Special Publication of Fisheries and Aquatic Sciences* **121**, 37–44.

Lehtonen, H. & Toivonen, J. (1988) Migration of pike-perch, *Stizostedion lucioperca* (L.), in different coastal waters in the Baltic Sea. *Finnish Fisheries Research* **1988** (7), 24–30.

Leino, R.L. (1982) Rodlet cells in the gill and intestine of *Catostomus commersoni* and *Perca flavescens*: a comparison of their light and electron microscopic cytochemistry with that of mucous and granular cells. *Canadian Journal of Zoology* **60** (11), 2768–82.

Leino, R.L. & McCormick, J.H. (1997) Histological evidence for repeat spawning of ruffe from the St Louis River estuary on western Lake Superior. In: *International Symposium on Biology and Management of Ruffe: Symposium Abstracts* (ed. D.A. Jensen), p. 10. *Minnesota Sea Grant Publication* **X48**.

Leino, R.L., McCormick, J.H. & Jensen, K.M. (1987) Changes in gill histology of fathead minnows and yellow perch transferred to soft water or acidified soft water with particular reference to chloride cells. *Cell Tissue Research* **250** (2), 389–99.

Leis, A.L. & Fox, M.G. (1994) Effect of mine tailings on the *in situ* survival of walleye (*Stizostedion vitreum*) eggs in a northern Ontario river. *Ecoscience* **1** (3), 215–22.

Leis, A.L. & Fox, M.G. (1996) Feeding, growth and habitat associations of young-of-the-year walleye (*Stizostedion vitreum*) in a river affected by mine tailings spill. *Canadian Journal of Fisheries and Aquatic Sciences* **35** (11), 2408–17.

Lelek, A. (1987) *The Freshwater Fishes of Europe 9. Threatened fishes of Europe.* Aula-Verlag, Wisbaden.

Lesser, W. & Vilstrup, R. (1979) The supply and demand for yellow perch, 1915–90. *University of Wisconsin College of Agricultural and Life Sciences Research Bulletin* **R3006.** Madison, Wisconsin.

Lessman, C.A. (1978) Effects of gonadotropin mixtures and two steroids on inducing ovulation in the walleye. *The Progressive Fish-Culturist* **40** (1), 3–5.

Letichevskii, M.A. (1946) K voprosu o plodovitosti ryb yuga Aralskogo morya. *Zoologichesky Zhurnal* **25** (4), 351–6.

Li, J., Cohen, Y., Schupp, D.H. & Adelman, I.R. (1996) Effects of walleye stocking on population abundance and fish size. *North American Journal of Fisheries Management* **16** (4), 830–39.

Li, S. & Mathias, J.A. (1982) Causes of high mortality among cultured larval walleyes. *Transactions of the American Fisheries Society* **111** (6), 710–21.

Lichtkoppler, F.R. (1992) *Ohio's fishing industry: a 1990 profile.* Factsheet. Ohio State University, Columbus, Ohio.

Lind, E.A., Ellonen, T., Keranen, M., Kukko, O. & Tenhunen, A. (1973) Advenen pyydystettävyyden vuorokauden ajat talvella ja Kesällä. *Kalamies* **3**, 3.

Lindesjoeoe, E. & Thulin, J. (1990) Fin erosion of perch *Perca fluviatilis* and ruffe *Gymnocephalus cernua* in a pulp mill effluent area. *Diseases of Aquatic Organisms* **8** (2), 119–26.

Lindesjoeoe, E., Thulin, J. Bengtsson, B.E., & Tjaernlund, U. (1994) Abnormalities of a gill cover bone, the operculum, in perch *Perca fluviatilis* from a pulp mill effluent area. *Aquatic Toxicology* **28** (3–4), 189–207.

Lindesjöö, E. (1994) Temporal variation and sexual dimorphism of the skin of perch *Perca fluviatilis* L.: a morphological study. *Journal of Applied Ichthyology* **10**, 154–66.

Lindquist, D.G., Shute, J.R. & Shute, P.W. (1981) Spawning and nesting behaviour of the Waccamaw darter, *Etheostoma perlongum. Environmental Biology of Fishes* **6** (2), 177–91.

Lindroth, A. (1947) Time of activity of freshwater fish spermatazoa in relation to temperature. *Zoologiska Bidrag fran Uppsala* **25**, 165–8.

Lindsay, H.L. & Adam, W.H. (1974) First published records of *Crystallaria asprella* in Oklahoma. *Southwestern Naturalist* **19**, 218–19.

Linfield, R.S.J. & Rickards, R.B. (1979) The zander in perspective. *Fisheries Management* **10** (1), 1–16.

Linlokken, A. (1984) Gill-net selectivity for perch, *Perca fluviatilis. Fauna Oslo* **37** (3), 112–16.

Loadman, N.L., Mathias, J.A. & Moodie, G.E.E. (1989) Methods for intensive culture of walleye. *The Progressive Fish-Culturist* **51**, 1–9.

Loew, E.L. & Wahl, C.M. (1991) A short-wavelength sensitive cone mechanism in juvenile yellow perch, *Perca flavescens. Vision Research* **31** (3), 353–60.

Loew, E.R., McFarland, W.N., Mills, E.L. & Hunter, D. (1993) A chromatic action spectrum for planktonic predation by juvenile yellow perch, *Perca flavescens*. *Canadian Journal of Zoology* **71**, 384–6.

Loftus, K.H. (1976) Science for Canada's fisheries rehabilitation needs. *Journal of the Fisheries Research Board of Canada* **33**, 1822–57.

Lom, J., Pike, A.W. & Dykova, I. (1991) *Myxobolus sandrae* Reuss, 1906, the agent of vertebral column deformities of perch *Perca fluviatilis* in northeast Scotland. *Diseases of Aquatic Organisms* **12** (1), 49–53.

Lom, J. & Vàvra, J. (1961) *Epistylis lowffi* from the skin of perches. *Acta Societatis Zoologicae Bohemosloven* **25** (4), 273–6.

Loos, J.J. & Woolcott, W.S. (1969) Hybridization and behavior in two species of *Percina* (Percidae). *Copeia* **1969**, 374–85.

Lott, J.P., Willis, D.W. & Lucchesi, D.O. (1996) Relationship of food habits to yellow perch growth and population structure in South Dakota lakes. *Journal of Freshwater Ecology* **11** (1), 27–37.

Lowenstein, O. (1971) The labyrinth. In: *Fish Physiology V* (eds W.S. Hoar & D.J.Randall), pp. 207–40. Academic Press, New York.

Ludwig, D., Hilborn, R. & Walters, C. (1993) Uncertainty, resource exploitation, and conservation: lessons from history. *Science* **260** No. 5104, pp. 17–18.

Lutterbie, G.W. (1979) Reproduction and age and growth in Wisconsin darters (Osteichthyes: Percidae). *Reports on the Fauna and Flora of Wisconsin* **15**.

Lutz, P.L. (1972) Ionic and body compartment responses to increasing salinity in the perch, *Perca fluviatilis*. *Comparative Biochemistry and Physiology* **42a**, 711–17.

Lynch, W.E., Jr., Johnson, D.L. & Schell, S.A. (1982) Survival, growth and food habits of walleye × sauger hybrids (saugeye) in ponds. *North American Journal of Fisheries Management* **2** (4), 381–7.

Lyons, J. (1983) Olfactory organ morphology and histology in the yellow perch *Perca flavescens*, and golden shiner, *Notemigonus crysoleucas*. *Canadian Journal of Zoology* **61** (12), 2987–90.

Lyons, J. & Magnuson, J.J. (1987) Effects of walleye predation on the population dynamics of small littoral-zone fishes in a northern Wisconsin lake. *Transactions of the American Fisheries Society* **116** (1), 29–39.

MacCrimmon, H.R., Gots, B.L. & MacGregor, R.B. (1989) A note on an atypical yellow perch, *Perca flavescens* Mitchill. *Journal of Fish Biology* **34** (3), 485–6.

Mackay, W.C. & Craig, J.F. (1983) A comparison of four systems for studying the activity of pike (*Esox lucius* L.) and perch (*Perca fluviatilis* L. and *Perca flavescens* (Mitchill)). In: *Proceedings of the Fourth International Wildlife Biotelemetry Conference* (ed. D.G. Pincock), pp. 22–30. Halifax, Canada.

Maclean, N.G., Teleki, G.C. & Polak, J. (1982) Ultrasonic telemetry studies of fish activity near the Nanticoke thermal generating station. *Journal of Great Lakes Research* **8** (3), 495–504.

Madenjian, C.P., Tyson, J.T., Knight, R.L., Kershner, M.W. & Hansen, M.J. (1996) First-year growth, recruitment, and maturity of walleyes in western Lake Erie. *Transactions of the American Fisheries Society* **125** (6), 821–30.

Madon, S.P. (1993) *Bioenergetics of young of year walleye and trophic interactions in experimental ponds*. Ph.D. Thesis, Ohio State University, Columbus, Ohio, USA.

Madon, S.P. & Culver, D.A. (1993) Bioenergetics model for larval and juvenile walleyes: an *in situ* approach with experimental ponds. *Transactions of the American Fisheries Society* **122**, 797–813.

Magnuson, J.J., Meisner, J.D. & Hill, D.K. (1990) Potential changes in the thermal habitat of Great Lakes fish after global climate warming. *Transactions of the American Fisheries Society* **119** (2), 254–64.

Maitland, P.S. (1977) *The Hamlyn Guide to Freshwater Fishes of Britain & Europe.* Hamlyn, London.

Malison, J.A., Best, C., Kayes, T.B., Amundson, C.H. & Wentworth, B. (1985) Hormonal growth promotion and evidence for a size-related difference in response to estradiol-17β in yellow perch (*Perca flavescens*). *Canadian Journal of Fisheries and Aquatic Sciences* **42** (10), 1627–33.

Malison, J.A. & Garcia-Abiado, M.A.R. (1996) Sex control and ploidy manipulations in yellow perch (*Perca flavescens*) and walleye (*Stizostedion vitreum*). *Journal of Applied Ichthyology* **12**, 189–94

Malison, J.A. & Held, J.A. (1992) Effects of fish size at harvest, initial stocking density and tank lighting conditions on the habituation of pond-reared yellow perch (*Perca flavescens*) to intensive culture conditions. *Aquaculture* **104**, 67–78.

Malison, J.A. & Held, J.A. (1996a) Reproduction and spawning in walleye (*Stizostedion vitreum*). *Journal of Applied Ichthyology* **12** (3–4), 153–6.

Malison, J.A. & Held, J.A. (1996b) Reproductive biology and spawning. In: *Walleye Culture Manual* (ed. R.C. Summerfelt). *NCRAC Culture Series* **101**. North Central Regional Aquaculture Center Publications Office, Iowa State University, Ames.

Malison, J.A. & Held, J.A. (1996c) Habituating pond-reared fingerlings to formulated feed. In: *Walleye Culture Manual* (ed. R.C. Summerfelt), pp. 199–204. *NCRAC Culture Series* **101**. North Central Regional Aquaculture Center Publications Office, Iowa State University, Ames.

Malison, J.A., Held, J.A., Procarione, L.S. & Garcia-Abiado, M.A.R. (1998b) Production of monosex female populations of walleye from intersex broodstock. *The Progressive Fish-Culturist* **60**, 20–24.

Malison, J.A., Kayes, T.B., Best, C.D., Amundson, C.H. & Wentworth, B. (1986) Sexual differentiation and use of hormones to control sex in yellow perch (*Perca flavescens*). *Canadian Journal of Fisheries and Aquatic Sciences* **43** (1), 26–35.

Malison, J.A., Kayes, T.B., Held, J.A., & Amundson, C.H. (1990) Comparative survival, growth, and reproductive development of juvenile walleye and sauger and their hybrids reared under intensive culture conditions. *The Progressive Fish-Culturist* **52**, 73–82.

Malison, J.A., Kayes, T.B., Held, J.A., Barry, T.P. & Amundson, C.H. (1993a) Manipulation of ploidy in yellow perch (*Perca flavescens*) by heat shock, hydrostatic pressure shock, and spermatozoa inactivation. *Aquaculture* **110**, 229–42.

Malison, J.A., Northey, D.L., Held, J.A. & Kuczynski, T.E. (1994b) Habituation of yellow perch (*Perca flavescens*) fingerlings to formulated feeds in ponds using lights and vibrating feeders. *Abstracts of the 29th Annual Meeting of the World Aquaculture Society*. New Orleans, Louisiana.

Malison, J.A., Procarione, L.S., Barry, T.P., Kapuscinski, T.P. & Kayes, T.B. (1994a) Endocrine and gonadal changes during the annual reproductive cycle of the freshwater teleost *Stizostedion vitreum* (walleye). *Fish Physiology and Biochemistry* **13**, 473–84.

Malison, J.A., Procarione, L.S., Held, J.A., Kayes, T.B. & Amundson, C.H. (1993b) The influence of triploidy and heat and pressure shocks on the growth and reproductive development of juvenile yellow perch (*Perca flavescens*). *Aquaculture* **116**, 121–33.

Malison, J.A., Procarione, L.S., Kayes, T.B., Hansen, J.F. & Held, J.A. (1998a) Induction of out-of-season spawning in walleye (*Stizostedion vitreum*). *Aquaculture* **163**, 151–61.

Mamcarz, A., Kucharczyck, D., Kujawa, R. & Skrzypczak, A. (1997) Influence of fish density on survival, growth and cannibalism development of pikeperch (*Stizostedion lucioperca* L.) larvae. In: *Island Aquaculture and Tropical Aquaculture* (eds L. Creswell & Y. Harache), pp. 343–6. Second International Workshop on Aquaculture of Percid Fish. *European Aquaculture Society Special Publication*, Oostende, Belgium.

Manci, W.E., Malison, J.A., Kayes, T.B. & Kuczynski, T.E. (1983) Harvesting photopositive juvenile fish from a pond using a lift net and light. *Aquaculture* **34**, 157–64.

Mansueti, A.J. (1964) Early development of the yellow perch, *Perca flavescens*. *Chesapeake Science* **5** (1–2), 46–66.

Manteifel, B.P., Girsa, I.I., Leshcheva, T.S. & Pavlov, D.S. (1965) Sutochnye ritmy pitaniya i dvigatelnoi aktivnosti nekotorykh presnovodnykh khishchnykh ryb. In: *Pitanie khishchnykh ryb* (ed. B.P. Manteifel), pp. 3–81. Nauka, Moscow.

Marcino, J. (1996) Observed diseases of walleye in Minnesota. In: *Walleye Culture Manual* (ed. R.C. Summerfelt), pp. 369–70. *NCRAC Culture Series* **101**. North Central Regional Aquaculture Center Publications Office, Iowa State University, Ames.

Margenau, T.L., Shram, S.T. & Blust, W.H. (1988) Lymphocystis in a walleye population. *Transactions of the American Fisheries Society* **117** (3), 308–10.

Margolis, L. & Arthur, J.R. (1979) Synopsis of the parasites of fishes of Canada. *Bulletin of the Fisheries Research Board of Canada* **199**, 1–269.

Margolis, L., Esch, G.W., Holmes, J.C., Kuris, A.M. & Shad, G.A. (1982) The use of ecological terms in parasitology. *Journal of Parasitology* **68**, 131–3.

Markevich, A.P. (1963) Parasitic fauna of freshwater fish of the Ukrainian SSR. *Israel Program for Scientific Translations* **884**.

Marsden, J.E., Gunderson, J.L., Klepinger, M.R. & Bronte, C.R. (eds) (1998) Special section on an international symposium on Eurasian ruffe (*Gymnocephalus cernuus*) biology, impacts, and control. *Journal of Great Lakes Research* **24** (2), 163–379.

Marsh, E. (1984) Egg size variation in central Texas populations of *Etheostoma spectabile* (Pisces: Percidae). *Copeia* **1984**, 291–301.

Marsh, E. (1986) Effects of egg size on offspring fitness and maternal fecundity in the orangethroat darter, *Etheostoma spectabile* (Pisces: Percidae). *Copeia* **1986**, 18–30.

Martineau, D., Bowser, P.R., Renshaw, R.R. & Casey, J.W. (1992) Molecular characterization of a unique retrovirus associated with a fish tumor. *Journal of Virology* **66** (1), 596–9.

Martineau, D., Renshaw, R., Williams, J.R., Casey, J.W. & Bowser, P.R. (1991) A large unintegrated retrovirus DNA species present in a dermal tumor of walleye *Stizostedion vitreum*. *Diseases of Aquatic Organisms* **10** (3), 153–8.

Mason, D.M. & Brandt, S.B. (1996) Effect of alewife predation on survival of larval yellow perch in an embayment of Lake Ontario. *Canadian Journal of Fisheries and Aquatic Sciences* **53** (7), 1609–17.

Mason, D.M., MacNeill, D.V., Brandt, S.B. & Gannon, J. (1986) Interaction between alewife and larval yellow perch in a Lake Ontario embayment. In: *Program and Abstracts, 29th*

Conference of the International Association for Great Lakes, p. 41. IAGLR, Buffalo, New York.

Matei, V.E. (1984) Comparative analysis of the gill epithelium ultrastructure in the perch, *Perca fluviatilis*, from basins with different ion composition. *Tsitologiya* **26** (7), 778–82. (In Russian.)

Mathews, S.H. & Dolan, D.M. (1983) Polychlorinated biphenyls (PCBs) in Saginaw Bay yellow perch, 1977–1980. In: *Program and Abstracts of the 26th Conference of the International Association for Great Lakes Research*, p. 36. IAGLR, Buffalo, New York.

Mathias, J.A., Franzin W.G., Craig, J.F. & Babaluk, J.A. (1992) Evaluation of stocking walleye fry to enhance a commercial fishery in a large, Canadian prairie lake. *North American Journal of Fisheries Management* **12**, 299–306.

Mathias, J.A. & Li, S. (1982) Feeding habits of walleye larvae and juveniles: comparative laboratory and field studies. *Transactions of the American Fisheries Society* **121** (6), 722–35.

Mathur, D. (1973a) Food habits and feeding chronology of the black-banded darter, *Percina nigrofasciata*, in Halawakee Creek, Alabama. *Transactions of the American Fisheries Society* **102** (1), 48–55.

Mathur, D. (1973b) Some aspects of life history of the blackbanded darter, *Percina nigrofasciata* (Agassiz), in Halawakee Creek, Alabama. *American Midland Naturalist* **89**, 381–93.

Matthews, W.J. (1985) Critical current speeds and microhabitats of the benthic fishes *Percina roanoka* and *Etheostoma flabellare*. *Environmental Biology of Fishes* **12**, 303–8.

Matthews, W.J. (1998) *Patterns in Freshwater Fish Ecology*. Chapman & Hall, New York.

Matthews, W.J., Bek, J.R. & Surat, E. (1982) Comparative ecology of the darters, *Etheostoma podostemone*, *E. flabellare*, and *Percina roanoka* in the upper Roanoke River dainage Virginia. *Copeia* **1982**, 805–14.

Matty, A.J. (1985) *Fish Endocrinology*. Croom Helm, London.

Mauck, W.L., Olson, L.E. & Hogan, J.W. (1977) Effects of water quality on deactivation and toxicity of mexacarbate (Zectran) to fish. *Archives of Environmental Contamination and Toxicology* **6** (4), 385–93.

May, B. (1969) Observations on the biology of the variegated darter, *Etheostoma variatum* (Kirtland). *Ohio Journal of Science* **69**, 85–92.

Mayden, R.L. (1985) Nuptial structures in the subgenus *Catonotus*, genus *Etheostoma* (Percidae). *Copeia* **1985**, 580–83.

Mayr, B., Kalat, M., Ráb, P. & Lambrou, M. (1987) Band karotypes and specific types of heterochromatins in several species of Europen percid fishes (Percidea, Pisces). *Genetica* **75**, 199–205.

McAllister, D.E. (1970) Rare or endangered Canadian fishes. *Canadian Field Naturalist* **84** (1), 124–5.

McAllister, P.E. (1996) Viruses associated with diseases of walleye, sauger, and walleye × sauger hybrids. In: *Walleye Culture Manual* (ed. R.C. Summerfelt), pp. 355–68. *NCRAC Culture Series* **101**. North Central Regional Aquaculture Center Publications Office, Iowa State University, Ames.

McCarthy, D.M. (1975) Fish furunculosis caused by *Aeromonas salmonicida* var. *achromogenes*. *Journal of Wildlife Diseases* **12**, 489–93.

McConville, D.R. & Fossum, J.D. (1981) Movement patterns of walleye (*Stizostedion v. vitreum*) in Pool 3 of the upper Mississippi River as determined by ultrasonic telemetry. *Journal of Freshwater Ecology* **1** (3), 279–85.

McCormick, J.H., Jensen, K.M. & Leino, R.L. (1989) Survival, blood osmolality, and gill morphology of juvenile yellow perch, rock bass, black crappie, and largemouth bass exposed to acidified soft water. *Transactions of the American Fisheries Society* **118** (4), 386–99.

McCormick, L.M. (1892) Descriptive list of the fishes of Lorain County, Ohio. *Laboratory Bulletin of Oberlin College* **2**, 1–33.

McDonald, D.G., Cavdek, V. & Ellis, R. (1991a) Gill design in freshwater fishes: interrelationships among gas exchange, ion regulation, and acid-base regulation. *Physiological Zoology* **64** (1), 103–23.

McDonald, D.G., Freda, J., Cavdek, V., Gonzalez, R. & Zia, S. (1991b) Interspecific differences in gill morphology of freshwater fish in relation to tolerance of low-pH environments. *Physiological Zoology* **64** (1), 124–44.

McElman, J.F. & Balon, E.K. (1979) Early ontogeny of walleye, *Stizostedion vitreum*, with steps of saltatory development. *Environmental Biology of Fishes* **4** (4), 309–48.

McWilliams, P.G. (1982) A comparison of physiological characteristics in normal and acid exposed populations of the brown trout, *Salmo trutta*. *Comparative Biochemistry and Physiology* **72a** (3), 515–22.

Mehner, T. & Wieser, W. (1994) Effects of temperature on allocation of metabolic energy in perch (*Perca fluviatilis*) fed submaximal rations. *Journal of Fish Biology* **45**, 1079–86.

Mélard, C., Baras, E., Mary, L. & Kestemont, P. (1996a) Relationships between stocking density, growth, cannibalism and survival rate in intensively cultured larvae and juveniles of perch (*Perca fluviatilis*). *Annales Zoologici Fennici* **33**, 643–51.

Mélard, C. & Kestemont, P. (1994) Diversification de la pisciculture en Région wallonne. Mise au point de l'élevage intensif de nouvelles espèces de consommation et de repeuplement – cyprinidés et percidés. *Rapport de recherches à la Région wallonne*, ULG, FUNDP, mars 1994, Belgium.

Mélard, C., Kestemont, P. & Baras, E. (1995) First results of European perch (*Perca fluviatilis*) intensive rearing in tank: effect of temperature and size grading on growth. *Bulletin Français de la Pêche et de Pisciculture* **336**, 19–27.

Mélard, C., Kestemont, P. & Grignard, J.C. (1996b) Intensive culture of juvenile and adult Eurasian perch (*P. fluviatilis*): effect of major biotic and abiotic factors on growth. *Journal of Applied Ichthyology* **12** (3–4), 175–80.

Mezhnin, F.I. (1972) Interrenal and chromaffin tissue of freshwater fish. *Journal of Ichthyology* **12**, 671–84.

Mezhnin, F.I. (1977) Interrenalovaya i suprarenalovaya zheleze i tel'tsa stanniusa okunya v nerestovyj period. *Biologicheskie Nauki. (Moscow)* **11**, 62–9.

Mezhnin, F.I. (1978) Development of the sex cells in the early ontogeny of the common perch, *Perca fluviatilis*. *Journal of Ichthyology* **18** (1), 71–86.

Mezhnin, F.I. (1979) Morphology and topography of interrenal and suprarenal glands of bony fishes. *Nauchnye Doklady Vysshei Shkoly, Biologicheskie Nauki* **3**, 33–8. (In Russian.)

Mikhailov, V. (1951) Stadialnosi rozwoju nictorych tasiemcow (Cestoda). *Annales Universitatis Mariae Curie-Skladowska (C)* **6** (3).

Mikhman, A.S. (1978) Pitanie i pishchevye potrebnosti lichinok perkariny *Percarina demidoffi* Nordmann v vostochnoj chasti Taganrogskogo zaliva. *Voprosy Ikhtiologii* **18** (2), 366–70.

Miller, L.W. (1967) The introduction, growth, diet and depth distribution of walleye,

Stizostedion vitreum (Mitchill), in El Capitan Reservoir, San Diego County. *Administrative Report California Resource Agency, Department of Fish and Game, Inland Fisheries Branch* **67–10**, 1–14.

Miller, P.J. (1979) Adaptiveness and implications of small size in teleosts. In: *Fish Phenology: Anabolic Adaptiveness in Teleosts* (ed. P. J. Miller), pp. 263–306. Academic Press, London.

Miller, T.J., Crowder, L.B. & Rice, J.A. (1993) Ontogenetic changes in behavioural and histological measures of visual acuity in three species of fish. *Environmental Biology of Fishes* **37**, 1–8.

Milliken, M.R. (1982) Qualitative and quantitative nutrient requirements of fishes: a review. *Fisheries Bulletin* **80** (4), 655–86.

Mills, C.A. & Hurley, M.A. (1990) Long-term studies on the Windermere populations of perch (*Perca fluviatilis*), pike (*Esox lucius*) and Arctic charr (*Salvelinus alpinus*). *Freshwater Biology* **23** (1), 119–36.

Mills, E.L., Confer, J.L. & Kretchmer, D.W. (1986) Zooplankton selection by young yellow perch: the influence of light, prey density, and predator size. *Transactions of the American Fisheries Society* **115**, 716–25.

Mills, E.L., Confer, J.L. & Ready, R.C. (1984) Prey selection of young yellow perch, *Perca flavescens*, the influence of capture success, visual acuity and prey choice. *Transactions of the American Fisheries Society* **113** (5), 579–87.

Mills, E.L. & Forney, J.L. (1981) Energetics, food consumption and growth of young yellow perch in Oneida Lake, New York. *Transactions of the American Fisheries Society* **110** (4), 479–88.

Mills, E.L., Sherman, R. & Robson, D.S. (1989) Effect of zooplankton abundance and body size on growth of age–0 yellow perch (*Perca flavescens*) in Oneida Lake, New York, 1975–86. *Canadian Journal of Fisheries and Aquatic Sciences* **46**, 880–86.

Mills, E.L., Widzowski, D.V. & Jones, S.R. (1987) Food conditioning and prey selection by young yellow perch (*Perca flavescens*). *Canadian Journal of Fisheries and Aquatic Sciences* **44**, 549–55.

Minns, C.K. & Hurley, D.A. (1988) Effects of net length and set time on fish catches in gill nets. *North American Journal of Fisheries Management* **8** (2), 216–223.

Minns, C.K. & Moore, J.E.. (1992) Predicting the impact of climate change on the spatial pattern of freshwater fish yield capability in eastern Canadian lakes. *Climatic Change* **22** (4), 327–46.

Minton, J.W.F. & McLean, R.B. (1982) Measurements of growth and consumption of sauger (*Stizostedion canadense*): implication for fish energetics studies. *Canadian Journal of Fisheries and Aquatic Sciences* **39**, 1396–1403.

Mitchill, S.L. (1818) Memoir on ichthyology. The fishes of New York described and arranged. *American Monthly Magazine Critical Review* **1817–1818** (2), 241–8 & 321–8.

Mittelbach, G.G. & Persson, L. (1998) The ontogeny of piscivory and its ecological consequences. *Canadian Journal of Fisheries and Aquatic Sciences* **55**, 1454–65.

Molnar, K., Hanek, G. & Fernando, C.H. (1974) Parasites of fishes from Laurel Creek, Ontario. *Journal of Fish Biology* **6**, 717–28.

Moodie, G.E.E., Loadman, N.L., Wiegand, M.D. & Mathias, J.A. (1989) Influence of egg characteristics on survival, growth and feeding in larval walleye (*Stizostedion vitreum*). *Canadian Journal of Fisheries and Aquatic Sciences* **46**, 516–21.

Mooij, W.M., Lammens, E.H.R.R. & Van Densen, W.L.T. (1994) Growth rate of 0+ fish in relation to temperature, body size, and food in shallow eutrophic Lake Tjeukemeer. *Canadian Journal of Fisheries and Aquatic Sciences* **51**, 516–26.

Moore, A. (1987) Short-term storage and cryopreservation of walleye semen. *The Progressive Fish-Culturist* **49**, 40–45.

Moore, A. (1996) Use of semen extenders for walleye. In: *Walleye Culture Manual* (ed. R.C. Summerfelt), pp. 51–3. *NCRAC Culture Series* **101**. North Central Regional Aquaculture Center Publications Office, Iowa State University, Ames.

Moore, A., Prange, M.A., Summerfelt, R.C. & Bushman, R.P. (1994) Evaluation of tank shape and a surface spray for intensive culture of larval walleyes fed formulated feed. *The Progressive Fish-Culturist* **56**, 100–110.

Moore, G.A. (1944) The retinae of two North American teleosts with special reference to their tapeta lucida. *Journal of Comparative Neurology* **80** (3), 369–79.

Moravec, F. (1979) Occurrence of the endoparasitic helminths in pike (*Esox lucius* L.) from the Mácha Lake fishpond system. *Věstnik Československe Spolecnosti Zoologicke* **43**, 174–93.

Mount, D.I. (1959) Spawning behavior of the bluebreast darter, *Etheostoma camurum* (Cope). *Copeia* **1959**, 240–43.

Müller, M. & Osse, J.W.M. (1984) Hydrodynamics of suction feeding in fish. *Transactions of the Zoological Society of London* **37** (2), 51–136.

Muncy, R.J. (1962) Life history of the yellow perch, *Perca flavescens*, in estuarine waters of Severn River, a tributary of Chesapeake Bay, Maryland. *Chesapeake Science* **3** (3), 143–59.

Murphy, B.R., Nielson, L.A. & Turner, B.J. (1983) Use of genetic tags to evaluate stocking success for reservoir walleyes. *Transactions of the American Fisheries Society* **122** (4), 457–63.

Muzik, V. (1996) Fish distribution of the middle part of the Turiec River. *Zivocisna Vuroba* **41** (11), 491–9.

Nagel, T.O. (1978) Walleye production in controlled environments. *Ohio Department of Natural Resources Division of Wildlife In-Service Note* **386**.

Nagel, T.O. (1996) Intensive culture of fingerling walleye on formulated feeds. In: *Walleye Culture Manual* (ed. R.C. Summerfelt), pp. 205–7. *NCRAC Culture Series* **101**. North Central Regional Aquaculture Center Publications Office, Iowa State University, Ames.

Nakashima, B.S. & Leggett, W.C. (1978) Daily ration of yellow perch (*Perca flavescens*) from Lake Memphremagog, Quebec-Vermont, with a comparison of methods for *in situ* determinations. *Journal of the Fisheries Research Board of Canada* **35** (12), 1597–1603.

Natochin, Y.V. & Lavrova, E.A. (1974) The influence of water salinity and stage of life history on ion concentration of fish blood serum. *Journal of Fish Biology* **6**, 545–55.

NCRAC (1994) *Annual Progress Report*. North Central Regional Aquaculture Center, East Lansing, Michigan.

Neacsu, I., Craciun, V. & Craciun, M. (1981) L'équilière hydro-mineral chez le sandre (*Stizostedion lucioperca* (L.)) transfere de 1'eau douce en milieux mixo-mesohalins et inversement. *Cercetari Marine (Recherches Marines)* **14**, 201–15.

Neill, W.H. & Magnuson, J.J. (1974) Distributional ecology and behavioural thermoregulation of fishes in relation to heated effluent from a power plant at Lake Monona, Wisconsin. *Transactions of the American Fisheries Society* **103**, 663–710.

Neja, Z. (1988) On some problems of reproduction of ruff, (*Gymnocephalus cernuus* L., 1758) in the Lake Dabie. *Acta Ichthyologica et Piscatoria* **18**, 33–50.

Nelson, J.A., Magnuson, J.J. & Chulakasem, W. (1988) Blood oxygen capacity differences in yellow perch (*Perca flavescens*) from northern Wisconsin lakes differing in pH. *Canadian Journal of Fisheries and Aquatic Sciences* **45** (10), 1699–1704.

Nelson, J.S. (1994) *Fishes of the World.* 3rd edn. John Wiley & Sons Inc., New York.

Nelson, W.R. (1968) Embryo and larval characteristics of sauger, walleye and their reciprocal hybrids. *Transactions of the American Fisheries Society* **97** (2), 167–74.

Nelson, W.R., Hines, N.H. & Bergman, L.G. (1965) Artificial propagation of saugers and hybridization with walleye. *The Progressive Fish-Culturist* **38**, 90–91.

Nepszy, S.J. (1977) Changes in percid populations and species interactions in Lake Erie. *Journal of the Fisheries Research Board of Canada* **34** (10), 1861–8.

Neuhaus, E. (1934) Studien über das Stettiner Haff und seine Nebengewaser. Untersuchungen über den Zander. *Zeitschrift für Fischerei* **32**, 599–634.

Neuman, E., Thoresson, G. & Sandström, O. (1996) Swimming activity of perch, *Perca fluviatilis*, in relation to temperature, day-length and consumption. *Annales Zoologici Fennici* **33**, 669–78.

New, J.G. (1966) Reproductive behavior of the shield darter, *Percina peltata peltata*, in New York. *Copeia* **1966**, 20–28.

Newsome, G.E. & Aalto, S.K. (1987) An egg-mass census method for tracking fluctuations in yellow perch (*Perca flavescens*) populations. *Canadian Journal of Fisheries and Aquatic Sciences* **44** (6), 1221–32.

Newsome, G.E. & Leduc, G. (1975) Seasonal changes of fat content in the yellow perch. *Journal of the Fisheries Research Board of Canada* **32**, 2214–21.

Newsome, G.E. & Tompkins, J. (1985) Yellow perch egg masses deter predators. *Canadian Journal of Zoology* **63** (12), 2882–4.

Newton, E.T. (1908) Note relative à des fragments fossiles de petite vertébrés trouvés dans les dépôts pliocènes de Tegelen sur Meuse. *Bulletin Société Belge de Géologie* **21**, 591–6

Ney, J.J. (1978) A synoptic review of yellow perch and walleye biology. In: *Selected Coolwater Fishes of North America* (ed. R.L. Kendall), pp. 1–12. *American Fisheries Society Special Publication* **11**.

Nickum, J.G. (1978) Intensive culture of walleyes: the state of the art. *American Fisheries Society Special Publication* **11**.

Niimi, A.J. & Morgan, S.L. (1980) Morphometric examination of the gills of walleye, *Stizostedion vitreum vitreum* (Mitchill), and rainbow trout, *Salmo gairdneri* Richardson. *Journal of Fish Biology* **16** (6), 685–92.

Noaillac-Depeyre, J. & Gas, N. (1978) Ultrastructural and cytochemical study of the gastric epithelium in a fresh water teleostean fish (*Perca fluviatilis*). *Tissue and Cell* **10** (1), 23–37.

Noaillac-Depeyre, J. & Gas, N. (1979) Structure and function of the intestinal epithelial cells in the perch (*Perca fluviatilis* L.). *Anatomical Record* **195** (4), 621–39.

Noaillac-Depeyre, J. & Gas, N. (1982) Ultrastructure of endocrine cells in the stomach of two teleost fish, *Perca fluviatilis* L. and *Ameiurus nebulosus* L. *Cell Tissue Research* **221** (3), 657–78.

Noble, R.L. (1970a) Parasites of yellow perch in Oneida Lake, New York. *New York Fish and Game Journal* **17** (2), 95–101.

Noble, R.L. (1970b) Evaluation of the Miller high-speed sampler for sampling yellow perch and walleye fry. *Journal of the Fisheries Research Board of Canada* **27** (6), 1033–44.

Noble, R.L. (1972) A method of direct estimation of total food consumption with application to young yellow perch. *The Progressive Fish-Culturist* **34**, 191–4.

Nonnotte, G. (1981) Cutaneous respiration in six freshwater teleosts. *Comparative Biochemistry and Physiology* **70A** (4), 541–3.

Nordstrom, P.R., Bailey, J.L. & Heaton, J.R. (1960) A bacterial disease of yellow perch (*Perca flavescens*). *Transactions of the American Fisheries Society* **89**, 310–12.

Nyberg, P. (1979) Production and food consumption of perch, *Perca fluviatilis* L., in two Swedish forest lakes. *Report Institute of Freshwater Research, Drottningholm* **58**, 140–57.

O'Neil, P.E. (1981) Life history aspects of *Etheostoma coosae* (Pisces: Percidae) in Barbaree Creek, Alabama. *Tulane Studies in Zoology and Botany* **23**, 75–83.

Ogden, J.C. (1970) Relative abundance, food habits, and age of the American eel, *Anguilla rostrata* (Lesueur), in certain New Jersey streams. *Transactions of the American Fisheries Society* **99**, 54–9.

Ogle, D.H., Selgeby, J.H., Newman, R.M. & Henry, M.G. (1995) Diet and feeding periodicity of ruffe in the St. Louis River estuary, Lake Superior. *Transactions of the American Fisheries Society* **124**, 356–69.

Oliver, D.R. (1960) The macroscopic bottom fauna of Lac la Ronge, Saskatchewan. *Journal of the Fisheries Research Board of Canada* **17** (5), 607–24.

Olson, D.E., Schupp, D.H. & Macins, V. (1978) An hypothesis of homing behavior of walleyes as related to observed patterns of passive and active movement. In: *Selected Coolwater Fishes of North America* (ed. R.L. Kendall). *American Fisheries Society Special Publication* **11**, 52–7.

Orlova, E.L. (1976) Feeding habits of predatory fishes in the outer delta of the Volga. *Hydrobiological Journal* **12** (2), 41–7.

Orr, J.W. & Ramsey, J.S. (1990) Reproduction in the greenbreast darter, *Etheostoma jordani* (Teleostei: Percidae). *Copeia* **1990**, 100–107.

Oseid, D.M. & Smith, L.L. Jr. (1971) Survival and hatching of walleye eggs at various dissolved oxygen levels. *The Progressive Fish-Culturist* **33** (2), 81–5.

Osse, J.W.M. (1969) Functional morphology of the head of the perch (*Perca fluviatilis* L.): an electromyographic study. *Netherlands Journal of Zoology* **19** (3), 289–392.

Page, L.M. (1974a) The subgenera of *Percina* (Percidae: Etheostomatini). *Copeia* **1974**, 6–86.

Page, L.M. (1974b) The life history of the spottail darter, *Etheostoma squamiceps*, in Big Creek, Illinois, and Ferguson Creek, Kentucky. *Illinois Natural History Survey Biological Notes* **89**, 1–20.

Page, L.M. (1975) The life history of the stripetail darter, *Etheostoma kennicotti*, in Big Creek, Illinois. *Illinois Natural History Survey Biological Notes* **93**, 1–15.

Page, L.M. (1976) The modified midventral scales of *Percina* (Osteichthyes; Percidae). *Journal of Morphology* **148**, 255–64.

Page, L.M. (1977) The lateralis system of darters (Etheostomatini). *Copeia* **1977**, 472–5.

Page, L.M. (1980). The life histories of *Etheostoma olivaceum* and *Etheostoma striatulum*, two species of darters in central Tennessee. *Illinois Natural History Survey Biological Notes* **113**, 1–14.

Page, L.M. (1981) The genera and subgenera of darters (Percidae, Etheostomatini). *Occasional Paper Museum of Natural History University of Kansas* **90**, 1–69.

Page, L.M. (1983) *Handbook of Darters.* T.F.H. Publications, Inc., Neptune City, New Jersey.

Page, L.M. (1985) Evolution of reproductive behaviors in percid fishes. *Bulletin of the Illinois Natural History Survey* **33**, 275–95.

Page, L.M. & Bart, H.L., Jr (1989) Egg-mimics in darters (Pisces: Percidae). *Copeia* **1989**, 514–17.

Page, L.M. & Burr, B.M. (1976) The life history of the slabrock darter, *Etheostoma smithi*, in Ferguson Creek, Kentucky. *Illinois Natural History Survey Biological Notes* **99**, 1–12.

Page, L.M. & Burr, B.M. (1979) The smallest species of darter (Pisces: Percidae). *American Midland Naturalist* **101**, 452–3.

Page, L.M., Keller, W.L. & Cordes, L.E. (1981) *Etheostoma* (*Boleosoma*) *longimanum* and *E.* (*Catontous*) *obeyense*, two more darters confirmed as egg-clusterers. *Transactions of the Kentucky Academy of Science* **42**, 35–6.

Page, L.M. & Mayden, R.L. (1979) Nesting site of the lollypop darter, *Etheostoma neopterum*. *Transactions of the Kentucky Academy of Science* **40**, 56–7.

Page, L.M. & Mayden, R.L. (1981) The life history of the Tennesse sunnose darter. *Etheostoma simoterum* in Brush Creek, Tennessee. *Illinois Natural History Survey Biological Notes* **117**, 1–11.

Page, L.M., Retzer, M.E. & Stiles, R.A. (1982) Spawning behaviour in seven species of darters (Pisces: Percidae). *Brimleyana* **8**, 135–43.

Page, L.M. & Sabaj, M.H. (1994) The function of the caudal keel in *Percina* (Percidae). *Environmental Biology of Fishes* **40**, 105–7.

Page, L.M. & Simon, T.P. (1988) Observations on the reproductive behavior and eggs of four species of darters, with comments on *Etheostoma tippecanoe* and *E. camurum*. *Transactions of the Illinois State Academy of Science* **81**, 205–10.

Page, L.M. & Smith, P.W. (1970) The life history of the dusky darter, *Percina sciera*, in the Embarras River, Illinois. *Illinois Natural History Survey Biological Notes* **69**, 1–15.

Page, L.M. & Smith, P.W. (1971) The life history of the slenderhead darter, *Percina phoxocephala*, in the Embarras River, Illinois. *Illinois Natural History Survey Biological Notes* **74**, 1–14.

Page, L.M. & Swofford, D.L. (1984) Morphological correlates of ecological specialization in darters. *Environmental Biology of Fishes* **11**, 139–59.

Page, L.M. & Whitt, G.S. (1973a) Lactate dehydrogenase isozymes, malate dehydrogenase isozymes and tetrazolium oxidase mobilities of darters (Etheostomatini). *Comparative Biochemistry and Physiology* **44B**, 611–23.

Page, L.M. & Whitt, G.S. (1973b) Lactate dehydrogenase isozymes of darters and the inclusiveness of the genus *Percina*. *Illinois Natural History Survey Biological Notes* **82**, 1–7.

Paine, M.D. (1990) Life history tactics of darters (Percidae: Etheostomatini) and their relationship with body size, reproductive behaviour, latitude and rarity. *Journal of Fish Biology* **37**, 473–88.

Paine, M.D., Dodson, J.J., & Power, G. (1982) Habitat and food resource partitioning among four species of darters Percidae: *Etheostoma* in a southern Ontario stream. *Canadian Journal of Zoology* **60**, 1635–41.

Pankhurst, N.W., Van Der Kraak, G. & Peter, R.E. (1986) Effects of human chorionic gonadotropin, des-Gly10 (D-Ala6)LHRH-ethylamide and pimozide on oocyte final maturation, ovulation, and levels of plasma sex steroids in the walleye (*Stizostedion vitreum*). *Fish Physiology and Biochemistry* **1**, 45–56.

Parsons, J.W. (1967) Contributions of year classes of blue pike to the commercial fishery of Lake Erie, 1943–59. *Journal of the Fisheries Research Board of Canada* **24** (5), 1035–66.

Patzner, R.A. & Glechner, R. (1996) Attaching structures in eggs of native fishes. *Limnologica* **26** (2), 179–82.

Patzner, R.A., Glechner, R. & Riehl, R. (1994) Die eier heimischer Fische 9. Streber, *Zingel streber* Siebold, 1863 (Percidae). *Österreichs Fischerei* **47**, 122–5.

Pauly, D. (1980) On the inter-relationships between natural mortality, growth parameters and mean environmental temperatures in 175 fish stocks. *Journal du Conseil internationale pour l'Exploration de la Mer* **39**, 175–92.

Pavlov, D.S., Lupandin, A.L., Taradina, D.G. & Khara, R. (1996) Diurnal buoyancy dynamics in young cyprinids and percids in relation to their downstream migration in the Volga Delta. *Voprosy Ikhtiologii* **36** (6), 847–50.

Pavlovskiy, S.L. & Sterligova, O.P. (1986) Predation of ruffe, *Gymnocephalus cernuus*, and benthic invertebrates on the eggs of Lake Syam whitefish, *Coregonus lavaretus palassi*. *Journal of Ichthyology* **26** (6), 80–86.

Payan, P., Girard, J.P. & Mayer-Gostan, N. (1984) Branchial ion movements in teleosts: the roles of respiratory and chloride cells. In: *Fish Physiology XB* (eds W.S. Hoar & D.J. Randall), pp. 39–64. Academic Press. Orlando.

Pearson, A.A. (1936) The acoustic-lateral nervous sytem in fishes. *Journal of Comparative Neurology* **64**, 235–73.

Pereira, D.L., Anderson, C.S., Radomski, P.J., Sak, T. & Stevens, A. (1995) Potential to index climate with growth and recruitment of temperate fish. *Canadian Special Publication of Fisheries and Aquatic Sciences* **121**, 709–17.

Persson, L. (1979) The effects of temperature and different food organisms on the rate of gastric evacuation in perch, *Perca fluviatilis*. *Freshwater Biology* **9** (2), 99–104.

Persson, L. (1981) The effects of temperature and meal size on the rate of gastric evacuation in perch (*Perca fluviatilis*) fed on fish larvae. *Freshwater Biology* **11** (2), 131–8.

Persson, L. (1983) Food consumption and competition between age classes in a perch, *Perca fluviatilis*, population in a shallow eutrophic lake. *Oikos* **40** (2), 197–207.

Persson, L. (1986) Effects of reduced interspecific competition on resource utilization in perch (*Perca fluviatilis*). *Ecology* **67** (2), 355–64.

Persson, L. (1987a) The effects of resource availability and distribution on size class interactions in perch, *Perca fluviatilis*. *Oikos* **48** (2), 148–60.

Persson, L. (1987b) Competition-induced switch in young-of-the-year perch, *Perca fluviatilis*: an experimental test of resource limitation. *Environmental Biology of Fishes* **19** (3), 235–9.

Persson, L. (1997) Competition, predation and environmental factors as structuring forces in freshwater fish communities: Sumari (1971) revisited. *Canadian Journal of Fisheries and Aquatic Sciences* **54**, 85–8.

Persson, L., Andersson, J., Wahlstroem, E. & Ekloev, P. (1996) Size-specific interactions in lake systems: predator gape limitation and prey growth rate and mortality. *Ecology* **77** (3), 900–911.

Persson, L. & Greenberg, L.A. (1990a) Juvenile competition bottlenecks: the perch (*Perca fluviatilis*)–roach (*Rutilus rutilus*) interaction. *Ecology* **71** (1), 44–56.

Persson, L. & Greenberg, L.A. (1990b) Interspecific and intraspecific size class competition affecting resource use and growth of perch, *Perca fluviatilis*. *Oikos* **59** (1), 97–106.

Peter, R.E. & Nagahama, Y. (1974) A light and electron microscope study of the structure of the nucleus preopticus and nucleus lateral tuberis of the goldfish, *Carassius auratus*. *Canadian Journal of Zoology* **54**, 1423–37.

Petravicz, J.J. (1936) The breeding habits of the least darter, *Microperca punctulata* Putnam. *Copeia* **1936**, 77–82.

Petravicz, W.P. (1938) The breeding habits of the black-sided darter, *Hadropterus maculatus* Girard. *Copeia* **1938**, 40–44.

Petrosky, B.R. & Magnuson, J.J. (1973) Behavioural responses of northern pike, yellow perch, and bluegill to oxygen concentrations under simulated winterkill conditions. *Copeia* **1973** (1), 124–33.

Petrushevskii, G.K. (ed.) (1963) Parasites and diseases of fish. *Israel Program for Scientific Translations 105*.

Pfeifer, G., Ponyi, J.E. & Nagy, Z. (1979) Pesticide residues in Lake Balaton. In: *Human impacts of life in freshwaters* (eds P. Biró & J. Salanki), pp. 21–6. Akademiai Kiado, Budapest.

Pfleiger, W.L. (1975) *The Fishes of Missouri*. Missouri Department of Conservation.

Pfleiger, W L. (1978) Distribution, status, and life history of the Niangua darter, *Etheostoma nianguae*. *Missouri Department of Conservation Aquatic Series* **16**, 1–25.

Pfleiger, W.L. (1984) Distribution, status, and life history of the bluestripe darter, *Percina cymatotaenia*. *Missouri Department of Conservation Aquatic Series* **18**, 1–22.

Pihu, E. & Maeemets, A. (1982) The management of fisheries in Lake Voertsjaerv. *Hydrobiologia* **86** (1–2), 207–10.

Piironen, J. & Hyvaerinen, H. (1983) Composition of the milt of some teleost fishes. *Journal of Fish Biology* **22** (3), 351–61.

Piller, K.R. & Burr, B.M. (1998) Distribution and population estimates of the federally endangered relict darter, *Etheostoma chienense*, Bayou du Chien, Kentucky. *Journal of the Kentucky Academy of Science* **59**, 64–75.

Piller, K.R. & Burr, B.M. (1999) Reproductive biology and spawning habitat supplementation of the relict darter, *Etheostoma chienense*, a federally endangered species. *Environmental Biology of Fishes* **55**, 145–55.

Pitcher, T.J. (in press) The use of ecosystem modelling in comparative policy analysis: maximising sustainable benefits from Lake Nasser's aquatic resources. In: *Sustainable Fish Production in Lake Nasser: ecological basis and management policy* (ed. J.F. Craig). Proceedings of the workshop, Aswan, Egypt, June 1998. *ICLARM Conference Proceedings*.

Pitlo, J. Jr. (1989) Walleye spawning habitat in Pool 13 of the upper Mississippi River. *North American Journal of Fisheries Management* **9** (3), 303–8.

Pitt, C.E. & Grundman, A.W. (1957) A study into the effects of parasitism on the growth of yellow perch produced by the larvae of *Ligula intestinalis* (Linnaeus, 1758) Gmelin, 1790. *Proceedings of the Helminthological Society of Washington* **24**, 73–80.

Pokrovskii, V.V. (1951) Materialy po issledovaniyu v nutrividovoi izmenchivosti okunya (*Perca fluviatilis* L.). *Trudy Kärelo-Finsk Vsesoyuznyi Nauchna-Issledovakl'shii Institut Ozernogo i Rechnogo Rybnogo Okeanografi Khozyaistva* **3**, 95–149.

Poole, B.C. & Dick, T.A. (1985) Parasite recruitment by stocked walleye, *Stizostedion vitreum vitreum* (Mitchill), fry in a small boreal lake in central Canada. *Journal of Wildlife Diseases* **21** (4), 371–6.

Poole, B.C. & Dick, T.A. (1986) *Raphidascaris acus* (Bloch, 1779) in northern pike, *Esox lucius* L., walleye, *Stizostedion vitreum vitreum* (Mitchill), and yellow perch, *Perca flavescens* (Mitchill), from central Canada. *Journal of Wildlife Diseases* **22** (3), 435–6.

Popova, O.A. & Sytina, L.A. (1977) Food and feeding relations of Eurasian perch (*Perca fluviatilis*) and pikeperch (*Stizostedion lucioperca*) in various waters of the USSR. *Journal of the Fisheries Research Board of Canada* **34** (10), 1559–70.

Porterfield, J.C. (1997) Separation of spawning habitat in the sympatric snubnose darters *Etheostoma flavum* and *E. simoterum* (Teleostei, Percidae). *Transactions of the Kentucky Academy of Science* **58**, 4–8.

Porterfield, J.C. (1998) Spawning behavior of snubnose darters (Percidae) in natural and laboratory environments. *Environmental Biology of Fishes* **53**, 413–19.

Post, J.R. (1988) Size-dependent process in yellow perch recruitment. *Dissertation Abstracts International B Science and Engineering* **49** (1).

Post, J.R. (1990) Metabolic allometry of larval and juvenile yellow perch (*Perca flavescens*): *in situ* estimates and bioenergetic models. *Canadian Journal of Fisheries and Aquatic Sciences* **47** (3), 554–60.

Post, J.R. & Evans, D.O. (1989a) Size-dependent overwinter mortality of young-of-the-year yellow perch (*Perca flavescens*): laboratory, *in situ* enclosure and field experiments. *Canadian Journal of Fisheries and Aquatic Sciences* **46** (11), 1958–68.

Post, J.R. & Evans, D.O. (1989b) Experimental evidence of size-dependent predation mortality in juvenile yellow perch. *Canadian Journal of Zoology* **67** (2), 521–3.

Post, J.R., Johannes, M.R.S. & McQueen, D.J. (1997) Evidence of density-dependent cohort splitting in age-0 yellow perch (*Perca flavescens*): potential behavioural mechanisms and population-level consequences. *Canadian Journal of Fisheries and Aquatic Sciences* **54** (4), 867–75.

Post, J.R. & McQueen, D.J. (1987) The impact of planktivorous fish on the structure of a plankton community. *Freshwater Biology* **17** (1), 79–89.

Post, J.R. & McQueen, D.J. (1988) Ontogenetic changes in the distribution of larval and juvenile yellow perch (*Perca flavescens*): a response to prey or predators? *Canadian Journal of Fisheries and Aquatic Sciences* **45**, 1820–26.

Post, J.R. & McQueen, D.J. (1994) Variability in first-year growth of yellow perch (*Perca flavescens*): predictions from a simple model, observations, and an experiment. *Canadian Journal of Fisheries and Aquatic Sciences* **51**, 2501–12.

Post, J.R. & Prankevicius, A.B. (1987) Size-selective mortality in young-of-the-year yellow perch (*Perca flavescens*): evidence from otolith microstructure. *Canadian Journal of Fisheries and Aquatic Sciences* **44** (11), 1840–47.

Post, J.R., Vandenbos, R. & McQueen, D.J. (1996) Uptake rates of food-chain and water-borne mercury by fish: field measurements, a mechanistic model, and an assessment of uncertainties. *Canadian Journal of Fisheries and Aquatic Sciences* **53** (2), 395–407.

Pratt, D.M., Blust, W.H. & Selgeby, J.H. (1992) Ruffe, *Gymnocephalus cernuus*: newly introduced in North America. *Canadian Journal of Fisheries and Aquatic Sciences* **49**, 1616–18.

Priegel, G.R. (1964) Early scale development in the walleye. *Transactions of the American Fisheries Society* **93** (2), 199–200.

Priegel, G.R. (1970) Reproduction and early life history of the walleye in the Lake Winnebago region. *Wisconsin Department of Natural Resources Technical Bulletin* **45**. Madison, Wisconsin.

Proteau, J.P., Schlumberger, O. & Albiges, C. (1993) Sandre: des efforts encore sur reproduction et élevage larvaire. *Aqua Revue* **47**, 23–6.

Puke, C. (1952) Pikeperch studies in Lake Vanern. *Report, Institute of Freshwater Research, Drottningholm* **33**, 168–78.

Purdom, C.E. (1983) Genetic engineering by the manipulation of chromosomes. *Aquaculture* **33**, 287–300.

Pycha, R.L. & Smith, L.L., Jr. (1955) Early life history of the yellow perch, *Perca flavescens*. (Mitchill) in the Red Lakes, Minnesota. *Transactions of the American Fisheries Society* **84**, 249–60.

Qin, J. & Culver, D.A. (1992) The survival and growth of larval walleye, *Stizostedion vitreum*, and trophic dynamics in fertilized ponds. *Aquaculture* **108**, 257–76.

Qin, J., Culver, D. A. & Yu, N. (1994) Comparative growth of larval walleye and saugeye, and their impact on zooplankton in experimental ponds. *The Progressive Fish-Culturist* **56**, 91–9.

Qin, J., Madon, S.P. & Culver, D.A. (1995) Effect of larval walleye (*Stizostedion vitreum*) and fertilization on the plankton community: implications for larval fish culture. *Aquaculture* **130**, 51–65.

Ráb, P., Roth, P. & Mayr, B. (1987) Karyotype study of eight species of European percid fishes (Pisces, Percidae). *Caryologia* **40** (4), 307–18.

Radke, R.J. & Eckmann, R. (1999) First-year overwinter mortality in Eurasian perch (*Perca fluviatilis* L.). Results from a field study and a simulation experiment. *Ecology of Freshwater Fish* **8** (2), 94–101.

Rahel, F.J. (1983) Population differences in acid tolerance between yellow perch, *Perca flavescens*, from naturally acidic and alkaline lakes. *Canadian Journal of Zoology* **61** (1), 147–52.

Rahkonen, R. (1994) Parasites of pike perch *Stizostedion lucioperca* (Linnaeus 1758) fry reared in two different types of natural food ponds in southern Finland. *Aquaculture* **122** (4), 279–93.

Raitaniemi, J. & Rask, M. (1990) Preliminary observations on the effects of liming to the fish populations of small acidic lakes in southern Finland. *Aqua Fennici* **20** (1), 115–23.

Ramsey, J.S. & Suttkus, R.D. (1965) *Etheostoma ditrema*, a new darter of the subgenus *Oligocephalus* (Percidae) from springs of the Alabama River basin in Alabama and Georgia. *Tulane Studies in Zoology* **12**, 65–77.

Ramseyer, L.J. (1992) *The optimal protein: metabolizable energy ratio and total protein requirement for yellow perch in a single-pass, flow-through system*. M.Sc. Thesis, Michigan State University, East Lansing, Michigan, USA.

Randall, D. & Daxboeck, C. (1984) Oxygen and carbon dioxide transfer across fish gills. In: *Fish Physiology XA* (eds W.S. Hoar & D.J. Randall), pp. 263–314. Academic Press, Orlando.

Raney, E.C. & Lachner, E.A. (1939) Observations on the life history of the spotted darter, *Poecilichthys maculatus* (Kirtland). *Copeia* **1939**, 157–65.

Raney, E.C. & Lachner, E.A. (1942) Studies of the summer food, growth, and movements of young yellow pike-perch, *Stizostedion v. vitreum*, in Oneida Lake, New York. *Journal of Wildlife Management* **6**, 1–16.

Raney, E.C. & Suttkus, R.D. (1966) *Etheostoma rubrum*, a new percid fish of the subgenus *Nothonotus* from Bayon Pierre, Mississippi. *Tulane Studies in Zoology* **13**, 95–102.

Rankin, J.C. & Bolis, L. (1984) Hormonal control of water movement across the gills. In:

Fish Physiology XB (eds W.S. Hoar & D.J. Randall), pp. 177–201. Academic Press, Orlando.

Rao, G.M.M. (1968) Oxygen consumption of rainbow trout (*Salmo gairdneri*) in relation to activity and salinity. *Canadian Journal of Zoology* **46**, 781–5.

Rask, M. (1984) The effect of low pH on perch, *Perca fluviatilis* L. 3. The perch population in a small, acidic, extremely humic forest lake. *Annales Zoologici Fennici* **21** (1), 15–22.

Rask, M., Heinaenen, A., Salonen, K., Arvola, L., Bergstroem, I., Liukkonen, M. & Ojala, A. (1986) The limnology of a small, naturally acidic, highly humic forest lake. *Archiv für Hydrobiologie* **106** (3), 351–71.

Rask, M. & Hiisivuori, C. (1985) The predation on *Asellus aquaticus* (L.) by perch, *Perca fluviatilis* (L.), in a small forest lake. *Hydrobiologia* **121** (1), 27–33.

Rask, M., Mannio, J., Forsius, M., Posch, M. & Vuorinen, P.J. (1995a) How many fish populations in Finland are affected by acid precipitation? *Environmental Biology of Fishes* **42** (1), 51–63.

Rask, M. & Raitaniemi, J. (1988) The growth of perch, *Perca fluviatilis* L., in recently acidified lakes of southern Finland – a comparison with unaffected waters. *Archiv für Hydrobiologie* **112**, (3), 387–97.

Rask, M., Raitaniemi, J., Mannio, J., Vuorenmaa, J. & Nyberg, K. (1995b) Losses and recoveries of fish populations in acidified lakes of southern Finland in the last decade. In: *Acid Reign '95? Proceedings of the Fifth International Conference on Acidic Deposition: Science and Policy* (eds P. Grennfelt, H. Rodhe, E. Thoerneloef & J. Wisniewski). *Water, Air and Soil Pollution* **85** (2), 315–20.

Rask, M. & Virtanen, E. (1986) Responses of perch, *Perca fluviatilis* L., from an acidic and a neutral lake to acidic water. In: *Acidic Precipitation. Proceedings of the International Symposium on Acidic Precipitation* (ed. H.C. Martin). *Water, Air and Soil Pollution* **30** (3–4), 537–44.

Rask, M., Vuorinen, P.J. & Vuorinen, M. (1990) Delayed spawning of the perch, *Perca fluviatilis* L., in acidified lakes. *Journal of Fish Biology* **36** (3), 317–25.

Rawson, D.S. (1946) Successful introduction of fish in a large saline lake. *Canadian Fish Culturist* **1** (1), 5–8.

Rawson, D.S. (1957) The life history and ecology of the yellow walleye, *Stizostedion vitreum*, in Lac la Ronge, Saskatchewan. *Transactions of the American Fisheries Society* **86**, 15–37.

Reed, R.J. (1968) Mark and recapture studies of eight species of darters (Pisces: Percidae) in three streams of northwestern Pennsylvania. *Copeia* **1968**, 172–5.

Reeves, C.D. (1907) The breeding habits of the rainbow darter (*Etheostoma coeruleum* Storer), a study in sexual selection. *Biological Bulletin (Woods Hole)* **14**, 35–59.

Regier, H.A., Applegate, V.C. & Ryder, R.A. (1969) The ecology and management of the walleye in western Lake Erie. *Technical Report Great Lakes Fisheries Commission* **15**.

Regier, H.A. & Hartman, W.L. (1973) Lake Erie's fish community: 150 years of cultural stresses. *Science* **180**, 1248–55.

Reighard, J. (1890) The development of the walleyed pike *Stizostedion vitreum*. A popular introduction to the development of bony fishes. *Michigan Fisheries Commission Bulletin* **1**.

Reyes, M.R.-delos, Arzbach, H.H. & Braum, E. (1992) *In situ* development of perch eggs, *Perca fluviatilis* L. (Pisces, Percidae) in a small eutrophic lake, Lake Plussee, Holstein, Germany. *Internationale Revue der Gesamten Hydrobiologie* **77** (3), 467–81.

Reynolds, W.W. & Casterline, M.E. (1979) Behavioural thermoregulation and locomotor activity of *Perca flavescens*. *Canadian Journal of Zoology* **57**, 2239–42.

Richter, R., Scheinert, P. & Hoffman, R.W. (1991) Food quality and development of digestive tract on larvae of perch (*Perca fluviatilis*). *Fischerei Forschung* **29** (3), 40–41.

Ricker, W.E. (1954) Stock and recruitment. *Journal of the Fisheries Research Board of Canada* **11**, 559–623.

Ricker, W.E. (1958) Handbook of computations for biological statistics of fish populations. *Bulletin of the Fisheries Research Board of Canada* **119**.

Ricker, W.E. (1975) Computation and interpretation of biological statistics of fish populations. *Bulletin of the Fisheries Research Board of Canada* **191.**

Roberts, N.J. & Winn, H.E. (1962) Utilization of the senses in feeding behavior of the johnny darter, *Etheostoma nigrum*. *Copeia* **1962**, 567–70.

Robillard, S.R. & Marsden, J.E. (1996) Comparison of otolith and scale ages for yellow perch from Lake Michigan. *Journal of Great Lakes Research* **22** (2), 429–35.

Romer, A.S. (1966) *Vertebrate Paleontology*, 3rd edn. University of Chicago Press, Chicago.

Romuk-Wodoracki, D. (1990) *Henneguya wolinensis* n. sp. (Myxosporea) from perch *Perca fluviatilis* L. from Szczecin Firth, Poland. *Acta Ichthyologica et Piscatoria* **20** (2), 91–6.

Ronneberger, D., Kasprzak, P. & Krienitz, L. (1993) Long-term changes in the rotifer fauna after biomanipulation in Haussee (Feldberg, Germany, Mecklenberg-Vorpommern) and its relationship to the crustacean and phytoplankton communities. *Rotifer Symposium VI* (eds J.J. Gilbert, E. Lubzens & M.R. Miracle). *Hydrobiologia* **255–256**, 297–304.

Rösch, R., Kangur, A., Kangur, K., Krämer, A., Ráb, R., Schlechta, V., Tapaninen, M. & Treasurer, J. (1996) Minutes of the Percis II working group. Ruffe (*Gymnocephalus cernuus*). *Annales Zoologici Fennici* **33**, 305–8.

Roseman, E.F., Mills, E.L., Forney, J.L. & Rudstam, L.G. (1996) Evaluation of competition between age-0 yellow perch (*Perca flavescens*) and gizzard shad (*Dorosoma cepedianum*) in Oneida Lake, New York. *Canadian Journal of Fisheries and Aquatic Sciences* **53** (4), 865–74.

Ross, J., Powles, P.M. & Berrill, M. (1977) Thermal selection and related behavior in larval yellow perch (*Perca flavescens*). *Canadian Field Naturalist* **91** (4), 406–10.

Ross, M.J. & Siniff, D.B. (1980) Spatial distribution and temperature selection of fish near the thermal outfall of a power plant during fall, winter and spring. *U.S. Environmental Protection Agency Ecological Research Series EPA-600/3-80-009*, (PB8O-148703).

Ross, M.J. & Siniff, D.B. (1982) Temperatures selected in a power plant thermal effluent by adult yellow perch (*Perca flavescens*) in winter. *Canadian Journal of Fisheries and Aquatic Sciences* **39** (2), 346–9.

Ross, S.T. & Wilkins, S.D. (1993) Reproductive behavior and larval characteristics of the threatened bayou darter (*Etheostoma rubrum*) in Mississippi. *Copeia* **1993**, 1127–32.

Rottiers, D.V. & Lemm, C.A. (1985) Movement of underyearling walleyes, *Stizostedion vitreum*, in response to odour and visual cues. *The Progressive Fish-Culturist* **47** (1), 34–41.

Runn, P., Johansson, N. & Milbrink, G. (1977) Some effects of low pH on the hatchability of eggs of perch, *Perca fluviatilis* L. *Zoon* **5** (2), 115–25.

Ruple, D.L., McMichael, R.H., Jr & Baker, J.A. (1984) Life history of the gulf darter, *Etheostoma swaini* (Pisces: Percidae). *Environmental Biology of Fishes* **11**, 121–30.

Ruuhijärvi, J & Hyvärinen, P. (1996) The status of pike-perch culture in Finland. *Journal of Applied Ichthyology* **12** (3–4), 185–8.

Ruuhijärvi, J, Virtanen, E., Salminen, M. & Myunda, M. (1991) The growth and survival of pikeperch *Stizostedion lucioperca* L. larvae fed on formulated feeds. In: *Larvi'91* (eds P. Lavens, P. Sorgeloos, E. Jaspers & F. Ollevier). Fish and Crustacean Larviculture Symposium, Gent, Belgium. *European Aquaculture Society Special Publication* **15**, 154–6.

Ryan, P.M. & Harvey, H.H. (1980) Growth responses of yellow perch, *Perca flavescens* (Mitchill), to lake acidification in the La Cloche Mountain lakes of Ontario. *Environmental Biology of Fishes* **5** (2), 97–108.

Ryder, R.A. (1968) Dynamics and exploitation of mature walleyes, *Stizostedion vitreum vitreum*, in the Nipigon Bay region of Lake Superior. *Journal of the Fisheries Research Board of Canada* **25** (7), 1347–76.

Ryder, R.A. (1977) Effects of ambient light variation on behaviour of yearling, subadult, and adult walleyes (*Stizostedion vitreum vitreum*). *Journal of the Fisheries Research Board of Canada* **34** (10), 1481–91.

Ryon, M.G. (1986) The life history and ecology of *Etheostoma trisella* (Pisces: Percidae). *American Midland Naturalist* **115**, 73–86.

Sachs, T.R. (1878) Transportation of live pikeperch *Lucioperca zanda* in Germany called zander. *Land and Water* **25**, 476.

Saerkkae, J., Jaemsae, A. & Luukko, A. (1995) Chernobyl-derived radiocaesium in fish as dependent on water quality and lake morphometry. *Journal of Fish Biology* **46** (2), 227–40.

Saerkkae, J., Keskitalo, A. & Luukko, A. (1996) Temporal changes in concentration of radiocaesium in lake sediment and fish of southern Finland as related to environmental factors. *Science of the Total Environment* **191** (1–2), 125–36.

Sági, K. (1974). A Balaton-vidék történelme és régeszeti emléki. In: *Balaton Monográfia* (ed. K. Toth), pp. 337–56. Panorama, Budapest.

Sakovich, I.G. (1986) Pike-perch, *Stizostedion lucioperca* from the Amur River. *Journal of Ichthyology* **25** (5), 161–2.

Saksgaard, R. & Hesthagen, T. (1995) Differences in response to liming in a lake-dwelling fish community. In: *Acid Reign '95? Proceedings of the Fifth International Conference on Acidic Deposition: Science and Policy* (eds P. Grennfelt, H. Rodhe, E. Thoerneloef & E. Wisniewski). *Water, Air and Soil Pollution* **85** (2), 973–8.

Salminen, M. & Ruuhijärvi, J. (1991) Production of pikeperch (*Stizostedion lucioperca*) fry: procedures and devices. In: *Larvi '91.* (eds P. Lavens, P. Sorgeloos, E. Jaspers & F. Ollevier). Fish and Crustacean Larviculture Symposium, Gent, Belgium. *European Aquaculture Society Special Publication* **15**, 287–9.

Salminen, M., Ruuhijärvi, J. & Ahlfors, P. (1993) Spawning of wild zander (*Stizostedion lucioperca*) in cages. In: *Fish Reproduction '92* (eds Z. Adamek & M. Flajshans), pp. 42–3. Research Institute of Fishculture and Hydrobiology, Vodnany, Czech Republic.

Samokhvalova, L.K. (1982) Osobennosti pitaniya sudaka v kurshskom zalive Baltijskogo Morya v osennij period. In: *Pitanie i pishchevye otnosheniya ryb i bespozvonochnykh Atlantischeskogo Okeana,* pp. 67–74. Atlantniro, Kaliningrad.

Sandström, O. (1983) Seasonal variations in the swimming performance of perch (*Perca fluviatilis* L.) measured with the rotatory-flow technique. *Canadian Journal of Zoology* **61** (7), 1475–80.

Satterfield, J.R. Jr & Flickinger, S.A. (1995) Factors influencing storage potential of preserved walleye semen. *The Progressive Fish-Culturist* **57**, 175–81.

Scalet, C.G. (1973) Reproduction of the orangebelly darter, *Etheostoma radiosum cyanorum* (Osteichthyes: Percidae). *American Midland Naturalist* **89**, 156–65.

Schaefer, M.B. (1954). Some aspects of the dynamics of populations important to the management of the commercial marine fisheries. *Bulletin of the Inter-American Tropical Tuna Commission* **1** (2), 27–56.

Schaefer, M.B. (1957) A study of the dynamics of the fishery for yellowfin tuna in the eastern tropical Pacific Ocean. *Bulletin of the Inter-American Tropical Tuna Commission* **2**, 247–68.

Schaeffer, J., Thomas, M. & Haas, R. (1994) Importance of *Hexagenia* in food webs of mesotrophic Great Lakes waters: evidence from recent and historical investigations. In: *Program and Abstracts, 37th Conference of the International Association of Great Lakes Research and Estuarine Research Federation*, p. 166. IAGLR, Buffalo, New York.

Schaeffer, J.S. & Margraf, F.J. (1987) Predation on fish eggs by white perch, *Morone americana*, in western Lake Erie. *Environmental Biology of Fishes* **18** (1), 77–80.

Schael, D.M., Rudstam, L.G. & Post, J.R. (1991) Gape limitations and prey selection in larval yellow perch (*Perca flavescens*), freshwater drum (*Aplodinotus grunniens*), and black crappie (*Pomoxis nigromaculatus*). *Canadian Journal of Fisheries and Aquatic Sciences* **48**, 1919–25.

Schäperclauss, W. (1961) *Lehrbuch der Teichwirtschaft*. Paul Parey, Berlin, Hamburg.

Schat, K.A. & Carlisle, J.C. (1980) Susceptibility of walleye fry to infection by IHN and IPN viruses. *Abstracts of the Joint Fourth Biennial Fish Health Section and Fish Disease Workshops*, p. 26, Seattle, Washington.

Schenck, J.R. & Whiteside, B.G. (1976) Distribution, habitat preference and population size estimate of *Etheostoma fonticola*. *Copeia* **1976**, 697–703.

Schenck, J.R. & Whiteside, B.G. (1977a) Food habits and feeding behavior of the fountain darter, *Etheostoma fonticola* (Osteichthyes: Percidae). *Southwestern Naturalist* **21**, 487–92.

Schenck, J.R. & Whiteside, B.G. (1977b) Reproduction, fecundity, sexual dimorphism, and sex ratio of *Etheostoma fonticola* (Osteichthyes: Percidae). *American Midland Naturalist* **98** (2), 365–75.

Scherer, E. (1971) Effects of oxygen depletion and of carbon dioxide buildup on the photic behaviour of the walleye (*Stizostedion vitreum vitreum*). *Journal of the Fisheries Research Board of Canada* **28** (9), 1303–7.

Scherer, E. (1976) Overhead-light intensity and vertical positioning of the walleye, *Stizostedion vitreum vitreum*. *Journal of the Fisheries Research Board of Canada* **33** (2), 289–92.

Schlosser, I.J. & Toth, L.A. (1984) Niche relationships and population ecology of rainbow (*Etheostoma caeruleum*) and fantail (*E. flabellare*) darters in a temporally variable environment. *Oikos* **42**, 229–38.

Schlueter, L.R. (1989) Simple removal of sagitta otoliths from walleyes. *North American Journal of Fisheries Management* **9** (1), 123.

Schlumberger, O. & Proteau, J.P. (1991) Production de juvéniles de sandre (*Stizostedion lucioperca*). *Aqua Revue* **36**, 25–8.

Schlumberger, O. & Proteau, J.P. (1996) Reproduction of pike-perch (*Stizostedion lucioperca*) in captivity. *Journal of Applied Ichthyology* **12** (3–4), 149–52.

Schlumpberger, W. & Schmidt, K. (1979) *Untersuchungen zur Entwicklung eines industriemäbigen verfahrens für die Prodution von vorgestrecken Zandern* (Stizostedion lucioperca L.). Dissertation, Humbold University, Berlin, Germany.

Schmitt, D.N. & Hubert, W.A. (1982) Comparison of cleithra and scales for age and growth analysis of yellow perch. *The Progressive Fish-Culturist* **44** (2), 87–8.

Schoener, T.W. (1971) Theory of feeding strategies. *Annual Review of Ecology and Systematics* **2**, 369–404.

Scholz, T. (1986) Observations on the ecology of five species of intestinal helminths in perch (*Perca fluviatilis*) from the Macha Lake fishpond system, Czechoslovakia. *Vestnik Ceskoslovenske Spolecnosti Zoologicke* **50** (4), 300–320.

Schott, E.F., Kayes, T.B. & Calbert, H.E. (1978) Comparative growth of male versus female yellow perch fingerlings under controlled environmental conditions. In: *Selected Coolwater Fishes of North America* (ed. R.L. Kendall). *American Fisheries Society Special Publication* **11**, 181–6.

Schwartz, F.J. (1965) Densities and ecology of the darters of the upper Allegheny River watershed. In: *Studies on the Aquatic Ecology of the Upper Ohio River System* (ed. C.A. Tryon Jr, R.T. Hartman & K.W. Cummins), pp. 95–103. *Pymatuning Laboratory of Ecology Special Publication* **3**. Pymatuning, Pennsylvania.

Schweigert, J.F., Ward, F.J. & Clayton, J.W. (1977) Effects of fry and fingerling introductions on walleye (*Stizostedion vitreum vitreum*) production in West Blue Lake, Manitoba. *Journal of the Fisheries Research Board of Canada* **34** (11), 2142–50.

Scott, D.B.C. (1979) Environmental timing and the control of reproduction in teleost fish. In: *Fish Phenology* (ed. P.J. Miller), pp. 105–32. *Symposia of the Zoological Society of London* **44**. Academic Press, London.

Scott, D.C. (1955) Activity patterns of perch, *Perca flavescens*, in Rondeau Bay of Lake Erie. *Ecology* **36** (2), 320–27.

Scott, W.B. & Crossman, E.J. (1973) Freshwater fishes of Canada. *Bulletin of the Fisheries Research Board of Canada* **184**.

Seesock, W.E., Ramsey, J.S. & Seesock, F.L. (1978) Life and limitation of the coldwater darter (*Etheostoma ditrema*) in Glencoe Spring, Alabama. *ASB (Association of Southeastern Biologists) Bulletin* **25**, 26.

Sephton, D.H. & Driedzic, W.R. (1991) Effect of acute and chronic temperature transition on enzymes of cardiac metabolism in white perch (*Morone americana*), yellow perch (*Perca flavescens*), and smallmouth bass (*Micropterus dolomieui*). *Canadian Journal of Zoology* **69**, 258–62.

Serns, S.L. (1978) Effects of a minimum size limit on the walleye population of a northern Wisconsin lake. In: *Selected Coolwater Fishes of North America* (ed. R.L. Kendall). *American Fisheries Society Special Publication* **11**, 390–97.

Serns, S.L. (1986) Cohort analysis as an indication of walleye year-class strength in Escanaba Lake, Wisconsin, 1956–74. *Transactions of the American Fisheries Society* **115** (6), 849–52.

Serns, S.L. & Kempinger, J.J. (1981) Relationship of angler exploitation to the size, age and sex of walleyes in Escanaba Lake, Wisconsin. *Transactions of the American Fisheries Society* **110** (2), 216–20.

Shirota, A. (1970) Studies on the mouth size of fish larvae. *Bulletin of the Japanese Society of Scientific Fisheries* **36**, 353–68.

Shuter, B.J. & Koonce, J.F. (1977) A dynamic model of the western Lake Erie walleye (*Stizostedion vitreum vitreum*) population. *Journal of the Fisheries Research Board of Canada* **34** (10), 1972–82.

Shuter, B.J., Koonce, J.F. & Regier, H.A. (1979) Modelling the western Lake Erie walleye population: a feasibility study. *Technical Report of the Great Lakes Fishery Commission* **32**.

Shuter, B.J. & Post, J.R. (1990) Climate, population viability, and the zoogeography of temperate fishes. *Transactions of the American Fisheries Society* **119**, 314–36.

Siefert, R.E. (1972) First food of larval yellow perch, white sucker, bluegill, emerald shiner, and rainbow smelt. *Transactions of the American Fisheries Society* **101** (2), 219–25.

Siefert, R.E. & Spoor, W.A. (1974) Effects of reduced oxygen on embryos and larvae of the white sucker, coho salmon, brook trout and walleye. In: *The Early Life History of Fish* (ed. J.H.S. Blaxter), pp. 487–95. Springer-Verlag, New York.

Simon, T.P. (1991) Startle response and causes of burying behavior in captive eastern sand darters, *Ammocrypta pellucida* (Putnam). *Proceedings of the Indiana Academy of Science* **100**, 155–60.

Simon, T.P. (1992) Description of eggs, larvae, and early juveniles of the bigscale logperch, *Percina macrolepida* (Girard), from the West Fork of the Trinity River basin, Texas. *Southwestern Naturalist* **37**, 28–34.

Simon, T.P. & Faber, D.J. (1987) Descriptions of eggs, larvae, and early juveniles of the Iowa darter, *Etheostoma exile* Stevenson, from Lac Heney, Quebec. *Canadian Journal of Zoology* **65**, 1264–9.

Simon, T.P. & Garcia, N.J. (1990) Descriptions of eggs, larvae, and early juveniles of the stippled darter, *Etheostoma punctulatum* (Agassiz), from a tributary of the Osage River, Missouri. *Southwestern Naturalist* **35**, 123–9.

Simon, T.P. & Vondruska, J.T. (1991) Larval identification of the ruffe, *Gymnocephalus cernuus* (Linnaeus) (Percidae: Percini) in the St. Louis River Estuary, Lake Superior drainage basin, Minnesota. *Canadian Journal of Zoology* **69**, 436–42.

Simons, A.M. (1991) Phylogenetic relationships of the crystal darter, *Crystallaria asprella* (Teleostei: Percidae). *Copeia* **1991**, 927–36.

Simons, A.M. (1992) Phylogenetic relationships of the *Boleosoma* species group (Percidae: *Etheostoma*). In: *Historical Ecology, Systematics, and North American Freshwater Fishes* (ed. R. L. Mayden), pp. 268–92. Stanford University Press.

Sjoebeck, M.-L., Haux, C., Larsson, A. & Lithner, G. (1984) Biochemical and haematological studies on perch, *Perca fluviatilis*, from the cadmium-contaminated River Emaan. *Ecotoxicology and Environmental Safety* **8** (3), 303–12.

Skaala, O., Dahle, G., Joerstad, K.E. & Naevdal, G. (1990) Interactions between natural and farmed fish populations: information from genetic markers. *Journal of Fish Biology* **36** (3), 449–60.

Skorping, A. (1980) Population biology of the nematode, *Camallanus lacustris*, in perch, *Perca fluviatilis* L., from an oligotrophic lake in Norway. *Journal of Fish Biology* **16** (5), 483–92.

Slechtová, V. & Slechta, V. (1996) *Gymnocephalus baloni* – species validity and the position within the genus *Gymnocephalus* (Pisces, Percidae): evidence from lactate dehydrogenase patterns. *Acta Universitatis Carolinae, Biologica* **39** (3–4), 243–50.

Smith, C.L. (1954) Pleistocene fishes of the Berends fauna of Beaver County, Oklahoma. *Copeia* **1954** (4), 282–9.

Smith, H.M. (1892) Report on an investigation of the fisheries of Lake Ontario. *Bulletin of the United States Fisheries Commission* **10** (1890), 177–215.

Smith, I.R., Johnson, A.F., MacLennan, D. & Manson, H. (1992) Chemical contaminants, lymphocystis, and dermal sarcoma in walleyes spawning in the Thames River, Ontario. *Transactions of the American Fisheries Society* **121** (5), 608–16.

Smith, L.L. & Moyle, J.B. (1943) Factors influencing production of yellow pike-perch, *Stizostedion vitreum vitreum*, in Minnesota rearing ponds. *Transactions of the American Fisheries Society* **73**, 243–61.

Smith, L.L., Jr, Broderius, S.J., Oseid, D.M., Kimball, G.L. & Koenst, W.M. (1978) Acute toxicity of hydrogen cyanide to freshwater fishes. *Archives of Environmental Contamination and Toxicology* **7** (3), 325–37.

Smith, L.L. Jr, Kramer, R.H. & Oseid, D.M. (1966) Long-term effects of conifer groundwood paper fiber on walleyes. *Transactions of the American Fisheries Society* **95** (1), 60–70.

Smith, L.L., Jr, & Koenst, J. (1975). Temperature effects on eggs and fry of percoid fishes. *United States Environmental Protection Agency Project 18050 PAB.*

Smith, L.L., Jr & Moyle, J.B. (1943) Factors influencing production of yellow pike-perch *Stizostedion vitreum vitreum*, in Minnesota rearing ponds. *Transactions of the American Fisheries Society* **73**, 243–61.

Smith, O.H. & Van Oosten, J. (1939) Tagging experiments with lake trout, whitefish and other species of fish from Lake Michigan. *Transactions of the American Fisheries Society* **69**, 63–84.

Smyly, W.J.P. (1952) Observations on the food of the fry of perch (*Perca fluviatitis* L.) in Windermere. *Proceedings of the Zoological Society of London* **122**, 407–16.

Solomon, D.J. & Brafield, A.E. (1972) The energetics of feeding, metabolism and growth of perch (*Perca fluviatilis* L.). *Journal of Animal Ecology* **41**, 699–719.

Song, C.B. (1995) Genetic divergence and speciation of Eurasian and American yellow perch based on the nucleotide sequence of cytchrome *b* gene. *Journal of the Korean Fisheries Society* **28** (6), 699–707.

Song, C.B., Near, T.J. & Page, L.M. (1998) Phylogenetic relation among percid fishes as inferred from mitochondrial cytochrome *b* DNA sequence data. *Molecular Phylogenetics and Evolution* **10** (3), 343–53.

Šorić, V. & Naumovski, M. (1991) The species of the genus *Zingel* in the waters of Yugoslavia. *Ichthyologia* **23** (1), 59–72.

Spanovskaya, V.D. & Grygorash, V.A. (1977) Development and food of age-0 Eurasian perch (*Perca fluviatilis*) in reservoirs near Moscow, USSR. *Journal of the Fisheries Research Board of Canada* **34**, 1551–8.

Speare, E.P. (1965) Fecundity and egg survival of the central johnny darter (*Etheostoma nigrum nigrum*) in southern Michigan. *Copeia* **1965**, 308–14.

Sprenger, M.D., McIntosh, A.W. & Hoenig, S. (1988) Concentrations of trace elements in yellow perch (*Perca flavescens*) from six acidic lakes. *Water, Air and Soil Pollution* **37** (3–4), 375–88.

Stacey, N.E. (1984) Control of the timing of ovulation by exogenous and endogenous factors. In: *Fish Reproduction: Strategies and Tactics* (eds G.W. Potts & R.J. Wooton), pp. 207–22. Academic Press, London.

Stanescu, G. (1971) *Romanichthys vansanicola*, Pisces, Percidae, and its distribution in Romania and the causes of its extinction. *Vestnik Ceskoslovenske Spolecnosti Zoologicke* **35** (2), 132–5.

Starnes, W.C. (1977) The ecology and life history of the endangered snail darter, *Percina* (*Imostoma*) *tanasi* Etnier. *Tennessee Wildlife Research Agency Technical Report* **77–52.**

Steffens, W. (1960) Ernahrung und Wachstum des jungen Zanders (*Lucioperca lucioperca* (L.)) in Teichen. *Zeitschrift für Fischerei* **9** (3–4), 161–272.

Steffens, W., Geldhauser, F., Gerstner, P. & Hilge, V. (1996) German experiences in the propagation and rearing of fingerling pikeperch (*Stizostedion lucioperca*). *Annales Zoologici Fennici* **33**, 627–34.

Steinberg, R., Page, L.M. & Porterfield, J.C. (in press) The spawning behavior of the harlequin darter, *Etheostoma histrio* (Osteichthyes: Percidae). *Ichthyological Exploration of Freshwaters.*

Steindachner, F. (1878) Ichthyologische Beitrage (VII). X. *Perca fluviatilis,* Lin. Sitzungsber. *Akademie Wissenschaften Wien* **78** (1), 398.

Sterba, G. (1962) *Freshwater Fishes of the World.* Vista Books, London.

Sterns, S.L. (1981) Occurrence of accessory checks on the scales of walleye fingerlings stocked in mid-August. *The Progressive Fish-Culturist* **43** (1), 46–7.

Stevens, C. (1996) Cage culture of walleye and its hybrids to food size. In: *Walleye Culture Manual* (ed. R.C. Summerfelt), pp. 273–4. *NCRAC Culture Series* **101**. North Central Regional Aquaculture Center Publications Office, Iowa State University, Ames.

Stiles, R.A. (1972) *The comparative ecology of three species of* Nothonotus (*Percidae – Etheostoma*) in *Tennessee's Little River*. Ph.D. Thesis, University of Tennessee, USA.

Strawn, K. (1956) A method of breeding and raising three Texas darters. Part II. *Aquarium Journal* **27**, 12–14, 17, 31–2.

Strawn, K. & Hubbs, C. (1956) Observations on stripping small fishes for experimental purposes. *Copeia* **1956**, 114–16.

Stripp, R.A., Heit, M., Bogen, D.C., Bidanset, J. & Trombetta, L. (1990) Trace element accumulation in the tissues of fish from lakes with different pH values. *Water, Soil and Air Pollution* **51** (1–2), 75–87.

Sulistyo, I. (1998) *Contribution à l'étude et à la maîtrise de la reproduction de la perche eurasienne* Perca fluviatilis. Ph.D. Thesis, Université H. Poincaré, Nancy, France.

Sulistyo, I., Rinchard, J., Fontaine, P., Gardeur, J.N., Capdeville, B. & Kestemont, P. (1998) Reproductive cycle and plasma levels of sex steroids in female Eurasian perch, *Perca fluviatilis*. *Aquatic Living Resources* **11** (2), 101–10.

Sullivan, K.M. (1986) Physiology of feeding and starvation tolerance in overwintering freshwater fishes. In: *Gutshop '84* (eds C.A. Simonstad & G.M. Cailliet) *Contemporary Studies on Fish Feeding* **7**, 259–68.

Summerfelt, R.C. (1996) Intensive culture of walleye fry. In: *Walleye Culture Manual* (ed. R.C. Summerfelt), pp. 161–85. *NCRAC Culture Series* **101**. North Central Regional Aquaculture Center Publications Office, Iowa State University, Ames.

Summerfelt, S.T., Holland, K.H., Hankins, J.A. & Durant, M.D. (1995) A hydroacoustical waste feed controller for tank systems. *Water Science and Technology* **31**, 123–9.

Summerfelt, S.T. & Summerfelt, R.C. (1996) Aquaculture of walleye as food fish. In: *Walleye Culture Manual* (ed. R.C. Summerfelt), pp. 215–30. *NCRAC Culture Series* **101**. North Central Regional Aquaculture Center Publications Office, Iowa State University, Ames.

Suns, K. & Hitchin, G. (1990) Interrelationships between mercury levels in yearling yellow perch, fish condition and water quality. *Water, Air and Soil Pollution* **50** (3–4), 255–65.

Suriano, D.M. & Beverley-Burton, M. (1981) *Urocleidus aculeatus* (Van Cleave and Mueller, 1932) (Monogenea: Ancyrocephalinae) from *Stizostedion vitreum* (Mitchill) (Pisces:

Percidae) in eastern North America: anatomy and systematic position. *Canadian Journal of Zoology* **59**, 240–45.

Suthers, I.M. & Gee, J.H. (1986) Role of hypoxia in limiting diel spring and summer distribution of juvenile yellow perch (*Perca flavescens*) in a prairie marsh. *Canadian Journal of Fisheries and Aquatic Sciences* **43** (8), 1562–70.

Svärdson, G. (1976) Interspecific population dominance in fish communities of Scandinavian lakes. *Report, Institute of Freshwater Research Drottningholm* **55.**

Svetovidov, A.N. & Dorofeeva, E.A. (1963) Systematics, origin and history of the distribution of the Eurasian and North American perches and pikeperches (Genera *Perca*, *Lucioperca* and *Stizostedion*). *Voprosy Ikhtiologii* **3** (4), 625–51. (In Russian.)

Svirskii, A.M., Malinin, L.K. & Ovchinnikov, V.I. (1976) On the diel rhythm of motor activity in perch. *Informatsionnyi Byulleten Biologiya Vnutrennikh Vod* **30**, 12–14. (In Russian.)

Swenson, W.A. & Smith, L.L., Jr (1973) Gastric digestion, food consumption, feeding periodicity and food conversion efficiency in walleye (*Stizostedion vitreum vitreum*). *Journal of the Fisheries Research Board of Canada* **30** (9), 1327–36.

Swenson, W.A. & Smith, L.L., Jr (1976) Influence of food, competition, predation, and cannibalism on walleye (*Stizostedion vitreum*) and sauger (*Stizostedion canadense*) populations in Lake of the Woods, Minnesota. *Journal of the Fisheries Research Board of Canada* **33** (9), 1946–54.

Swift, D.R. & Pickford, G.E. (1965) Seasonal variations in the hormone content of the pituitary gland of the perch, *Perca fluviatilis* L. *General and Comparative Endocrinology* **5**, 354–65.

Sychevskaya, E.K. & Devyatkin, E.V. (1960) Pervyje nachodky ryb iz neogenovych i nizne cetverticnych otlozenij. *Gornogo Altaja Doklady Akademie Nauk SSSR* **142** (1), 175.

Szalai, A.J. & Dick, T.A. (1991) Role of predation and parasitism in growth and mortality of yellow perch in Dauphin Lake, Manitoba. *Transactions of the American Fisheries Society* **120**, 739–51.

Talikina, M.G., Komov, V.T. & Zelenetskij, N.M. (1996) Histophysiological characteristics of gonads in the perch *Perca fluviatilis* from acidified lakes in the Darwin Reservation. *Voprosy Ikhtiologii* **36** (3), 400–406.

Tamazouzt, L. (1995) *L'alimentation artificielle de la perche*, Perca fluviatilis, *en milieux confinés (eau recyclée, cage flottante)*. Ph.D. Thesis, Université H. Poincaré, Nancy, France.

Tamazouzt, L., Dubois, J.P. & Fontaine, P. (1993) Enquête sur la production et les marchés européens de la perche. *La Pisculture française* **114**, 4–8.

Tanasichuk, R.W. & Mackay, W.C. (1989) Quantitative and qualitative characteristics of somatic and gonadal growth of yellow perch (*Perca flavescens*) from Lake Ste Anne, Alberta. *Canadian Journal of Fisheries and Aquatic Sciences* **46** (6), 989–94.

Taylor, M.W. (1981) A generalized inland fishery simulator for management biologists. *North American Journal of Fisheries Management* **1** (1), 60–72.

Tedla, S. & Fernando, C.H. (1969) Observations on the biology of *Ergasilus* spp. (Cyclopoidea: Copepoda) infesting North American freshwater fishes. *Canadian Journal of Zoology* **47** (3), 405–8.

Temte, J., Allen-Rentmeester, Y., Luecke, C. & Vanni, M. (1988) Effects of a fish summerkill on zooplankton and phytoplankton populations in Lake Mendota. *Eighth Annual International Symposium on Lake and Watershed Management*, p 8. St Louis, Montana.

Theofan, G. & Goetz, F.W. (1983) The *in vitro* synthesis of final maturational steroids by ovaries of brook trout (*Salvelinus fontinalis*) and yellow perch (*Perca flavescens*). *General and Comparative Endocrinology* **51** (1), 84–95.

Thomas, D.L. (1970) An ecological study of four darters of the genus *Percina* (Percidae) in the Kaskaskia River, Illinois. *Illinois Natural History Survey Biological Notes* **70**, 1–18.

Thomas, K. & Ollevier, F. (1992) Paratenic hosts of the swimbladder nematode *Anguillicola crassus*. *Diseases of Aquatic Organisms* **13** (3), 165–74.

Thorgaard, G.H. (1983) Chromosome set manipulation and sex control in fish. In: *Fish Physiology IX B* (eds W.S. Hoar, D.J. Randall & E.M. Donaldson), pp. 405–34. Academic Press, New York.

Thorpe, J. (1977) Synopsis of biological data on the perch, *Perca fluviatilis* (Linnaeus, 1758) and *Perca flavescens* (Mitchill, 1814). *FAO Fisheries Synopsis* **113**.

Todd, T.N. & Haas, R.C. (1993) Genetic and tagging evidence for movements of walleyes between Lake Erie and Lake St Clair. *Journal of Great Lakes Research* **19**, 445–52.

Todd, T.N. & Hatcher, C.O. (1993) Genetic variability and glacial origins of yellow perch (*Perca flavescens*) in North America. *Canadian Journal of Fisheries and Aquatic Sciences* **50** (9), 1828–34.

Toivonen, J. (1991) The fish populations and fisheries in small forest lakes in Finland. *Finnish Fisheries Research* **12**, 19–24.

Tonn, W.M. & Paskowski, C.A. (1986) Size-limited predation, winterkill, and the organization of *Umbra–Perca* fish assemblages. *Canadian Journal of Fisheries and Aquatic Sciences* **43**, 194–202.

Tonn, W.M. & Paskowski, C. A. (1987) Habitat use of the central mudminnow (*Umbra limi*) and yellow perch (*Perca flavescens*) in *Umbra–Perca* assemblages: the roles of competition, predation, and the abiotic environment. *Canadian Journal of Zoology* **65**, 862–70.

Tonn, W.M., Paskowski, C.A. & Holopainen, I.J. (1992) Piscivory and recruitment: mechanisms structuring prey populations in small lakes. *Ecology* **73** (3), 951–8.

Tonn, W.M., Paskowski, C.A. & Moermond, T.C. (1986) Competition in *Umbra–Perca* fish assemblages: experimental and field evidence. *Oecologia* **69** (1), 126–33.

Tramer, E.J. (1977) Catastrophic mortality of stream fishes trapped in shrinking pools. *American Midland Naturalist* **97**, 469–78.

Trautman, M.B. (1957) *The Fishes of Ohio*. Ohio State University Press, Columbus.

Treasurer, J.W. (1981) Some aspects of the reproductive biology of perch *Perca fluviatilis* L. Fecundity, maturation and spawning behaviour. *Journal of Fish Biology* **18**, 729–40.

Treasurer, J.W. (1983) Estimates of egg and viable embryo production in a lacustrine perch, *Perca fluviatilis*. *Environmental Biology of Fishes* **8** (1), 3–16.

Treasurer, J.W. (1990) The food and daily food consumption of lacustrine 0+ perch, *Perca fluviatilis* L. *Freshwater Biology* **24** (2), 361–74.

Treasurer, J.W. & Holliday, F.G.T. (1981) Some aspects of the reproductive biology of perch, *Perca fluviatilis* L. A histological description of the reproductive cycle. *Journal of Fish Biology* **18** (3), 359–76.

Troitskiy, S.K. & Tsunikova, Ye. P. (1976) The time of spawning migrations and spawning of the pikeperch, *Lucioperca lucioperca*, in the Azov–Kuban Region. *Journal of Ichthyology* **26** (4), 592–8.

Tsepkin, E.A. (1984a) Fish remains from ancient settlements on the Baltic Sea coast. *Byulleten Moskovskogo Obshcestva Ispytatelei Prirody, Otdel Geologicheskii* **89** (2), 50–58. (In Russian.)

Tsepkin, E.A. (1984b) History of the commercial ichthyofauna in the upper Dnieper River USSR. *Biologicheski Nauki (Moscow)* **10** (7), 48–52. (In Russian.)

Turner, C.L. (1919) The seasonal cycle in the spermary of the perch. *Journal of Morphology* **32**, 681–711.

Turner, C.L. (1927) A case of hermaphroditism in the perch. *Anatomical Record* **37** (2), 186.

Turner, T.F. & Trexler, J.C. (1998) Ecological and historical associations of gene flow in darters (Teleostei: Percidae). *Evolution* **52**, 1781–1801.

Tuuha, H., Valtonen, E.T. & Taskinen, J. (1992) Ergasilid copepods as parasites of perch *Perca fluviatilis* and roach *Rutilus rutilus* in central Finland: seasonality, maturity and environmental influence. *Journal of Zoology* **228** (3), 405–22.

Twibell, R.G., Cartwright, D., Wilson, K.A. & Brown, P.B. (1998) Dietary lysine requirement of juvenile yellow perch. In: World Aquaculture Society symposium, *Aquaculture'98*, p. 553, Las Vegas, Nevada, 15–19 February 1998.

Tyutyunov, Y., Arditi, R., Buettiker, B., Dombrovsky, Y. & Straub, E. (1993) Modelling fluctuations and optimal harvesting in perch populations. *Ecological Modelling* **69** (1–2), 19–42.

Urho, L. (1996) Identification of perch (*Perca fluviatilis*), pikeperch (*Stizostedion lucioperca*) and ruffe (*Gymnocephalus cernuus*) larvae. *Annales Zoologici Fennici* **33**, 659–67.

Uthe, J.F., Roberts, E., Clarke, L.W. & Tsuyuki, H. (1966) Comparative electropherograms of representatives of the families Petromyzontidae, Esocidae, Centrarchidae and Percidae. *Journal of the Fisheries Research Board of Canada* **23** (11), 1663–71.

Uthe, J.F. & Ryder, R.A. (1970) Regional variation in muscle polymorphism in walleye (*Stizostedion vitreum vitreum*) as related to morphology. *Journal of the Fisheries Research Board of Canada* **27** (5), 923–7.

Valtonen, E.T. & Julkunen, M. (1995) Influence of the transmission of parasites from prey fishes on the composition of the parasite community of a predatory fish. *Canadian Journal of Fisheries and Aquatic Sciences* **52** (Suppl. 1), 233–45.

Vandenbyllaardt, L., Ward, F.J., Braekevelt, C.R. & McIntyre, D.B. (1991) Relationship between turbidity, piscivory and development of the retina in juvenile walleye. *Transactions of the American Fisheries Society* **120**, 382–90.

Van Densen, W.L.T. (1985a) Piscivory and the development of biomodality in the size distribution of 0+ pikeperch (*Stizostedion lucioperca* L.). *Journal of Applied Ichthyology* **1** (3), 119–31.

Van Densen, W.L.T. (1985b) Feeding behaviour of major 0+ fish species in a shallow, eutrophic lake (Tjeukemeer, The Netherlands). *Journal of Applied Ichthyology* **1** (2), 49–70.

Van Densen, W.L.T., Ligvoet, W. & Roozen, R.W.M. (1996) Intra-cohort variation in the individual size of juvenile pikeperch, *Stizostedion lucioperca*, and perch, *Perca fluviatilis*, in relation to the size spectrum of their food items. *Annales Zoologici Fennici* **33**, 495–506.

Vander-Zanden, M.J., Cabana, G. & Rasmussen, J.B. (1997) Comparing trophic position of freshwater fish calculated using stable nitrogen isotope ratios (delta super (15)N) and literature dietary data. *Canadian Journal of Fisheries and Aquatic Sciences* **54** (5), 1142–58.

Van Leeuwen, J.L. & Müller, M. (1984a) Optimum sucking techniques for predatory fish. *Transactions of the Zoological Society of London* **37** (2), 137–70.

Van Leeuwen, J.L. & Müller, M. (1984b) The recording and interpretation of pressures in prey sucking fish. *Netherlands Journal of Zoology* **33** (4), 425–75.

Verreth, J. (1984) Manipulation of the zooplankton population in nursing ponds of pikeperch fry (*Stizostedion lucioperca* L.). *Proceedings of the International Association of Limnology* **22** (3), 1672–80.

Verreth, J. & Kleyn, K. (1987) The effects of biomanipulation of the zooplankton on the growth, feeding, and survival of pike-perch (*Stizostedion lucioperca*) in nursing ponds. *Journal of Applied Ichthyology* **3**, 13–23.

Vetemaa, M. & Saat, T. (1996) Effects of salinity on the development of fresh-water and brackish-water ruffe *Gymnocephalus cernuus* (L.) embryos. *Annales Zoologici Fennici* **33**, 687–91.

Vinogradov, G.A. & Komov, V.T. (1985) Ion regulation in the perch, *Perca fluviatilis*, in connection with the problem of acidification of water bodies. *Journal of Ichthyology* **25** (2), 53–61.

Vivier, P. (1951) Poissons et crustacés d'eau douce acclimatés en France en eaux libres depuis le debut du siècle. *Terre Vie* **2**, 57–82.

Vlavonou, R., Masson, G. & Moretcau, J.C. (1995) Use of Artemia as unique starting food for cultured perch, *perca fluviatilis*, larvae. In: *Abstracts, Percis II, Second International Percid Fish Symposium*, p.79. FGFRI, Helsinki, Finland.

Volodin, V.M. (1980) Plodovitost, okunya *Perca fluviatilis* L., Rybinskogo Vodokhrani-lishcha. *Voprosy Ikhtiologii* **19** (4), 672–9.

Vuorinen, P.J., Vuorinen, M., Peuranen, S., Rask, M., Lappalainen, A. & Raitaniemi, J. (1992) Reproductive status, blood chemistry, gill histology and growth of perch (*Perca fluviatilis*) in three acidic lakes. *Environmental Pollution* **78** (1–3), 19–27.

Wahl, C.M. (1994) Periodic cone cell twists in the walleye, *Stizostedion vitreum*; a new type of retinomotor activity. *Vision Research* **34** (1), 11–18.

Wahl, D.H. (1995) Effect of habitat selection and behavior on vulnerability to predation of introduced fish. *Canadian Journal of Fisheries and Aquatic Sciences* **52** (11), 2312–19.

Walker, R. (1969a) Virus associated with epidermal hyperplasia in fish. *National Cancer Institute Monograph* **31**, 195–207.

Walker, R. (1969b) Epidermal hyperplasia in fish: two types without visible virus. *National Cancer Institute Monograph* **31**, 209–13.

Walker, R. & Weissenberg, R. (1965) Conformity of light and electron microscopic studies on virus particle distribution in lymphocystis tumor cells of fish. *Annals of the New York Academy of Sciences* **126** (1), 375–85.

Walters, C.J. & Punt, A. (1994) Placing odds on sustainable catch using virtual population analysis and survey data. *Canadian Journal of Fisheries and Aquatic Sciences* **51**, 946–58.

Walters, J.P. (1994) Spawning behavior of *Etheostoma zonale* (Pisces: Percidae). *Copeia* **1994**, 818–21.

Wang, N. (1994) Food and feeding of young perch (*Perca fluviatilis* L.) in Lake Constance. *Verhandlungen der Internationalen Vereinigung für Limnologie* **25**, 2148–52.

Wang, N. & Eckmann, R. (1994) Effects of temperature and food density on egg development, larval survival and growth of perch (*Perca fluviatilis*). *Aquaculture* **122**, 323–33.

Ward, F.J. & Robinson, G.G.C. (1974) A review of the research on the limnology of West Blue Lake, Manitoba. *Journal of the Fisheries Research Board of Canada* **31**, 977–1005.

Warren, M.L., Jr, Burr, B.M. & Kuhajda, B.R. (1986) Aspects of the reproductive biology of *Etheostoma tippecanoe* with comments on egg-burying behavior. *American Midland Naturalist* **116**, 215–18.

Warren, M.L., Jr, Burr, B.M. & Taylor, C.A. (1994) The relict darter, *Etheostoma chienense* (Percidae): status review of a Kentucky endemic. *Transactions of the Kentucky Academy of Science* **55**, 20–27.

Weatherley, A.H. (1963) Thermal stress and interrenal tissue in the perch, *Perca fluviatilis* (Linnaeus). *Proceedings of the Zoological Society of London* **141**, 527–55.

Webb, P.W. (1975) Hydrodynamics and energetics of fish propulsion. *Bulletin of the Fisheries Research Board of Canada* **90**, 1–159.

Webb, P.W. (1978) Fast-start performance and body form in seven species of teleost fish. *Journal of Experimental Biology* **74**, 211–26.

Weddle, G.K. & Burr, B.M. (1991) Fecundity and the dynamics of multiple spawning in darters: an in-stream study of *Etheostoma rafinesquei*. *Copeia* **1991**, 419–33.

Wehnes, R.A. (1973) *The food and feeding interrelationships of five sympatric darter species (Pisces: Percidae) in Salt Creek, Hocking County, Ohio*. M.Sc. Thesis, Ohio State University, USA

Weiler, W. (1933) Die Fischreste aus dem Oberpliocän von Willershausen. *Archiv für Hydrobiologie* **25**, 291–304.

Weinbauer, J.D., Thiel, D.A., Kaczynski, V.W. & Martin, C.S. (1980) Receiving stream fisheries studies relative to secondary treated pulp and paper mill effluents. *Tappi (Technical Association of the Pulp and Paper Industry)* **63** (10), 121–5.

Weinfurter, E. (1950) Die oberpannonische Fischfauna vom Eichkogel bei Mödling. Sitzungsber. *Österreichs Akademie Wissenschaften Mathematisch-Naturwissenschaftliche Klasse* **159** (1–5), 37–50.

Weissenberg, R. (1965) Fifty years of research on the lymphocystis virus disease of fishes (1914–1964). *Annals of the New York Academy of Sciences* **126** (1), 362–74.

Welsh, S.A. & Perry, S.A. (1998) Habitat partitioning in a community of darters in the Elk River, West Virginia. *Environmental Biology of Fishes* **51**, 411–19.

West, G.R. (1977) Intensive culture of yellow perch (*Perca flavescens*). In: *Perch Fingerling Production for Aquaculture* (ed. R.W. Soderberg). *University of Wisconsin Sea Grant Program Advisory Report 421*, 41–54.

Wheeler, A. & Maitland, P.S. (1973) The scarcer freshwater fishes of the British Isles. I. Introduced species. *Journal of Fish Biology* **5**, 49–68.

Whitear, M. (1986) The skin of fishes including cyclostomes. In: *The Biology of the Integument, Vertebrates, Vol. 2* (eds J. Bereiter-Hahn, A.G. Matoltsy & K.S. Richards), 8–77. Springer-Verlag, Berlin.

Whiteside, M.C., Swindoll, C.M. & Doolittle, W.L. (1985) Factors affecting the early life history of yellow perch, *Perca flavescens*. *Environmental Biology of Fishes* **12** (1), 47–56.

Whittington, R.J., Kearns, C., Hyatt, A.D., Hengstberger, S. & Rutzou, T. (1996) Spread of epizootic haematopoietic necrosis virus (EHNV) in redfin perch (*Perca fluviatilis*) in southern Australia. *Australian Veterinary Journal* **73** (3), 112–14.

Whoriskey, F.G. & Brown, K. (1994) Under-ice observations of behavior of selected native and exotic fishes of the St Lawrence River system. *Zoo Biology* **13** (6), 545–55.

Wickliff, E.L. (1941) Natural productivity of fish and crayfish in riffles. *Transactions of the North American Wildlife Conference* **5** (1940), 149–53.

Widlak, J.C. & Neves, R.J. (1985) Age, growth, food habits, and reproduction of the redline darter *Etheostoma rufilineatum* (Cope) (Perciformes: Percidae) in Virginia. *Brimleyana* **11**, 69–80.

Wieser, W. & Medgyesy, N. (1991) Metabolic rate and cost of growth in juvenile pike (*Esox lucius* L.) and perch (*Perca fluviatilis* L.): the use of energy budgets as indicators of environmental change. *Oecologia* **87** (4), 500–505.

Wiley, E.O. (1992) Phylogenetic relationships of the Percidae (Teleostei: Perciformes): A preliminary hypothesis. In: *Systematics, Historical Ecology, and North American Freshwater Fishes* (ed. R.L. Mayden), pp. 247–67. Stanford University Press, Stanford.

Wiley, M.L. & Collette, B.B. (1970) Breeding tubercles and contact organs in fishes: their occurrence, structure, and significance. *Bulletin of the American Museum of Natural History* **143**, 143–216.

Willemsen, J. (1977a) Influence of temperature on feeding, growth and mortality of pike-perch and perch. *Verhandlungen der Internationalen Vereinigung für Limnologie* **20** (3), 2127–33.

Willemsen, J. (1977b) Population dynamics of percids in Lake Ijssel and some smaller lakes in The Netherlands. *Journal of the Fisheries Research Board of Canada* **34** (10), 1710–19.

Williams, J.D. (1975) Systematics of the percid fishes of the subgenus *Ammocrypta,* genus *Ammocrypta,* with descriptions of two new species. *Bulletin of the Alabama Museum of Natural History* **1**, 1–56.

Williams, J.S. (1976) Spawning behavior of *Etheostoma edwini* (Pisces: Percidae). *ASB (Association of Southeastern Biologists) Bulletin* **23**, 107.

Willoughby, L.G. (1970) Mycological aspects of a disease of young perch in Windermere. *Journal of Fish Biology* **2**, 113–16.

Willoughby, L.G. & Roberts, R.J. (1991) Occurrence of the sewage fungus *Leptimitus lacteus*, a necrotroph on perch (*Perca fluviatilis*) in Windermere. *Mycological Research* **95** (6), 755–8.

Wilson, R.P. & Poe, W.E. (1985) Relationship of whole body and egg essential amino acid patterns to amino acid requirement patterns in channel catfish, *Ictalurus punctatus*. *Comparative Biochemistry and Physiology* **80B**, 385–8.

Winberg, G.G. (1956) Rate of metabolism and food requirements of fishes. Belorussian State University, Minsk. *Fisheries Research Board of Canada Translation Series* **194**, 1–253.

Winfield, I.J., Dodge, D.P. & Rösch, R. (1998a) Introductions of the ruffe, *Gymnocephalus cernuus*, to new areas of Europe and to North America: history, the present situation and management implications. In: *Stocking and Introduction of Fish in Freshwater and Marine Ecosystems* (ed. I.G. Cowx), pp. 191–201. Fishing News Books, Blackwell Scientific Publications, Oxford.

Winfield, I.J., Fletcher, J.M. & Cubby, P.R. (1998c) The threat to vendace (*Coregonus albula*) eggs from introduced ruffe (*Gymnocephalus cernuus*) in Bassenthwaite Lake, U.K. *Archiv für Hydrobiologie. Special Issues Advances in Limnology* **50**, 171–7.

Winfield, I.J., George, D.G., Fletcher, J.M. & Hewitt, D.P. (1998d) Environmental factors influencing the recruitment and growth of underyearling perch (*Perca fluviatilis*) in Windermere North Basin, UK, from 1966 to 1990. In: *Management of Lakes and Reservoirs during Global Climate Change* (eds D.G. George *et al.*), pp. 245–61. Kluwer Academic Publishers, The Netherlands.

Winfield, I.J. & McCulloch, I.D. (1995) An indexed bibliography of the ruffe *Gymnocephalus cernuus* (L.) (Percidae): a carnivorous freshwater fish threatening rare fishes following recent introductions in Britain, Europe and North America. *Freshwater Biological Association Occasional Publication* **31**.

Winfield, I.J., Rösch, R., Appelberg, M., Kinnerbäck, A. & Rask, M. (1998b) Recent introductions of the ruffe (*Gymnocephalus cernuus*) to *Coregonus* and *Perca* lakes in Europe and an analysis of their natural distributions in Sweden and Finland. *Journal of Great Lakes Research* **24** (2), 235–48.

Winn, H.E. (1953) Breeding habits of the percid fish *Hadropterus copelandi* in Michigan. *Copeia* **1953**, 26–30.

Winn, H.E. (1958a) Observations on the reproductive habits of darters (Pisces-Percidae). *American Midland Naturalist* **59**, 190–212.

Winn, H.E. (1958b) Comparative reproductive behavior and ecology of fourteen species of darters (Pisces-Percidae). *Ecological Monographs* **28**, 155–91.

Winn, H.E. & Picciolo, A.R. (1960) Communal spawning of the glassy darter *Etheostoma vitreum* (Cope). *Copeia* **1960**, 186–92.

Wolfert, D.R. (1969) Maturity and fecundity of walleyes from the eastern and western basins of Lake Erie. *Journal of the Fisheries Research Board of Canada* **26** (7), 1877–88.

Wolfert, D.R. (1981) The commercial fishery for walleyes in New York waters of Lake Erie, 1959–1978. *North American Journal of Fisheries Management* **1** (2), 112–26.

Wolfert, D.R. & Van Meter, H.D. (1978) Movements of walleyes tagged in eastern Lake Erie. *New York Fish and Game Journal* **25** (1), 16–22.

Wong B. & Ward, F.J. (1972) Size selection of *Daphnia pulicaria* by yellow perch (*Perca flavescens*) fry in west Blue Lake, Manitoba. *Journal of the Fisheries Research Board of Canada* **29** (12), 1761–4.

Wood, R.M. & Mayden, R.L. (1997) Phylogenetic relationships among selected darter subgenera (Teleostei: Percidae) as inferred from analysis of allozymes. *Copeia* **1997**, 265–74.

Woodbury L.A. (1942) A sudden mortality of fishes accompanying a supersaturation of oxygen in Lake Waubesa, Wisconsin. *Transactions of the American Fisheries Society* **71**, 112–17.

Worthington, E.B. (1950) An experiment with populations of fish in Windermere, 1939–48. *Proceedings of the Zoological Society of London* **120** (1), 113–49.

Woynarovich, E. (1960) Aufzucht der Zanderlarven bis zum Raubfischalter. *Zeitschrift für Fischerei* **9** (1–2), 73–83.

Woynarovich, E. (1962) Die kunstliche Erbrutung des Zanders. *Zeitschrift für Fischerei* **10** (8–10), 677–80.

Wu, L. & Culver, D.A. (1992) Ontogenetic diet shift in Lake Erie age-0 yellow perch (*Perca flavescens*): a size-related response to zooplankton density. *Canadian Journal of Fisheries and Aquatic Sciences* **49** (9), 1932–7.

Yakovlev, V.N. (1960) O sistematicheskom polozhenii presnovodnykh ryb is zeogena Zapadnoi Sibiri. *Paleontologicheskii Zhurnal* **3**, 102–8.

Yakovlev, V.N. (1961) Rasprostranenie presnovodnykh ryb neogena Golarktiki i zoogeographicheskoe raionirovanie. *Voprosy Ikhtiologie* **1** (2), 209–20.

Yakovleva, A.S., Amstislavskiy, A.Z. & Baymuratov, A. (1976) Features of the relative growth of some internal organs in the perch, *Perca fluviatilis*. *Journal of Ichthyology* **16** (3), 419–26.

Yamamoto, T., Kelly, R.K. & Nielsen, O. (1985) Morphological differentiation of virus associated skin tumors of walleye (*Stizostedion vitreum vitreum*). *Fish Pathology* **20** (2–3), 361–72.

Yamamoto, T., MacDonald, R.D., Gillespie, D.C. & Kelly, R.K. (1976) Viruses associated with lymphocystis disease and dermal sarcoma of walleye. *Journal of the Fisheries Research Board of Canada* **33** (11), 2408–19.

Zalewski, M., Brewinska-Zaras, B. & Frankiewicz, P. (1990a) Fry communities as a biomanipulating tool in a temperate lowland reservoir. *Proceedings of the International Conference on Reservoir Limnology and Water Quality. 3 Key Factors of Reservoir Limnology, Eutrophication, Water Quality and its Prediction* (eds O.T. Lind, V. Sladecek & V. Straskrabova). *Ergebnisse der Limnologie: Advances in Limnology* **33**, 763–74.

Zalewski, M., Brewinska-Zaras, B., Frankiewicz, P. & Kalinowski, S. (1990b) The potential for biomanipulation using fry communities in a lowland reservoir: concordance between water quality and optimal recruitment. In: *Biomanipulation, tool for Water Management* (eds R.D. Gulati, E.H.R.R. Lammens, M.L. Meijer & E. van Donk). *Hydrobiologia* **200–201**, 549–56.

Zalewski, M., Kapala, J., Tomczak, M. & Mnich, Z. (1996) Radioactive caesium in Sawinda Wielka Lake fish meat. *Bromatologia i Chemia Toksykologiczna* **29** (3), 221–4.

Zeh, M., Ritter, E. & Ribi, G. (1989) Spawning and egg development of *Perca fluviatilis* in Lake Zurich. *Aquatic Sciences* **51** (2), 100–107.

Zelenkov, V.M. (1982) Early gametogenesis and sex differentiation in the perch, *Perca fluviatilis*. *Journal of Ichthyology* **21** (2), 124–30.

Zhmurova, Ye. K.H. & Somkina, N.V. (1976) The effect of salinity on the early development stanzas of the walleye (*Lucioperca lucioperca*). *Journal of Ichthyology* **16** (3), 511–14.

Zyznar, E.S. & Ali, M.A. (1975) An interpretative study of the organization of the visual cells and tapetum lucidum of *Stizostedion*. *Canadian Journal of Zoology* **53**, 180–96.

Author Index

Fish Species Index

For species of the Percidae please see subject index and relevant chapter.

Geographic Index

Subject Index

CPSIA information can be obtained
at www.ICGtesting.com
Printed in the USA
LVHW101237130820
662941LV00007B/242